Lecture Notes in Mathematics

A collection of informal reports and seminars
Edited by A. Dold, Heidelberg and B. Eckmann, Zürich

T0184012

255

Conference in Mathematical Logic - London '70

Edited by Wilfrid Hodges, Bedford College, London/G. B.

Springer-Verlag
Berlin · Heidelberg · New York 1972

AMS Subject Classifications (1970): 02 A 05, 02 B 25, 02 C 10, 02 F 27, 02 F 30, 02 G 05, 02 G 20, 02 H xx, 02 K xx, 06 A 40

ISBN 3-540-05744-7 Springer-Verlag Berlin · Heidelberg · New York
ISBN 0-387-05744-7 Springer-Verlag New York · Heidelberg · Berlin

Offsetdruck: Julius Beltz, Hemsbach/Bergstr.

PREFACE

This volume represents the Proceedings of the Conference in Mathematical Logic, held at Bedford College, London on 24th-28th August 1970. The organising committee was Imre Lakatos (Chairman), Robin Gandy, Moshé Machover, Frederick Rowbottom and Wilfrid Hodges (Secretary). Thanks are due to the following bodies for their generous support at rather short notice: the International Union of History and Philosophy of Science, the London Mathematical Society, Bedford College, and the British Logic Colloquium.

WILFRID HODGES

Bedford College, November 1971

CONTENTS

LIST OF INVITED ADDRESSES

The invited addresses at the Conference were as follows:

M. A. DICKMANN
Languages with expressions of large cardinality.

SOLOMON FEFERMAN
Systems of ordinal functions and functionals.

HAIM GAIFMAN
Probabilities for logical calculi.

MIHÀLY MAKKAI
Preservation theorems concerning binary relations between structures.

GERT H. MÜLLER
Extension and intension in set theory.

HARTLEY ROGERS JR.
Topics in generalized recursion theory.

GERALD E. SACKS
The 1-section of a type 2 object.

ROBERT SOLOVAY
Large cardinals and projective determinacy.

YOSHINDO SUZUKI
Non-standard models for set theory.

C. E. M. YATES
Embedding lattices as initial segments of the degrees.

Some of these talks appear in this volume; the contents of others have appeared or will appear in print elsewhere. Gert H. Müller and Hartley Rogers Jr. had hoped to write up their talks for this volume, but were unfortunately prevented by various unavoidable circumstances.

LIST OF REGISTERED PARTICIPANTS

S. M. Abramsky
P. H. G. Aczel
R. D. Adams
P. Axt
P. D. Bacsich
H. P. Barendregt
K. J. Barwise
P. Bastable
D. Becchio
J. L. Bell
L. Blum
J. Bridge
D. Bryars
A. R. Bundy
A. J. Carpenter
C. C. Chang
P. J. Clark
T. Climo
P. M. Cohn
J. C. Cole
D. Coles
S. B. Cooper
R. H. Cowen
R. Cusin
N. J. Cutland
E. A. Davies
D. H. J. de Jongh
H. de Swart
K. J. Devlin
M. A. Dickmann
F. R. Drake
D. Edgington
S. Feferman
K. Fine
M. C. Fitting
P. Foulkes
E. Fredriksson
M. Frické
J. I. Friedman
D. M. Gabbay
H. Gaifman
R. O. Gandy
J. R. Geiser
J. Gielen
P. W. Grant
H. Gudjónsson
B. Harle
J. Heidema
P. Henrard
A. Heyting
R. Hindley
H. B. Hingert
A. Hirschelmann
H. Hiż
W. A. Hodges
G. Hunter
H. Ishiguro
D. Jack
S. C. Jackson
D. G. Jones
J. W. Kamp
H. J. Keisler
A. A. Khoury
G. T. Kneebone
F. Kriwaczek
N. S. Kroonenberg
J. C. Lablanquie
A. H. Lachlan
I. Lakatos
M. Lapscher
T. Larsen
D. K. Lewis
S. Lewis
J. A. T. Lorman
F. Lowenthal
T. Lucas
M. Machover
A. J. Macintyre
M. Makkai
J. A. Makowsky
A. R. D. Mathias
D. Miller
G. P. Monro
B. P. Moss
J. M. B. Moss
G. H. Müller
Muir
R. D. Nelson
A. Ostoja-Ostaszewski
Owen
J. F. Pabion
J. B. Paris
Y. Paul
B. J. Philp
A. Prestel
W. G. Raines
K. A. Rasmussen
H. Rogers Jr.
P. K. Rogers
H. E. Rose
S. B. Russ
G. Sabbagh
G. E. Sacks
H. Simmons
W. E. Singletary
M. Sintzoff
A. B. Slomson
R. C. Solomon
R. M. Solovay
G. Stahl
L. A. Steen
Y. Suzuki
M. Takahashi
M. M. Thomas
S. Thompson
S. A. Tracey
A. A. Treherne
S. Tsou
J. Tucker
R. Turner
D. van Dalen
W. P. van Stigt
S. C. van Westrhenen
A. Verbeek
F. Y. Villemin
S. S. Wainer
R. M. Whitehouse
D. R. P. Wiggins
A. J. Wilkie
G. Wilmers
C. E. M. Yates

131 people

INDUCTIVE DEFINITIONS AND ANALOGUES OF LARGE CARDINALS

Peter Aczel
Manchester University
and
Wayne Richter[1]
University of Minnesota

1. Introduction

An operation Γ on $P(A)$, the power set of A, determines a transfinite sequence $<\Gamma^{\xi} : \xi \in On>$ of subsets of A, where $\Gamma^{\lambda} = \cup\{\Gamma(\Gamma^{\xi}) : \xi < \lambda\}$. Let $|\Gamma|$, the <u>closure ordinal</u> of Γ, be the smallest ordinal α such that $\Gamma^{\alpha+1} = \Gamma^{\alpha}$. It is clear that $|\Gamma|$ is less than the first regular cardinal greater than $\bar{A}$. Γ is <u>monotone</u> if $X \subseteq Y$ implies $\Gamma(X) \subseteq \Gamma(Y)$ for all $X, Y \subseteq A$.

Given a set $\mathcal{C}$ of operations on $P(A)$ we may wish to characterize $|\mathcal{C}| = \sup\{|\Gamma| : \Gamma \in \mathcal{C}\}$. Interesting results concerning monotone operations on $P(\omega)$ were first obtained by Spector [13]. Let Π^n_m (resp. Π^n_m-mon) be the set of Π^n_m (resp. monotone Π^n_m) operations on $P(\omega)$. Similarly for Σ^n_m and Δ^n_m. Let ω_1 be the first non-recursive ordinal. Spector showed that $|\Pi^1_1\text{-mon}| = |\Pi^0_1\text{-mon}| = \omega_1$. A generalization of Spector's result has been obtained recently by Barwise, Gandy and Moschovakis [2]. Putnam [10] essentially showed that $|\Delta^1_2|$ is the first non-Δ^1_2 ordinal, and Gandy (unpublished) observed that $|\Sigma^1_2\text{-mon}| = |\Delta^1_2|$.

---

[1]Research supported in part by the U.S. National Science Foundation under Grant GP-20846.

Aczel [1] has obtained a characterization of $|\Sigma_1^1\text{-mon}|$.[2] Gandy (unpublished) showed that $|\Pi_1^0| = \omega_1$. Richter [11] obtained characterizations of certain natural extensions of Π_1^0 in terms of recursive analogues of large cardinals. In particular it was shown that even $|\Pi_2^0|$ is much larger than the first recursively Mahlo ordinal, the first recursively hyper-Mahlo ordinal, etc.

In this paper we provide characterizations of $|\Pi_n^0|$ for $n < \omega$. Theorem 3 below characterizes $|\Pi_n^0|$ in terms of reflection principles analogous to those used in defining the various families of indescribable cardinals (see Lévy [8]). Theorem 10 below shows how these reflection principles characterize recursive analogues of large cardinals, so that each $|\Pi_n^0|$ is characterized as a recursive analogue of a large cardinal.

We also obtain characterizations of $|\Delta_1^1|$, $|\Pi_1^1|$ and $|\Sigma_1^1|$. It turns out that $|\Delta_1^1|$ is strictly less than both $|\Pi_1^1|$ and $|\Sigma_1^1|$ ($|\Delta_1^1|$ is not even admissible), and $|\Pi_1^1| \neq |\Sigma_1^1|$, but the order relation between $|\Pi_1^1|$ and $|\Sigma_1^1|$ is unknown.

The results of section 2 use the techniques of [11]. In most cases only the briefest sketch of proofs is given here. A full treatment will be published elsewhere.

2. Reflecting ordinals

Let L_κ be the set of constructible sets of order less than κ.

DEFINITION. Let X be a class of ordinals. κ is Π_m^n-*reflecting on* X if for every Π_m^n sentence ϕ (of the Lévy [7] hierarchy of formulas of set theory) with parameters in L_κ,

---

[2] T. Grilliot has pointed out in conversation that $|\Sigma_1^1\text{-mon}| = |\Sigma_1^1|$.

if $L_\kappa \models \phi$ then for some $\alpha \in X \cap \kappa$, $L_\alpha \models \phi$.

κ is Π^n_m-<u>reflecting</u> if κ is Π^n_m-reflecting on On. Σ^n_m-<u>reflecting</u> ordinals are defined similarly.

The proof of the following is straightforward.

PROPOSITION 1.

(i) κ is Π^0_0-reflecting iff κ is Π^0_1-reflecting iff κ is a limit ordinal.

(ii) κ is Π^0_2-reflecting iff κ is admissible and $\kappa > \omega$.

Let $X = \{\alpha : \alpha$ is admissible$\}$.

(iii) $\kappa \in X$ is Π^0_1-reflecting on X iff κ is recursively inaccessible.

(iv) $\kappa \in X$ is Π^0_2-reflecting on X iff κ is recursively Mahlo.

Let π^n_m be the least Π^n_m-reflecting ordinal. σ^n_m is defined similarly. To understand the relative magnitude of these ordinals we proceed as follows. For $n > 0$ and $A \subseteq$ On, let

$$M_n(A) = \{\alpha \in A : \alpha \text{ is } \Pi^0_n\text{-reflecting on } A\}.$$

Let $M^0_n =$ On; for $\lambda > 0$ let $M^\lambda_n = \cap\{M_n(M^\xi_n) : \xi < \lambda\}$, and let $M^\Delta_n = \{\alpha : \alpha \in M^\alpha_n\}$. Using methods analogous to those of Lévy [8] we obtain:

PROPOSITION 2. $M^1_{n+1} \subset M^\Delta_n$ for $n > 0$.

As in [8] even stronger results may be proved. It follows from Proposition 2 that π^0_{n+1} is the π^0_{n+1}-th Π^0_n-reflecting ordinal, but it is far from being the first ordinal with this fixed-point property.

THEOREM 3. $|\Pi^0_n| = \pi^0_{n+1}$.

If $n = 0$ this is trivial since $\pi^0_1 = \omega = |\Pi^0_0|$. The case $n = 1$, due to Gandy, is proved in [11]. For $n > 1$ the proof that $|\Pi^0_n| \leq \pi^0_{n+1}$

involves little more than the definition of Π^0_{n+1}-reflection and the fact that for Π^0_n operations Γ, Γ^λ is Σ_1 definable on L_λ, for admissible ordinals λ, uniformly in λ. To show $|\Pi^0_n| \geq \pi^0_{n+1}$, a Π^0_n operation Γ is found such that $|\Gamma| = \pi^0_{n+1}$. This requires the techniques of [11]. In particular, the concept of an acceptable ordinal system, and Theorem 4.1 of [11] relating acceptable ordinal systems and admissible ordinals, play a basic role.

THEOREM 4.

(i) $$|\Pi^1_1| = \pi^1_1;$$

(ii) $$|\Sigma^1_1| = \sigma^1_1;$$

(iii) $$\pi^1_1 \neq \sigma^1_1.$$

Parts (i) and (ii) are proved as in the proof of Theorem 3. Let κ^+ be the first admissible ordinal greater than κ. Part (iii) may be proved by following the proof of Lévy [8] that the first Σ^2_1-indescribable cardinal is different from the first Π^2_1-indescribable cardinal, and using the fact [2] that for countable κ, Π^1_1 statements over L_κ may be written as Σ_1 statements over L_{κ^+}. The order relationship between π^1_1 and σ^1_1 is unknown and the same is true for the two indescribable cardinals. A solution to either of these problems will probably suggest a solution to the other. An upper bound for both π^1_1 and σ^1_1 is the first non-Δ^1_2 ordinal. A much sharper upper bound is the first non-projectible ordinal. For if Γ is Π^1_1 such that $|\Gamma| = \pi^1_1$ then Γ will define a system of notations for the ordinals $< \pi^1_1$ which may be used to project each admissible ordinal $\leq \pi^1_1$ down to ω. Similarly for σ^1_1.

Following Kripke [6] and Platek [9], we say that α is β-stable if $\alpha < \beta$ and $L_\alpha \prec_{\Sigma_1} L_\beta$. This notion is closely related to certain types of reflection properties. For example:

THEOREM 5. κ is $\kappa+1$-stable iff κ is Π^0_n-reflecting for all $n < \omega$.

Using stronger reflection properties we may also characterize the ordinals α that are $\alpha+\beta$-stable for $0 < \beta < \omega_1$, etc. Let ρ_β be the first α that is $\alpha+1+\beta$-stable. Thus, for example, $\rho_0 > \sup_{n<\omega} \pi^0_n$. It seems probable that $|\Delta^1_1| = \sup_{\alpha<\omega_1} \rho_\alpha$. Whether this is the case or not we get:

PROPOSITION 6. $|\Delta^1_1|$ is not admissible.

For if Σ^0_α, for each $\alpha < \omega_1$, is the family of operations that are definable in Σ^0_α form in the hyperarithmetical hierarchy (see e.g. [12] page 443) then $\Delta^1_1 = \cup_{\alpha<\omega_1} \Sigma^0_\alpha$. Now if $\kappa \geqslant |\Delta^1_1| = \sup_{\alpha<\omega_1} |\Sigma^0_\alpha|$ is admissible and $f : \kappa \rightarrow \kappa$ is given by $f(\alpha) = |\Sigma^0_\alpha|$ for $\alpha < \omega_1$ and $f(\alpha) = 0$ otherwise, then f is κ-recursive, so that $|\Delta^1_1| = \sup_{\alpha<\omega_1} f(\alpha) < \kappa$.

Since σ^1_1 and π^1_1 are admissible we have:

PROPOSITION 7. $|\Delta^1_1| < \sigma^1_1, \pi^1_1$.

Using results of Barwise, Gandy and Moschovakis [2] relating Π^1_1 definitions on L_κ, for countable admissible κ, to Σ_1 definitions on L_{κ^+} we obtain:

THEOREM 8. For countable κ, κ is Π^1_1-reflecting iff κ is κ^+-stable.

3. n-regular cardinals

One of the promising features of the study of closure ordinals of operations is the relationship between these ordinals and recursive analogues of large cardinals in set theory. For example $|\Pi^0_1| = \omega_1$ is a recursive analogue of the first uncountable ordinal. The main result of [11] is that the ordinals $|\mathcal{a}_2|$, $|\mathcal{a}_3|$, ... defined there are respectively, the first recursively Mahlo ordinal, the first recursively hyper-Mahlo ordinal, etc. One suspects that by considering larger natural classes $\mathcal{C}$ of operations, recursive analogues of larger cardinals will be obtained. Since $\mathcal{a}_n \subseteq \Pi^0_2$ for all $n < \omega$ (and indeed by

Proposition 2 much more is true) one suspects that even $|\Pi^0_3|$ should be the recursive analogue of some rather large cardinal. Since such large ordinals are obtained by such simple operations it seems quite possible that one will be led to interesting new axioms of infinity by the study of closure ordinals of operations. At present we do not have such results. In our characterization of arithmetic operations, however, we are led to a new and natural characterization of the Π^1_n-indescribable cardinals. In this section we define the n-regular cardinals and show that for $n > 0$, κ is n+1-regular iff it is strongly Π^1_n-indescribable. In the next section we define the n-admissible ordinals as the recursive analogue of the n-regular cardinals and show that κ is n-admissible iff κ is Π^0_{n+1}-reflecting. Thus by Theorem 10 $|\Pi^0_{n+1}|$ is a recursive analogue of the first Π^1_n-indescribable cardinal.

Certain properties of infinity can be stated in terms of fixed points of operations. For example, let $\kappa > 0$. Then $\kappa > \omega$ and κ is regular iff:

(1) for every $f : \kappa \longrightarrow \kappa$ there is some $0 < \alpha < \kappa$ such that $f''\alpha \subseteq \alpha$. (We say α is a <u>witness</u> for f.)

There are two natural ways of modifying (1) to imply existence of much larger cardinals. The first method used originally by Mahlo [4] consists of placing requirements on the witness. If we require in (1) that the witness be regular we obtain the class of (weakly) Mahlo cardinals. If we require the witness to be Mahlo we obtain the hyper-Mahlo cardinals, etc.

The second method which we investigate here consists of modifying (1) by using higher type functionals on κ.

Let $P^{(0)}_\kappa = \kappa$ and let $P^{(n+1)}_\kappa$ be the set of operations on $P^{(n)}_\kappa$ which are <u>bounded</u>; where every $f : \kappa \longrightarrow \kappa$ is bounded, and if

$F : {}^{\kappa}\kappa \longrightarrow {}^{\kappa}\kappa$ then F is <u>bounded</u> if for every $f : \kappa \longrightarrow \kappa$ and $\xi < \kappa$, the value $F(f)(\xi)$ is determined by less than κ values of f. More precisely, F is bounded if

$$(\forall f \epsilon {}^{\kappa}\kappa)(\forall \xi < \kappa)(\exists \gamma < \kappa)(\forall g \epsilon {}^{\kappa}\kappa)[g \upharpoonright \gamma = f \upharpoonright \gamma \Rightarrow F(g)(\xi) = F(f)(\xi)].$$

For $n > 1$ the bounded $F : P_{\kappa}^{(n)} \longrightarrow P_{\kappa}^{(n)}$ are defined in a similar spirit. α is a <u>witness</u> for $\beta < \kappa$ if $\beta < \alpha < \kappa$. α is a <u>witness</u> for $F \in P_{\kappa}^{(n+1)}$ if $0 < \alpha < \kappa$ and for all $f \in P_{\kappa}^{(n)}$, α a witness for f implies α is a witness for F(f). For example, α is a witness for $F \in P_{\kappa}^{(2)}$ if:

$$\forall f \in {}^{\kappa}\kappa \; [f''\alpha \subseteq \alpha \Rightarrow F(f)''\alpha \subseteq \alpha]. \tag{2}$$

Finally, $\kappa > 0$ is n+1-<u>regular</u> if every $F \in P_{\kappa}^{(n+1)}$ has a witness.

Following Lévy [8] we distinguish between two notions of indescribability.

DEFINITION.

(i) κ is <u>Π_n^1-indescribable</u> if for every structure of the form $\langle \kappa, U_1, \ldots, U_k \rangle$ and every Π_n^1 sentence ϕ for this structure, if $\langle \kappa, U_1, \ldots, U_k \rangle \models \phi$ then there is some $\alpha < \kappa$ such that $\langle \alpha, U_1 \upharpoonright \alpha, \ldots, U_k \upharpoonright \alpha \rangle \models \phi$.

(ii) κ is <u>strongly Π_n^1-indescribable</u> if for every structure of the form $\langle R(\kappa), \epsilon, X \rangle$, where $X \subseteq R(\kappa)$, and Π_n^1-sentence ϕ for this structure, if $\langle R(\kappa), \epsilon, X \rangle \models \phi$ then for some $\alpha < \kappa$, $\langle R(\alpha), \epsilon, X \cap R(\alpha) \rangle \models \phi$.

THEOREM 9. (i) κ is 1-regular iff κ is Π_0^1-indescribable;
(ii) for $n > 0$, κ is n+1-regular iff κ is strongly Π_n^1-indescribable.

4. n-admissible ordinals

Roughly speaking, the notion of n-admissible is obtained from that of n-regular by replacing in the definition of the latter, bounded by recursive and replacing the functions by their Gödel numbers. For any κ (not necessarily admissible) and $\xi < \kappa$, let $\{\xi\}_\kappa$ be the κ-partial recursive function of one variable with Gödel number ξ. The notation $\{\xi\}_\kappa : A \longrightarrow A$ implies $\{\xi\}_\kappa$ is defined at least on A. Let $Q_\kappa^{(0)} = \kappa$ and $Q_\kappa^{(n+1)} = \{\xi < \kappa \mid \{\xi\}_\kappa : Q_\kappa^{(n)} \longrightarrow Q_\kappa^{(n)}\}$. $\alpha < \kappa$ is a witness for $\xi \in Q_\kappa^{(n)}$ if $\xi \in Q_\alpha^{(n)}$. For example, (compare with (2)) α is a witness for $\xi \in Q_\kappa^{(2)}$ iff $\xi < \alpha < \kappa$ and

$$\forall \beta < \alpha\ [\{\beta\}_\alpha : \alpha \longrightarrow \alpha \Rightarrow \{\{\xi\}_\alpha(\beta)\}_\alpha : \alpha \longrightarrow \alpha].$$

Let κ be closed under primitive recursive functions of ordinals (cf. [9]). This implies in particular, that every κ-partial recursive function has a Gödel number less than κ. Such a κ is called n-admissible if every $\xi \in Q_\kappa^{(n)}$ has a witness.

THEOREM 10. For $n \geq 1$, κ is n-admissible iff κ is Π_{n+1}^0-reflecting.

In view of Theorems 3, 9, 10 we regard $|\Pi_{n+1}^0|$ as a recursive analogue of the first Π_n^1-indescribable cardinal. It is interesting to note that the first Π_1^1-indescribable cardinal appears to be the first natural stopping place when one iterates into the transfinite the process: Mahlo, hyper-Mahlo, ... Correspondingly, from [11] it appears that $|\Pi_2^0|$ is the first natural stopping place when one iterates the process: recursively Mahlo, recursively hyper-Mahlo, ...

REFERENCES

[1] P. Aczel, The ordinals of the superjump and related functionals, Abstract, this volume pp. 336f.

[2] J. Barwise, R. Gandy and Y. Moschovakis, The next admissible set, J. Symb. Logic 36 (1971), 108-120.

[3] R. B. Jensen and C. Karp, Primitive recursive set functions, in *Axiomatic Set Theory, Part I*, ed. Dana S. Scott, Amer. Math. Soc. Proceedings of Symposia in Pure Maths. (1971).

[4] P. Mahlo, Über lineare transfinite Mengen, Berichte über die Verhandlungen der Königlich Sächsischen Gesellschaft der Wissenschaften zu Leipzig, Mathematisch-Physische Klasse 63 (1911), 187-225.

[5] S. Kripke, Transfinite recursions on admissible ordinals, I, II, J. Symb. Logic 29 (1964), 161-162 (abstracts).

[6] S. Kripke, Transfinite recursion, constructible sets, and analogues of cardinals, in: *Lecture notes prepared in connection with the Summer Institute on Axiomatic Set Theory held at University of California, Los Angeles, California, July-August, 1967*.

[7] A. Lévy, A hierarchy of formulas in set theory, Memoirs of the Amer. Math. Soc. 57, Amer. Math. Soc., Providence, 1965.

[8] A. Lévy, The sizes of the indescribable cardinals, in *Axiomatic Set Theory, Part I*, ed. Dana S. Scott, Amer. Math. Soc. Proceedings of Symposia in Pure Maths. (1971).

[9] R. A. Platek, Foundations of recursion theory, Ph. D. Thesis and Supplement, Stanford University, 1966.

[10] H. Putnam, On hierarchies and systems of notations, Amer. Math. Soc. Proc. 15 (1964), 44-50.

[11] W. Richter, Recursively Mahlo ordinals and inductive definitions, in *Logic Colloquium '69*, R. O. Gandy, C. E. M. Yates (eds.), North-Holland (1971), 273-288.

[12] H. Rogers, Jr., *Theory of recursive functions and effective computability*, McGraw-Hill (1967).

[13] C. Spector, Inductively defined sets of natural numbers, in *Infinitistic methods*, New York-Oxford-London-Paris and Warsaw (1961), 97-102.

COMPACT INJECTIVES AND NON-STANDARD ANALYSIS

Paul D. Bacsich[1]

Mathematical Institute, Oxford

The notion of compact object in a category is introduced, and within this framework general theorems are given which deal uniformly with most algebraic or combinatorial consequences of the Prime Ideal Theorem. In fact all known proofs of Extension Theorems using the Principle of Consistent Choices, the method of ultrapowers, or Non-Standard Analysis, can be reduced to category-theoretic consideration of certain inverse limits of compact spaces. Further, three other methods of building up injectives using the Prime Ideal Theorem are given, one of which is the dual of Non-Standard Analysis in a certain sense. The techniques apply to certain non-axiomatic classes, for which the classical Compactness Theorem does not hold.

1. Preliminaries

We recall that the Prime Ideal Theorem (BPI) is the following statement: every nontrivial boolean algebra has an ultrafilter. Throughout this paper the system of Set Theory used is GB + BPI, where GB is the set of axioms in groups A to D of Gödel's monograph [9] (i.e. excluding the Axiom of Choice).

Our category-theoretic terminology is based on Freyd [8] and

---

[1]Research supported by an SRC NATO Studentship at the University of Bristol and subsequently by an SRC Research Fellowship at the Mathematical Institute, Oxford. This paper is a revised version of some of the author's Ph.D. Thesis.

Semadeni [19]. In particular (A,B) denotes the set of morphisms from A to B, while the ordered pair of A and B is denoted by <A,B>.

One important consequence of BPI (i.e. theorem of our system) that we shall need is the following result of [12], which we shall call TPT: if $(X_i : i \in I)$ is a family of nonvoid compact T_2 spaces, then the product topology on $\Pi_{i \in I} X_i$ is compact T_2 and the space is nonvoid.

Let $\underset{\sim}{K}$ be the category of compact T_2 spaces and continuous maps, M the underlying set functor on $\underset{\sim}{K}$. We shall require two results on inverse limits in $\underset{\sim}{K}$: these can be extracted from Section 1.9.6 of Bourbaki [6] and it is not hard to check that the proofs given there are valid in our system.

LEMMA 1.1. Let X_i, d_{ij}, $i \leq j \in D$ be an inverse system of nonvoid spaces in $\underset{\sim}{K}$. Let $X^* = \Pi_{i \in D} X_i$, and X be the set of $x \in X^*$ for which $d_{ij}(x(j)) = x(i)$ whenever $i \leq j$. For $i \in D$, let $d_i : X \longrightarrow X_i$ be given by $d_i(x) = x(i)$. Then X, (d_i) is an inverse limit in $\underset{\sim}{K}$ of the above system, and has the following properties:

(1) X is nonvoid

(2) for each $i \in D$, if d_{ij} is surjective for all $j \geq i$ then d_i is surjective

(3) whenever $f : A \longrightarrow X$ is a morphism such that $d_i f$ is surjective for each $i \in D$, then f is surjective.

From this, the next result is easily deduced.

THEOREM 1.2. With the hypotheses of 1.1, let Y_i, e_{ij}, $i \leq j \in D$ be another inverse system, and let Y, (e_i) be an inverse limit of this. Let $f_i : X_i \longrightarrow Y_i$, $i \in D$ be a compatible family of morphisms (i.e. $e_{ij} f_j = f_i d_{ij}$ whenever $i \leq j$) with limit $f : X \longrightarrow Y$. If all d_{ij} and f_i are surjective, then f is surjective.

2. Compact objects and injectivity

DEFINITION 2.1. A *compact object* over a category $\mathbf{C}$ is a pair $\langle D,T\rangle$ where D is a $\mathbf{C}$-object and $T : \mathbf{C} \longrightarrow \mathbf{K}$ a contravariant functor such that $MT = (-,D)$. The *category* $\mathbf{KC}$ *of* compact objects over $\mathbf{C}$ is given as follows: the objects of $\mathbf{KC}$ are the compact objects $\langle D,T\rangle$ over $\mathbf{C}$, and $\mathbf{u} : \langle D,T\rangle \longrightarrow \langle D',T'\rangle$ is a morphism of $\mathbf{KC}$ if $\mathbf{u} : T \longrightarrow T'$ is a natural transformation. The $\mathbf{KC}$-morphisms will be called *compact morphisms* over $\mathbf{C}$.

Although the number of objects in $\mathbf{KC}$ may be unmanageably large, the Hom-sets of $\mathbf{KC}$ are in fact *sets*. This follows from the following results.

LEMMA 2.2. Let D, D' be $\mathbf{C}$-objects. Then there is a bijection $Y : \mathrm{Hom}((-,D), (-,D')) \longrightarrow (D,D')$.

Proof. See e.g. Corollary 1.7 of [7].

THEOREM 2.3. There is a faithful functor $M' : \mathbf{KC} \longrightarrow \mathbf{C}$.

Proof. M' is defined by $M'(\langle D,T\rangle) = D$ and $M'(\mathbf{u}) = Y(M\mathbf{u})$. M' is faithful by 2.2 and since M is faithful.

The other key notion is that of injectivity.

DEFINITION 2.4. Let $\mathbf{C}$ be a category, $u : A \longrightarrow B$ a morphism of $\mathbf{C}$, D an object of $\mathbf{C}$. We say that D is u-*injective*, or that u is a D-*submorphism*, if $(u,D) : (B,D) \longrightarrow (A,D)$ is surjective. If X is a class of morphisms, D is said to be X-*injective* if D is u-injective for each $u \in X$.

With these two concepts we can now prove the basic result of our investigation.

THEOREM 2.5. (*The Product Theorem*) Let $\mathbf{C}$ be a category admitting products, $u : A \longrightarrow B$ a morphism of $\underset{\sim}{C}$. Let $\langle D_i, T_i \rangle$, $i \in I$ be a family of compact objects over $\underset{\sim}{C}$ such that D_i is u-injective for $i \in I$. Then $\Pi_{i \in I} D_i$ is u-injective.

Proof. Let D, (p_i) be a product of $(D_i : i \in I)$, and let $f \in (A,D)$. Since $(u,D_i) : (B,D_i) \longrightarrow (A,D_i)$ is continuous (with respect to the topologies $T_i(B)$ and $T_i(A)$), $H_i = (u,D_i)^{-1} \{p_i f\}$ is compact, as it is a closed subset of (B,D_i), and nonvoid, as D_i is u-injective. By TPT, there is a member g of $\Pi_{i \in I} H_i$. Let $h \in (B,D)$ be defined by $p_i h = g(i)$. Then $p_i f = g(i)u = p_i hu$, and so $f = hu = (u,D)(h)$.

By similar means we can establish:

THEOREM 2.6. Let $\underset{\sim}{C}$ be a category admitting sums, $\langle D,T \rangle$ a compact object over $\underset{\sim}{C}$. Then any sum of a family of D-submorphisms is a D-submorphism.

While 2.5 has many direct applications, it is necessary to join 2.6 with the following Adjointness Theorem in order to be useful.

THEOREM 2.7. (The Adjointness Theorem) Let $\underset{\sim}{C}$ be a category admitting multiples and powers, X an abstract class of morphisms closed under multiples. Then every power of an X-injective is X-injective.

Proof. Let E be X-injective, I a nonvoid set, $\Pi_I E$, (p_i) an I'th power of E. We shall show that $\Pi_I E$ is X-injective. Let $e : A \longrightarrow B$ be an X-morphism, $f \in (A, \Pi_I E)$. Let $\Sigma_I A$, (u_i) and $\Sigma_I B$, (v_i) be I'th multiples of A and B respectively. Define $m : \Sigma_I A \longrightarrow \Sigma_I B$ by $mu_i = v_i e$. Then m is a multiple of e and so $m \in X$. Let $f' \in (\Sigma_I A, E)$ be given by $f'u_i = p_i f$. As E is m-injective, there is $h' \in (\Sigma_I B, E)$ with $h'm = f'$. Define $h \in (B, \Pi_I E)$ by $p_i h = h'v_i$. Then $p_i f = f'u_i = h'mu_i = h'v_i e = p_i he$, and so $he = f$.

Hence we obtain:

THEOREM 2.8. Let $\underset{\sim}{C}$ be a category admitting multiples and powers, $<D,T>$ a compact object over $\underset{\sim}{C}$, X the class of D-submorphisms. Then every power of an X-injective is X-injective.

The last three results are of course without interest if the Axiom of Choice (AC) is assumed. However, the next is useful even in GB + AC: it deals with the preservation of injectivity under inverse limits, and is in a sense dual to 3.4.

THEOREM 2.9. (The Inverse Limit Theorem) Let $<P,\leq>$ be an up-directed poset, $<D_i,T_i>$, $\underset{\sim}{u}_{ij}$, $i \leq j \in P$ be an inverse system in $\underset{\sim}{KC}$. Let $u_{ij} = M'(\underset{\sim}{u}_{ij})$ for $i \leq j$. Suppose that the inverse system D_i, u_{ij}, $i \leq j \in P$ has a limit D, (u_i) in $\underset{\sim}{C}$. Let $e : A \longrightarrow B$ be a $\underset{\sim}{C}$-morphism. If D_i is e-injective for all $i \in I$, then D is e-injective.

Proof. Let $f \in (A,D)$ and let $H_i = \{g \in (B,D_i) : ge = u_i f\}$. Then H_i is nonvoid as D_i is e-injective. Now let us consider the topologies. H_i is a compact subset of (B,D_i) (with the topology $T_i(B)$), as $(e,D_i) : (B,D_i) \longrightarrow (A,D_i)$ is continuous and $H_i = (e,D_i)^{-1}\{u_i f\}$. For all pairs $i \leq j$, define $d_{ij} : H_j \longrightarrow H_i$ by $d_{ij}(g) = u_{ij}g$ for $g \in H_j$ (note that if $ge = u_j f$ then $u_{ij}ge = u_{ij}u_j f = u_i f$). Now $(B,u_{ij}) : (B,D_j) \longrightarrow (B,D_i)$ is continuous (in the topologies $T_j(B)$ and $T_i(B)$), as $\underset{\sim}{u}_{ij} : T_j \longrightarrow T_i$ is a compact morphism. Thus d_{ij} is continuous, as $d_{ij} = (B,u_{ij})|H_j$. Hence H_i, d_{ij}, $i \leq j \in P$ is an inverse system of nonvoid compact spaces and continuous maps, and so the inverse limit H is nonvoid (where H is given by the standard construction as in 1.1). Let $k \in H$. Then $k(i) \in H_i$ and $d_{ij}(k(j)) = k(i)$ if $i \leq j$. Thus $k(i) \in (B,D_i)$ and $u_{ij}k(j) = k(i)$ if $i \leq j$. Let $h \in (B,D)$ be such that $u_i h = k(i)$. Then $u_i he = k(i)e = u_i f$, and so $he = f$.

3. The Construction Theorems

In this section we show that if D is a compact object, the class of D-submorphisms is closed under certain powerful direct limit operations. We shall express this closure by means of a definition.

DEFINITION 3.1. Let $\underset{\sim}{C}$ be a category, D an object of $\underset{\sim}{C}$. D is called <u>metacompact</u> if whenever A_i, d_{ij}, $i \leqslant j \in P$ is a direct system, with direct limit A, (d_i), then

(1) if $(A_i,D) \neq \emptyset$ for all $i \in P$, then $(A,D) \neq \emptyset$

(2) for each $i \in P$, if d_{ij} is a D-submorphism for all $j \geqslant i$ then d_i is a D-submorphism

(3) if $u : A \longrightarrow B$ is such that ud_i is a D-submorphism for all $i \in P$, then u is a D-submorphism.

THEOREM 3.2. If <D,T> is a compact object then D is metacompact.

<u>Proof</u>. We verify conditions (1) - (3) of Definition 3.1, the notation of which we use. Note first that it is easy to check that $T(A)$, $(T(d_i))$ is the inverse limit in $\underset{\sim}{K}$ of the inverse system $T(A_i)$, $T(d_{ij})$, $i \leqslant j \in P$. Now (1) follows easily from 1.1(1). As regards (2), by hypothesis (d_{ij},D) is surjective for all $j \geqslant i$, and so (d_i,D) is surjective by 1.1(2). Lastly, assume the hypotheses of 3.1(3). Then $(ud_i,D) : (B,D) \longrightarrow (A_i,D)$ is surjective for all $i \in P$. But $(ud_i,D) = (d_i,D)(u,D)$, and so (u,D) is surjective by 1.1(3).

DEFINITION 3.3. Let X be a class of morphisms. A morphism $u : A \longrightarrow B$ is a <u>direct limit</u> of X-<u>morphisms</u> if there are direct systems A_i, d_{ij}, $i \leqslant j \in P$ with limit A, (d_i) and B_i, e_{ij}, $i \leqslant j \in P$ with limit B, (e_i) and a direct system of morphisms $u_i : A_i \longrightarrow B_i$ with limit u with respect to the given direct systems, and u_i, d_{ij}, $e_{ij} \in X$ for all $i, j \in P$ with $i \leqslant j$.

COROLLARY 3.4. Let D be metacompact. Then

(1) the class of A with $(A,D) \neq \emptyset$ is closed under direct limits

(2) the class of D-submorphisms is closed under direct limits.

Proof. (1) follows by 3.1(1). For (2), assume the notation of 3.3. Note that we need only assume that the u_i and e_{ij} are D-submorphisms, as $u_j d_{ij} = u_i$ is a D-submorphism and so d_{ij} is a D-submorphism whenever $i \leq j$. Now e_i is a D-submorphism by 3.1(2), and so $e_i u_i = ud_i$ is. Hence u is, by 3.1(3).

This result is the category-theoretic core of many proofs in Non-Standard Analysis.

4. Examples of compact objects

A. Algebras

We recall that a topological algebra is a pair $\langle A,t\rangle$ where A is an algebra and t is a T_2 topology on the underlying set of A for which the operations of A are continuous. $\langle A,t\rangle$ is a compact topological algebra if in addition t is compact. We shall establish the following result.

THEOREM 4.1A. Let $\underset{\sim}{C}$ be an equational class of (similar) algebras, $\langle A,t\rangle$ a compact topological $\underset{\sim}{C}$-algebra. Then there is a functor $T : \underset{\sim}{C} \longrightarrow \underset{\sim}{K}$ such that $\langle A,T\rangle$ is a compact object over $\underset{\sim}{C}$, and T(E) is homeomorphic to $\langle A,t\rangle$ where E is the free $\underset{\sim}{C}$-algebra on 1 generator.

The proof depends on the following easy lemma.

LEMMA. Let B, A be algebras of $\underset{\sim}{C}$, A admit a T_2 topology, A^B have the product topology. Let $s(u_1,\dots,u_k)$, $s'(u_1,\dots,u_m)$ be terms in the language of $\underset{\sim}{C}$ (with free variables $u_1, u_2, \dots$). Let $x_1, \dots, x_k$, $x_1', \dots, x_m' \in B$. Then the set of $f \in A^B$ with $s_A(f(x_1),\dots,f(x_k)) = s_A'(f(x_1'),\dots,f(x_m'))$ is a closed subset of A^B (where s_A is the interpretation of s in A).

Proof of 4.1A. Let $B \in \underset{\sim}{C}$. Now A^B is compact by TPT. Also (B,A) is a closed subset of A^B, as it is the intersection of a family of sets of the type considered in the Lemma. Thus let $T(B)$ be the compact space (B,A). Now let $f : B \longrightarrow C$ be a $\underset{\sim}{C}$-morphism, and let $f^* = (f,A) : (C,A) \longrightarrow (B,A)$. Since the projections $p_b : A^B \longrightarrow A$, $q_c : A^C \longrightarrow A$ are continuous, and $p_b f^*(g) = p_b(gf) = g(f(b)) = q_{f(b)}(g)$ for $g \in (C,A)$, $p_b f^*$ is continuous for $b \in B$, and so f^* is continuous. Let $T(f)$ be this continuous map. We have now constructed the compact object $\langle A,T\rangle$. Lastly let E have free generator x. Then the projection $(E,A) \longrightarrow A$ onto the x'th coordinate is a continuous bijection and thus a homeomorphism.

THEOREM 4.2A. Let $\langle B,t'\rangle$ be another compact topological algebra in $\underset{\sim}{C}$, $u : A \longrightarrow B$ a continuous homomorphism. Then for any algebra C of $\underset{\sim}{C}$, $(C,u) : (C,A) \longrightarrow (C,B)$ is continuous in the appropriate topologies. Thus u induces a compact morphism $\underset{\sim}{u}$.

Proof. Let $\underset{\sim}{u}(C) = (C,u)$. Now the projections $p_x : (C,A) \longrightarrow A$ and $q_x : (C,B) \longrightarrow B$ are continuous by construction. Since $(q_x \underset{\sim}{u}(C))(f) = q_x(uf) = u(f(x)) = (up_x)(f)$ for $f \in (C,A)$, $q_x \underset{\sim}{u}(C) = up_x$ and so is continuous, for $x \in C$. Hence $\underset{\sim}{u}(C)$ is continuous.

B. Partially ordered sets (posets)

Let $\underset{\sim}{P}$ be the category of posets and order-preserving maps. Now we define a compact topological poset to be a pair $\langle A,t\rangle$ where $A = \langle A_0, \leq_A\rangle$ is a poset and t is a compact T_2 topology on A_0 such that $\leq_A$ is a closed subset of $A_0{}^2$ (with the product topology).

THEOREM 4.1B. Let $\langle A,t\rangle$ be a compact topological poset. Then there is a functor $T : \underset{\sim}{P} \longrightarrow \underset{\sim}{K}$ such that $\langle A,T\rangle$ is a compact object over $\underset{\sim}{P}$ and $T(1)$ is homeomorphic to $\langle A_0,t\rangle$. Further, any continuous morphism between compact topological posets induces a compact morphism.

Proof. Let B be a poset, b, b' $\in$ B. Then $\{f \in A^B : f(b) \leq_A f(b')\}$ is a closed subset of A^B. Hence (B,A) is a closed subset of A^B and so is compact. The rest of the proof is similar to 4.1A and 4.2A.

C. Metric spaces

Let $\underset{\sim}{M}$ be the category of metric spaces and contractions. Every metric space has a natural topology, the metric topology, and a metric space is called compact if this is compact. Then the next result is not surprising.

THEOREM 4.1C. Let A be a compact metric space (with metric d_A). Then there is a functor $T : \underset{\sim}{M} \longrightarrow \underset{\sim}{K}$ such that $\langle A,T\rangle$ is a compact object over $\underset{\sim}{M}$ and T(1) is homeomorphic to A (where 1 is the one-element metric space). Further, any $\underset{\sim}{M}$-morphism between compact metric spaces induces a compact morphism.

Proof. Let B be a metric space, x, x' $\in$ B. Then $\{f \in A^B : d_A(f(x),f(x')) \leq k\}$, where $k = d_B(x,x')$, is a closed subset of A^B. Hence (B,A) is a closed subset of A^B and thus is compact. The rest of the proof is similar to that in Part A.

D. Normed spaces

Let $\underset{\sim}{N}$ be the category of normed linear spaces and linear contractions, R be the real line with the usual norm. The following result is basic to our proof of the Hahn-Banach Theorem.

THEOREM 4.1D. R is a compact object over $\underset{\sim}{N}$.

Proof. Let $\langle A,p\rangle$ be a normed linear space. Let A^* be the set of linear maps $A \longrightarrow R$. Now A^* is a closed subset of R^A, by the Lemma in Part A applied to the topological linear space R, and (A,R) = $A^* \cap \Pi_{x \in A} [-p(x),p(x)]$. It now follows that (A,R) is compact, as the product of closed intervals is compact by TPT. The rest is easy.

E. Measures

Let I be the closed unit interval. A subset J of I is called <u>e-closed</u> if J is closed and 0, 1 $\in$ J. If A is a boolean algebra, a function $m : A \longrightarrow J$ is called a <u>measure</u> if $m(0_A) = 0$, $m(1_A) = 1$, and $m(a \vee b) = m(a) + m(b)$ whenever a, b $\in$ A are such that $a \wedge b = 0_A$. Let $\underset{\sim}{B}$ be the category of nontrivial boolean algebras and homomorphisms, and $\underset{\sim}{B}^*$ the category defined as follows:

(1) A is an object of $\underset{\sim}{B}^*$ if $A \in \underset{\sim}{B}$ or $A \subseteq I$ is e-closed.

(2) for A, B $\in \underset{\sim}{B}^*$, (A,B) is the set of homomorphisms from A to B if A, B $\in \underset{\sim}{B}$, the set of measures from A to B if $A \in \underset{\sim}{B}$ and $B \notin \underset{\sim}{B}$, the set consisting of the inclusion of A in B if $A \subseteq B \subseteq I$ are e-closed, and the null set otherwise.

To see that $\underset{\sim}{B}^*$ is indeed a category it is enough to remark that if $m : A \longrightarrow J$ is a measure and $f : B \longrightarrow A$ a boolean homomorphism then $mf : B \longrightarrow J$ is a measure.

THEOREM 4.1E. Every e-closed set J in $\underset{\sim}{B}^*$ is a compact object.

<u>Proof</u>. J is a compact space. Now let A $\in \underset{\sim}{B}$. Then J^A is a compact space (in the product topology), and the following sets are closed in J^A: $K_0 = \{m \in J^A : m(0_A) = 0\}$, $K_1 = \{m \in J^A : m(1_A) = 1\}$, and $H(x,y) = \{m \in J^A : m(x \vee y) = m(x) + m(y)\}$, for all disjoint x, y $\in$ A. But (A,J) is the intersection of these sets and so is compact. The rest is easy.

F. Graphs

It may be thought more natural in certain cases to consider

compact objects in certain <u>small</u> categories. If these are suitably chosen, our compactness arguments reduce to classical compactness arguments, as we shall see later.

We take a <u>graph</u> to be a pair $\langle G,R\rangle$ where G is a set and R is a symmetric irreflexive relation on G. We can regard a k-<u>colouring</u> of G (where k is a nonzero finite ordinal) as a function $f : G \longrightarrow k$ such that $f(x) \neq f(y)$ whenever xRy. From a graph G and its colourings we construct a category $\underset{\sim}{C}(G)$ as follows:

(1) the objects of $\underset{\sim}{C}(G)$ are G and all finite subgraphs of G, and the nonzero finite ordinals,

(2) the morphisms of $\underset{\sim}{C}(G)$ are all inclusions between graphs and between ordinals, together with for each graph H and ordinal k, all k-colourings of H.

THEOREM 4.1F. Each ordinal k is a compact object of $\underset{\sim}{C}(G)$.

<u>Proof</u>. Let k have the discrete topology, H be a graph in $\underset{\sim}{C}(G)$. Then k^H is compact. Let $F(x,y) = \{f \in k^H : f(x) \neq f(y)\}$, if xRy in H. This is closed in k^H as k is <u>finite</u>. (H,k) is the intersection of all F(x,y), and so is compact.

G. <u>Co-compactness</u>

It is of course possible to dualise Definition 2.1, and the dual notion has natural examples. Thus a <u>co-compact object</u> over $\underset{\sim}{C}$ is a pair $\langle D,T\rangle$ where D is a $\underset{\sim}{C}$-object and $T : \underset{\sim}{C} \longrightarrow \underset{\sim}{K}$ is a covariant functor with $MT = (D,-)$. We shall show that $\underset{\sim}{K}$ itself has a proper class of co-compact objects.

LEMMA 4.3. The underlying set functor M on $\underset{\sim}{K}$ has a left adjoint β.

Proof. The following standard proof is in fact valid in our system. For each set I let βI be the set of ultrafilters on I. Then βI has a natural compact T_2 topology. Let $f : I \longrightarrow X$ be any function, where $X \in \underset{\sim}{K}$. For $D \in \beta I$ let D' be the ultrafilter on X given by $A \in D'$ iff $f^{-1}A \in D$, and let g(D) be the unique limit point in X of D'. This process defines a continuous function $g : \beta I \longrightarrow X$ with $ge_I = f$, where $e_I : I \longrightarrow \beta I$ is the obvious embedding. The other verifications are easy.

THEOREM 4.4. Let I be a nonvoid set. Then βI is a co-compact object over $\underset{\sim}{K}$.

Proof. Let $u : (\beta I,-) \longrightarrow (I,M(-))$ be a natural equivalence. Since $(I,M(A)) = M(A^I)$ let $(\beta I,A)$ have the topology T(A) such that $u(A) : \langle(\beta I,A),T(A)\rangle \longrightarrow A^I$ is a homeomorphism. Clearly $\langle\beta I,T\rangle$ is a co-compact object.

5. Applications of the Construction Theorems

To begin with, let us consider only those categories introduced in A - D of Section 4. In each of these there is a natural notion of injection, defined as follows:

(a) in a variety, u is an injection iff u is a 1-1 homomorphism.

(b) in $\underset{\sim}{P}$, $u : A \longrightarrow B$ is an injection if u is an order-isomorphism of A onto a subset of B.

(c) in $\underset{\sim}{M}$, u is an injection iff u is an isometrical embedding.

(d) in $\underset{\sim}{N}$, u is an injection iff u is a linear isometrical embedding.

An injection $u : A \longrightarrow B$ is called small if A and B are finitely generated. An object C is called injective if it is u-injective for all injections u, and a local injective if it is u-injective for all small injections u.

The terms 'injection' and 'injective' are from the Theory of Bicategories as outlined in [19] - the above all form bicategories with the prescribed notion of injection.

Let $\underset{\sim}{C}$ be one of the above types of category. Let X be the class of small injections, S the class of finitely generated objects. Then the following two lemmas hold.

LEMMA 5.1. Every object is a direct limit of S-objects.

Proof. Let A be an object, D the set of finite subsets of A, ordered by inclusion. For $d \in D$ let A_d be the substructure of A generated by d. Then A is the directed union of $(A_d : d \in D)$ and so clearly the direct limit of this family with the obvious inclusion morphisms.

Note that 5.1 clearly applies also to the category in Part F of Section 4.

LEMMA 5.2. Every injection is a direct limit of X-injections.

Proof. Let $u : A \longrightarrow B$ be an injection. Without loss of generality we can assume that u is in fact an inclusion. Let D be the set of finite subsets of B, ordered by inclusion. For $d \in D$ let B_d be the substructure of B generated by d and A_d the substructure of A generated by $d \cap A$. Then u is the direct limit of the family of inclusions $A_d \longrightarrow B_d$, $d \in D$.

By applying 5.1 and 5.2 to (1) and (2) of 3.4 respectively we obtain the following two Compactness Theorems.

THEOREM 5.3. Let D be a metacompact object such that (A,D) is nonvoid for all finitely generated A. Then (A,D) is nonvoid for all A.

THEOREM 5.4. A metacompact local injective is injective.

EXAMPLES 5.5. As applications of 5.3 we have the following:

(1) Prime Ideal Theorem. For consider the variety $\underset{\sim}{B}$ of boolean algebras. Now an algebra B has an ultrafilter iff (B,2) is nonvoid, and 2 is compact. The proof is completed by noting that every finitely generated algebra is finite and so atomic, and thus has an ultrafilter.

(2) Prime Filter Theorem for distributive lattices. For consider the variety $\underset{\sim}{E}$ of distributive lattices with 0 and 1. Now an algebra L has a prime filter iff (L,2) is nonvoid, and 2 is compact. Lastly note that every finitely generated algebra is finite and so contains a meet-irreducible element, and the principal filter generated by this is prime (see Chapter 9 of [5] for details).

(3) Graph Colouring Theorem. If every finite subgraph of a graph G is k-colourable then G is k-colourable. For consider the category $\underset{\sim}{C}(G)$ of 4.F: the result then follows as k is metacompact.

We note that (2) was first proved in Scott [18] using a metamathematical approach, and that (3) was proved in Luxemburg [15] by means of ultrapowers. It is interesting to observe that if our method of proving (3) is translated into concrete terms the classical proof of (3) by means of TPT is obtained.

EXAMPLES 5.6. We now discuss some applications of 5.4.

(1) Let $\underset{\sim}{B}$ be the variety of boolean algebras. Now every algebra of $\underset{\sim}{B}$ is a local injective. For if $u : A \longrightarrow B$ is a small inclusion we may assume by induction that A is a maximal proper subalgebra of B: then some atom of A splits into two atoms of B and it is then clear how to extend a morphism on A to one on B. Hence any compact algebra, and in particular 2, is injective.

(2) Let $\underset{\sim}{D}$ be the variety of distributive lattices. Following [4], it is easy to check that 2 is injective in $\underset{\sim}{D}$ iff Stone's theorem

(see [22]) on the separation of an ideal and filter is true. But 2 is compact and Stone's theorem clearly holds for finite algebras (by a simple combinatorial argument). Hence 2 is injective in $\underset{\sim}{D}$.

(3) Let $\underset{\sim}{A}$ be the variety of abelian groups. It is easy to see that a group G is a local injective iff G is divisible. Hence any compact divisible group, and in particular the compact circle group, is injective in $\underset{\sim}{A}$.

(4) Let us consider posets. A poset P is said to have the ω-Interpolation Property (ω-IP) if whenever X, Y are finite subsets of P such that $X \leq Y$, there is $z \in P$ with $X \leq z \leq Y$. By considering cuts and simple extensions, it is easy to see that P is a local injective iff P has ω-IP, and that an injective poset is complete. Hence any compact poset with ω-IP is injective and so complete. This is related to the classical result that a lattice compact in its interval topology is complete (see Section IV.8 of [5]). For a lattice clearly has ω-IP and if $\langle P,t \rangle$ is a compact topological poset then t extends the interval topology.

(5) As regards $\underset{\sim}{M}$, it was proved in Theorems 2.2 and 2.3 of [1] that a metric space is a local injective iff it is ω-hyperconvex. Clearly the real line R is ω-hyperconvex and so the closed unit interval I is. Thus I is injective, as it is compact.

(6) The Hahn-Banach Theorem. By results of Banach [2], R is a local injective in $\underset{\sim}{N}$. As R is compact by 4.1D, it follows that R is injective.

(7) Tarski's Measure Extension Lemma: If B is a boolean algebra, A a subalgebra of B, $m : A \longrightarrow R$ a (finitely additive) measure on A, then there is a measure $m' : B \longrightarrow R$ with the range of m' contained in the closure of the range of m. In the category $\underset{\sim}{B}^*$ of 4.E let V be the class of embeddings between boolean algebras. Now by normalising all

measures, it is clear that the above Lemma is equivalent to: every e-closed interval J is V-injective. If X is the set of V-morphisms with finite domain and codomain, it is clear from the proof of 5.2 that every V-morphism is a direct limit of X-morphisms. But it is easy to check that J is X-injective (see (1) above). Now J is compact, and so V-injective.

Other applications of these techniques are to the Ordering Principle (see [12]) and to Kelley's Measure Extension Lemma (see [11]). Other proofs of (1) can be found in [16], of (6) in [12] and [14], and of (7) in [13] and [21]. In practice, (4) is not very useful as there is a direct proof in [3] that every complete poset is injective.

6. Applications of the Product Theorems

It follows from the Axiom of Choice that every complete boolean algebra is injective in $\underset{\sim}{B}$ (see 33.1 of [20]). However, it is not known whether this result can be proved from the Prime Ideal Theorem. Thus it is natural to ask what closure conditions the class of injective boolean algebras satisfies, and we shall give an answer by proving the following result.

THEOREM 6.1. The class X of injective boolean algebras has the following properties.

(1) $2 \in X$

(2) X is closed under retracts

(3) X is closed under powers

(4) X is closed under restrictions.

Proof. (1) follows from 5.6 (1), and (2) is an easy argument (it is 4.1 of Semadeni [19]). However, (3) and (4) are harder to establish. It is easy to see that u is a 2-submorphism iff u is an injection. Also note that $\underset{\sim}{B}$ has sums - this follows from the Stone duality as explained in Halmos [10]. Then we can apply 2.8 (with X the class of 2-submorphisms) to establish (3). For (4) we consider the variety $\underset{\sim}{D}$. Let $A \in X$, $a \in A$ be nonzero. Then the algebra A/a with underlying set $\{x \in A : x \leq a\}$, sup and inf as in A, and complement relative to a, is called the <u>restriction</u> of A by a. Now A/a is a retract of A in $\underset{\sim}{D}$, and A is a retract of some 2^I in $\underset{\sim}{B}$ and so in $\underset{\sim}{D}$. But 2^I is injective in $\underset{\sim}{D}$, by 5.6 (2) and 2.5. Hence A/a is injective in $\underset{\sim}{D}$ and so in $\underset{\sim}{B}$.

A direct application of the <u>dual</u> of the Product Theorem gives the following result.

THEOREM 6.2. In $\underset{\sim}{K}$, every space βI is projective.

<u>Proof</u>. Let 1 denote the 1-point space. Then 1 is clearly projective. Now βI is the sum of I copies of 1, and 1 is co-compact, so that βI is projective by the dual of 2.5.

7. <u>Concluding remarks</u>

The techniques developed here can also be applied to <u>equational compactness</u> (see [17] for the definition), in view of the following result.

THEOREM 7.1. Let $\underset{\sim}{C}$ be an equational class (or the category of posets). Then any equationally compact object is metacompact.

Finally we remark that our techniques do not seem applicable to $\underset{\sim}{K}$ itself, as we can establish:

THEOREM 7.2. Assume the Axiom of Choice, and let I be the closed unit interval. Then I is not metacompact.

<u>Proof</u>. By 11.2 of [19], u is an injection iff u is an I-submorphism. Now consider the direct system $1 \to 2 \to \ldots$ with inclusion maps,

and note that $\beta\omega$ is the direct limit. Let ω^* be the 1-point compactification of ω with the discrete topology, $u_n : n \longrightarrow \omega^*$ be the obvious inclusion for $n < \omega$. But the direct limit $u : \beta\omega \longrightarrow \omega^*$ of the system $(u_n : n < \omega)$ cannot be an injection. This contradicts 3.1(3) and so I is not metacompact.

REFERENCES

[1] N. Aronszajn and P. Panitchpakdi, Extension of uniformly continuous transformations and hyperconvex metric spaces, Pac. J. Math. 6 (1956) 405-439.

[2] S. Banach, Théorie des opérations linéaires, Warsaw (1932).

[3] B. Banaschewski and G. Bruns, Categorical characterisation of the MacNeille completion, Archiv der Math. Basel 18 (1967) 369-377.

[4] ---- and ---- , Injective hulls in the category of distributive lattices, J.f. reine u. ange. Math. 232 (1968) 102-109.

[5] G. Birkhoff, Lattice Theory (2nd edition), Amer. Math. Soc. Colloq. Pub. 25 (1948).

[6] N. Bourbaki, Topologie Générale, Hermann, Paris (1961).

[7] I. Bucur and A. Deleanu, Introduction to the theory of categories and functors, Wiley, London (1968).

[8] P. Freyd, Abelian categories, Harper and Row, New York (1966).

[9] K. Gödel, The consistency of the Continuum Hypothesis, Annals of Math. Studies 3 (1940).

[10] P. Halmos, Lectures on Boolean algebras, Van Nostrand, Princeton (1963).

[11] J. Kelley, Measures in boolean algebras, Pac. J. Math. 9 (1959) 1165-1177.

[12] J. Łoś and C. Ryll-Nardzewski, On the application of Tychonoff's theorem in mathematical proofs, Fund. Math. 38 (1951) 233-237.

[13] ---- and ---- , Effectiveness of the representation theorem for Boolean algebras, Fund. Math. 41 (1954) 49-56.

[14] W. Luxemburg, Two applications of the method of construction by ultrapowers to analysis, Bull. Amer. Math. Soc. 68 (1962) 416-419.

[15] ---- , A remark on a paper by N. G. de Bruijn and P. Erdos, Nederl. Akad. Wetensch. Proc. Ser. A 65 (1962) 343-345.

[16] W. Luxemburg, A remark on Sikorski's extension theorem for homomorphisms in the theory of Boolean algebras, Fund. Math. 55 (1964) 239-247.

[17] J. Mycielski, Some compactifications of general algebras, Coll. Math. 13 (1964) 1-9.

[18] D. Scott, Prime Ideal Theorems for rings, lattices, and boolean algebras (abstract), Bull. Amer. Math. Soc. 60 (1954) 388.

[19] Z. Semadeni, Projectivity, injectivity, and duality, Rozprawy Matematiczne 35 (1963).

[20] R. Sikorski, Boolean algebras (2nd edition), Springer-Verlag, Berlin (1964).

[21] J. Staples, A Non-Standard representation of Boolean algebras, and applications, Bull. Lond. Math. Soc. 1 (1969) 315-320.

[22] M. Stone, Topological representation of distributive lattices and Brouwerian logics, Casopis Pest. Mat. 67 (1937) 1-25.

NON-AXIOMATIZABILITY RESULTS IN INFINITARY LANGUAGES FOR HIGHER-ORDER STRUCTURES

J. C. Cole and M. A. Dickmann

Matematisk Institut, Aarhus

This paper clarifies to some extent the question of whether infinitary quantifications are first or second order notions. The guiding principle seems to be that a second-order quantifier which need only range over sets of some bounded cardinality is properly an infinitary first-order notion. For example, the topological axiom that an arbitrary union of open sets is open is strictly second-order when one is considering all spaces, but in separable spaces, this quantifier over collections of open sets need only range over subsets of the countable base, so we obtain an axiomatization in $L_{\omega_1\omega_1}$. This principle may be viewed as a generalization of Kleene's Normal Form Theorem which tells us that a quantifier over, say, recursive sets can be reduced to a number quantifier.

We show that many classes of second-order structures are not axiomatizable in any $L_{\kappa\lambda}$, for example topological spaces, compact spaces, complete Boolean algebras, and complete linear orderings. The method used in all these cases was inspired by a paper of Kopperman [1]. One supposes that the class is an $RPC_\Delta(L_{\kappa\lambda})$ for some κ, λ. One chooses a structure of this class of sufficiently large cardinality with the property that it is "generated" by a set of smaller cardinality. Applying the Downward Löwenheim-Skolem Theorem for $L_{\kappa\lambda}$ quickly gives a contradiction. It is interesting to remark that in each case the "generating" process is by means of the power-set operation - the only properly non-first-order notion known to us.

For a pair of cardinals, κ, λ, with $\kappa \geq \lambda$, the language $L_{\kappa\lambda}(\mu)$ is the first-order language of similarity type μ obtained by allowing conjunctions (disjunctions) of any set of formulae of cardinality $< \kappa$, and allowing simultaneous universal (existential) quantifications of any set of variables of cardinality $< \lambda$. Thus the usual (finitary) first-order language in this notation is $L_{\omega\omega}$.

A class $\mathbb{K}$ of structures of similarity type μ is an $RPC_{\Delta}(L_{\kappa\lambda}(\mu))$-class if there exist

> a similarity type $\nu \supseteq \mu$,
> a set Σ of sentences of $L_{\kappa\lambda}(\nu)$, and
> a formula ϕ of $L_{\kappa\lambda}(\nu)$ with one free variable,
> such that $\mathfrak{A} \in \mathbb{K}$ iff there exists $\mathfrak{A}' \in Mod_{\nu}(\Sigma)$
> such that $\mathfrak{A} = (\mathfrak{A}' \restriction \phi^{\mathfrak{A}'}) \restriction \mu$.

Thus an $RPC_{\Delta}(L_{\kappa\lambda})$-class is obtained from the class of models of some theory in $L_{\kappa\lambda}$, by cutting down the domain of each model to the extension of a fixed formula, and then forgetting some of the relation and function symbols of the similarity type. Then this notion is the most general known notion of the axiomatizability of a class of structures.

If $\mathfrak{A}$, $\mathfrak{B}$ are structures of similarity type μ, we say that $\mathfrak{B}$ is a κ-λ-elementary substructure of $\mathfrak{A}$ ($\mathfrak{B} \prec_{\kappa\lambda} \mathfrak{A}$) iff $\mathfrak{B}$ is a substructure of $\mathfrak{A}$ such that for every formula ϕ of $L_{\kappa\lambda}(\mu)$, with a set $Fv(\phi)$ of free variables, and for every $f \in |\mathfrak{B}|^{Fv(\phi)}$ we have

$$\mathfrak{B} \models \phi[f] \quad \text{iff} \quad \mathfrak{A} \models \phi[f].$$

We obtain the <u>Downward Löwenheim-Skolem Theorem</u> for $L_{\kappa\lambda}$:

If $\mathfrak{A}$ is an infinite structure of similarity type μ, $X \subseteq |\mathfrak{A}|$, and τ is a cardinal such that

$$\max\{\overline{\overline{X}},\overline{\overline{\mu}}\} \leq \tau = \tau^{\kappa} \leq \overline{\overline{\mathfrak{A}}},$$

then there is a structure $\mathfrak{B}$ such that

$$\mathfrak{B} \prec_{\kappa\lambda} \mathfrak{A},\ X \subseteq |\mathfrak{B}| \text{ and } \overline{\overline{\mathfrak{B}}} = \tau.$$

Note that there are arbitrarily large τ for which $\tau = \tau^{\kappa}$, for example any τ of the form α^{κ} for some cardinal α.

We begin by proving a slight generalization of Kopperman's result [1].

We shall say that $\mathbb{K}$ is a class of topological spaces (with property P) if $\mathbb{K}$ is a class closed under isomorphism of structures of similarity type $\{Pt, \mathcal{O}, E\}$, of form $\langle X \cup T, X, T, E \rangle$ each of which is isomorphic to a structure $\langle Y \cup \mathcal{O}(Y), Y, \mathcal{O}(Y), \epsilon_Y \rangle$ where $\mathcal{O}(Y)$ is a topology on Y and ϵ_Y is the standard membership relation (and with this topology the space Y has property P). For example, in this definition, the class of all discrete spaces is the class of all structures isomorphic to

$$\langle X \cup \mathcal{P}(X), X, \mathcal{P}(X), \epsilon_X \rangle$$

for some set X.

An examination of Kopperman's proof shows that he actually obtained the result:

THEOREM 1. Let $\mathbb{K}$ be a class of topological spaces which contains discrete spaces of arbitrarily large cardinality. Then $\mathbb{K}$ is not an $RPC_{\Delta}(L_{\kappa\lambda})$ for any κ, λ.

Proof. Suppose $\mathbb{K} \in RPC_{\Delta}(L_{\kappa\lambda})$ for some κ, λ, $\kappa \geq \lambda$. Then there are $\mu \supseteq \nu$, $\Sigma \subseteq Sent(L_{\kappa\lambda}(\mu))$ and $\phi \in Form_1(L_{\kappa\lambda}(\mu))$ such that $\mathbb{K} = (Mod_{\mu}(\Sigma))^{\phi} \upharpoonright \nu$; i.e., for every structure $\mathfrak{A}$ of type ν:

$$(*) \qquad \mathfrak{A} \in \mathbb{K} \iff \text{there is } \mathfrak{A}' \text{ of type } \mu \text{ such that } \mathfrak{A}' \in Mod_{\mu}(\Sigma) \text{ and } \mathfrak{A} = (\mathfrak{A}' \upharpoonright \phi^{\mathfrak{A}'}) \upharpoonright \nu.$$

Let τ be a cardinal such that $\tau \geq \overline{\overline{\mu}}$ and $\tau^{\kappa} = \tau$ (e.g., $\tau = 2^{\alpha}$, where $\alpha = \max\{\overline{\overline{\mu}}, \kappa\}$). Let $\mathfrak{A} = \langle Y \cup \mathcal{P}(Y), Y, \mathcal{P}(Y), \epsilon \rangle$ be a discrete space of cardinality $\geq \tau$ belonging to $\mathbb{K}$; such $\mathfrak{A}$ exists by hypothesis. Also, we demand that $Y \cap \mathcal{P}(Y) = \emptyset$.

By (*) there is $\mathfrak{A}' \in Mod(\Sigma)$ such that $\mathfrak{A} = (\mathfrak{A}' \upharpoonright \phi^{\mathfrak{A}'}) \upharpoonright \nu$. Let $Y' \subseteq Y$ be such that $\overline{\overline{Y'}} = \tau$, and let $X = Y' \cup \{\{y\} \mid y \in Y'\}$; then $\overline{\overline{X}} = \tau$. By the Downward Löwenheim-Skolem theorem there is $\mathfrak{B}' \prec_{\kappa\lambda} \mathfrak{A}'$ such that $X \subseteq |\mathfrak{B}'|$ and $\overline{\overline{\mathfrak{B}'}} = \tau$.

Therefore $\mathfrak{B}' \in Mod(\Sigma)$, and $\mathfrak{B} = (\mathfrak{B}' \upharpoonright \phi^{\mathfrak{B}'}) \upharpoonright \nu \in \mathbb{K}$; then $\mathfrak{B} \upharpoonright \{\mathcal{O}, E\} \cong \langle Z, \mathcal{T}, \epsilon \rangle$ for some set Z and topology $\mathcal{T}$ on Z. Clearly we can (and do) assume $Z \subseteq Pt^{\mathfrak{B}}$, $\mathcal{T} \subseteq \mathcal{O}^{\mathfrak{B}}$ and $Z \cup \mathcal{T} = |\mathfrak{B}|$.

$\mathfrak{B}' \prec_{\kappa\lambda} \mathfrak{A}'$ implies $\phi^{\mathfrak{B}'} = \phi^{\mathfrak{A}'} \cap |\mathfrak{B}'|$; hence $X \subseteq \phi^{\mathfrak{B}'} = |\mathfrak{B}|$ and we get: $Y' \subseteq Z$ and $\{\{y\} \mid y \in Y'\} \subseteq \mathcal{T}$. Since $\mathcal{T}$ is a topology, it is closed under arbitrary unions of its members sets; then $\mathcal{P}(Y') \subseteq \mathcal{T}$; hence

$$\overline{\overline{\mathfrak{B}'}} \geq \overline{\overline{\mathfrak{B}}} = \overline{\overline{Z \cup \mathcal{T}}} \geq 2^{\tau},$$

contradicting the choice of $\mathfrak{B}'$. This contradiction shows that $\mathbb{K} \notin RPC_{\Delta}(L_{\kappa\lambda})$.

For topological algebraic structures, we can easily generalize the proof. By a class of <u>topological universal algebras</u>, we understand a class of structures of similarity type $\{Pt, \mathcal{O}, E\} \cup F$, where F is a collection of finitary function symbols, such that the reduct of each such structure is a topological space in the above formulation, and the functions of each structure are continuous and make the underlying set of the space into a universal algebra of a fixed class. Examples are given by topological groups, rings, modules, etc. Since any universal algebra can be given the discrete topology, we obtain by a slight modification the

COROLLARY TO PROOF. Let $\mathbb{K}$ be a class of topological universal algebras containing discrete algebras of arbitrarily large cardinality. Then $\mathbb{K}$ is not an $RPC_{\Delta}(L_{\kappa\lambda})$ for any κ, λ.

Since infinite discrete spaces are not compact, an immediate question arises: are classes of compact spaces non-axiomatizable? We use the same general method, letting the Stone space of a power-set algebra play the crucial role. We recall some facts about Boolean algebras and their Stone spaces.

(i) A Stone space is a compact Hausdorff space with a base of clopen sets.

(ii) A Boolean algebra has a unique Stone space associated to it, namely the set of ultrafilters, with the product topology. The algebra of clopen sets of this space is isomorphic to the given Boolean algebra.

(iii) A Boolean algebra is complete iff its Stone space is extremally disconnected (i.e. the closure of an open set is open).

(iv) A Boolean algebra is atomic iff the set of isolated points of the Stone space is dense.

Thus a Boolean algebra is isomorphic to a power-set algebra iff its Stone space satisfies (iii) and (iv).

(v) The power-set algebra of a set of cardinality τ has 2^{2^τ} ultrafilters (Tarski and others). Hence for each τ, there is a Stone space of cardinality 2^{2^τ} with clopen base of cardinality 2^τ.

THEOREM 2. Let $\mathbb{K}$ be a class of compact topological spaces, containing Stone spaces of arbitrarily large power-set algebras. Then $\mathbb{K}$ is not an $RPC_\Delta(L_{\kappa\lambda})$ for any κ, λ.

Proof. Assume $\mathbb{K} \in RPC_\Delta(L_{\kappa\lambda})$ for some κ, λ, $\kappa \geq \lambda$. With the same notation as in the proof of Theorem 1, let $\mu \supseteq \nu$, $\Sigma \subseteq Sent(L_{\kappa\lambda}(\mu))$ and $\phi \in Form_1(L_{\kappa\lambda}(\mu))$ be the objects that make $\mathbb{K}$ an $RPC_\Delta(L_{\kappa\lambda})$.

By assumption there is a $\tau \geq \max\{\kappa, \bar{\bar{\mu}}\}$ such that the Stone space $\mathfrak{A}$ of the Boolean algebra $\mathcal{P}(\tau)$ is in $\mathbb{K}$. By the remarks above, $\bar{\bar{\mathfrak{A}}} = 2^{2^\tau}$ and $\mathfrak{A}$ has a base $\mathcal{Y}$ of clopen sets; also $\bar{\bar{\mathcal{Y}}} = 2^\tau$ (even more, $\mathcal{Y}$ with set-theoretic operations is a Boolean algebra isomorphic to $\mathcal{P}(\tau)$). By hypothesis $\mathfrak{A} = (\mathfrak{A}' \upharpoonright \phi^{\mathfrak{A}'}) \upharpoonright \nu$ for some $\mathfrak{A}' \in Mod_\mu(\Sigma)$. Enlarge $\mathfrak{A}'$ to $\mathfrak{A}''$ by adding a predicate Clb whose extension is $\mathcal{Y}$; i.e., $\mathfrak{A}'' = \langle \mathfrak{A}', \mathcal{Y} \rangle$. The following hold in $\mathfrak{A}''$:

(i) $\forall v[Clb(v) \rightarrow \mathcal{O}(v)]$ (every member of Clb is open);

(ii) $\forall v_0[Clb(v_0) \rightarrow \exists v_1(\mathcal{O}(v_1) \wedge \forall z(zEv_1 \leftrightarrow \neg z \epsilon v_0))]$ (the complement of every member of Clb is open);

(iii) $\forall v[Clb(v) \rightarrow \phi(v)]$;

(iv) $\forall v_0 v_1[\phi(v_0) \wedge \phi(v_1) \wedge v_0 E v_1 \rightarrow \exists v_2[Clb(v_2) \wedge v_0 E v_2 \wedge \forall z(zEv_2 \rightarrow zEv_1)]]$ (Clb is a base);

(v) $\forall v_0 v_1[Pt(v_0) \wedge Pt(v_1) \wedge \phi(v_0) \wedge \phi(v_1) \rightarrow \exists w_0 w_1[\mathcal{O}(w_0) \wedge \mathcal{O}(w_1) \wedge \phi(w_0) \wedge \phi(w_1) \wedge v_0 E w_0 \wedge v_1 E w_1 \wedge \neg \exists z(zEw_0 \wedge zEw_1)]]$ ($\mathfrak{A}$ is a Hausdorff space).

We observe that $(2^\tau)^\kappa = 2^\tau$. By the Downward Löwenheim-Skolem theorem, there is $\mathfrak{B}'' \prec_{\kappa\lambda} \mathfrak{A}''$ such that $\mathcal{Y} \subseteq |\mathfrak{B}''|$ and $\bar{\bar{\mathfrak{B}''}} = 2^\tau$.

Let $\mathfrak{B}' = \mathfrak{B}''\restriction\mu$ and $\mathfrak{B} = (\mathfrak{B}'\restriction\phi^{\mathfrak{B}'})\restriction\nu$. Then $\mathfrak{B} \in \mathbb{K}$ and we can assume that it is a compact space. By (v) $\mathfrak{B}$ is also Hausdorff. Let $\mathcal{X} = \mathrm{Clb}^{\mathfrak{B}''}$; by (iii), $\mathcal{X} \subseteq \phi^{\mathfrak{B}'}$; by (i), (ii) and (iv), $\mathcal{X}$ is a clopen base for $\mathfrak{B}$. Since $\mathfrak{B}'' \prec_{\kappa\lambda} \mathfrak{A}''$, for every $a \in |\mathfrak{B}''|$:

$$\mathfrak{B}'' \models \mathrm{Clb}[a] \iff \mathfrak{A}'' \models \mathrm{Clb}[a];$$

therefore:

$$\mathcal{X} = \mathrm{Clb}^{\mathfrak{B}''} = \mathrm{Clb}^{\mathfrak{A}''} \cap |\mathfrak{B}''| = \mathcal{Y}.$$

Conclusion: $\mathfrak{B}$ is a Stone space with $\mathcal{Y}$ as a clopen base. We now prove:

<u>Claim</u>: $\overline{\overline{\mathfrak{B}}} \geq 2^{2^{\tau}}$.

By the remarks preceding the theorem, $\mathfrak{B}$ is the Stone space of the Boolean algebra $\mathcal{A}$ of its clopen subsets with the usual set-theoretic operations as Boolean operations. Moreover, $\mathcal{Y} \subseteq \mathcal{A}$ (as sets) and $\mathcal{Y}$ is isomorphic to the power-set algebra $\mathcal{P}(\tau)$.

First we show that $\mathcal{Y}$ is a Boolean subalgebra of $\mathcal{A}$; the verification of this fact is a matter of routine except, possibly, for the complement operation; thus we prove:

$$(*) \qquad -_{\mathcal{Y}}\, x = -_{\mathcal{A}}\, x \quad \text{for every } x \in \mathcal{Y}.$$

Let $\psi(v_0, v_1)$ be the formula: $\forall w[wEv_0 \leftrightarrow \neg(wEv_1)]$. Then

$$\mathfrak{A}'' \models \psi[x, -_{\mathcal{Y}}\, x]$$

and

$$\mathfrak{B}'' \models \psi[x, -_{\mathcal{A}}\, x].$$

Since x, $-_{\mathcal{Y}}\, x \in \mathcal{Y} \subseteq |\mathcal{L}''|$, we get:

$$\mathcal{L}'' \models \psi[x, -_{\mathcal{Y}}\, x];$$

on the other hand we know:

$$\mathcal{L}'' \models \forall v_0 v_1 v_2 [\psi(v_0, v_1) \wedge \psi(v_0, v_2) \longrightarrow v_1 \approx v_2];$$

hence (*) holds; so $\mathcal{Y}$ is a Boolean subalgebra of $\mathcal{A}$.

Each ultrafilter $\mathcal{F}$ of $\mathcal{Y}$ is a subset of $\mathcal{A}$ with the finite intersection property; let * be a map choosing an ultrafilter $\mathcal{F}^*$ of $\mathcal{A}$ containing $\mathcal{F}$.

We show that * is a one-one map. If $\mathcal{F}, \mathcal{G}$ are ultrafilters on $\mathcal{Y}$, $\mathcal{F} \neq \mathcal{G}$, then there is $x \in \mathcal{Y}$ such that $x \in \mathcal{F}$ and $-_{\mathcal{Y}}\, x \in \mathcal{G}$; therefore $x \in \mathcal{F}^*$ and $-_{\mathcal{Y}}\, x = -_{\mathcal{A}}\, x \in \mathcal{G}^*$; hence $\mathcal{F}^* \neq \mathcal{G}^*$.

Since $\mathcal{L}$ is (homeomorphic to) the set of all ultrafilters of $\mathcal{A}$ and $\mathcal{Y}$ has 2^{2^τ} distinct ultrafilters, we conclude that $\overline{\overline{\mathcal{L}}} \geqslant 2^{2^\tau}$; this proves the claim.

Hence we have on the one hand $\overline{\overline{\mathcal{L}}} \leqslant 2^\tau$, by Downward Löwenheim-Skolem theorem, and on the other $\overline{\overline{\mathcal{L}}} \geqslant 2^{2^\tau}$. Thus we have reached a contradiction, which proves that $\mathbb{K}$ cannot be an $RPC_\Delta(L_{\kappa\lambda})$.

Theorems 1 and 2 cover a very wide collection of classes of spaces, including metrizable spaces, contrary to a remark of Kopperman [1], p. 266. This latter seems to contradict the $L_{\omega_1\omega}$ axiomatization of metric spaces. But the metric topology of a space is radically different from the metric, a first-order notion. What these results do show is that the metric topology is not (first-order) definable from the metric in any $L_{\kappa\lambda}$.

We turn now to another widely occurring class of second-order structures. By a complete partial ordering we mean a partial order $\langle A,\leq\rangle$, such that any linearly ordered subset with an upper bound (resp. lower bound) has a supremum (resp. infimum). The class of complete partial orderings is extremely wide: it contains complete linear orderings, complete lattices, complete Boolean algebras. We give two criteria for non-axiomatizability of such classes. In the first, we use the inclusion ordering of a power set as the special object. In the second, we use a certain linear ordering.

THEOREM 3. Let $\mathbb{K}$ be a class of complete partial orderings containing orderings isomorphic to $\langle \mathcal{P}(\alpha),\subseteq\rangle$ for arbitrarily high cardinals, α. Then $\mathbb{K}$ is not an $RPC_{\Delta}(L_{\kappa\lambda})$ for any κ, λ.

Proof. (In sketch) Suppose $\mathbb{K} \in RPC_{\Delta}(L_{\kappa\lambda})$ for some κ, λ, $\kappa \geq \lambda$. Let μ, Σ, ϕ have the same meaning as in Theorem 1. Choose τ as in Theorem 1. By assumption there is $\alpha \geq \tau$ such that $\langle \mathcal{P}(\alpha),\subseteq\rangle \in \mathbb{K}$; hence $\langle \mathcal{P}(\alpha),\subseteq\rangle = (\mathfrak{A}' \restriction \phi^{\mathfrak{A}'}) \restriction \{\leq\}$ for some $\mathfrak{A}' \in Mod_{\mu}(\Sigma)$.

Let X be a set of atoms of $\mathcal{P}(\alpha)$ of power τ. Notice that the set of all atoms is defined by the following formula $\psi(v_0)$:

$$\forall v_1[(\phi(v_1) \wedge v_1 \leq v_0) \rightarrow (v_1 \approx v_0 \vee (\neg(v_1 \approx v_0) \wedge \forall w(\phi(w) \rightarrow v_1 \leq w)))].$$

By the Downward Löwenheim-Skolem Theorem there is $\mathfrak{B}' \prec_{\kappa\lambda} \mathfrak{A}'$ such that $X \subseteq |\mathfrak{B}'|$ and $\overline{\overline{\mathfrak{B}'}} = \tau$. Then:

$$\text{(i)}\quad \mathfrak{B}' \in Mod_{\mu}(\Sigma);$$
$$\text{(ii)}\quad X \subseteq \psi^{\mathfrak{B}'} \subseteq \psi^{\mathfrak{A}'}.$$

Putting these together we have:

$$\mathfrak{B} = (\mathfrak{B}' \restriction \phi^{\mathfrak{B}'}) \restriction \{\leq\} \in \mathbb{K}$$

and $x \in X \Rightarrow \mathcal{B} \models$ "x is an atom".

Therefore $\mathcal{B}$ is a complete partially ordered structure with a set of atoms of power $\geq \tau$. Notice that the Boolean algebra axioms can be written in terms of $\leq$; hence, from $\mathcal{B}' \prec_{\kappa\lambda} \mathfrak{A}'$ we also get:

(iii) $\mathcal{B}$ is a Boolean algebra.

By well-ordering the set, X, of atoms, one can express each subset of X as the supremum of a linearly ordered subset of $\mathcal{B}$, and hence show that $\mathcal{B}$ contains a subalgebra isomorphic to $\mathcal{P}(X)$; hence $\overline{\overline{\mathcal{B}}} \geq 2^{\tau}$, contradicting the choice of $\mathcal{B}'$.

A precisely similar argument for complete lattices, in the similarity type $\{\wedge, \vee\}$, gives a similar result. We give a slight generalization, easily proved by noting that a complete Boolean algebra contains a subalgebra isomorphic to the power-set algebra of any set of atoms.

COROLLARY TO PROOF. Let $\mathbb{K}$ be a class of complete lattices containing Boolean algebras with sets of atoms of arbitrarily high cardinality. Then $\mathbb{K}$ is not an $RPC_{\Delta}(L_{\kappa\lambda})$ for any κ, λ.

The second criterion is proved using the following lemma:

LEMMA. (Implicitly contained in Sierpiński [3], [4].) For any limit ordinal, ξ, there is a linearly ordered set of cardinality $\beth_{\xi}$, whose order-completion has cardinality $\beth_{\xi+1}$. (The cardinals $\beth_{\alpha}$ are defined by $\beth_0 = \aleph_0$, $\beth_{\alpha+1} = 2^{\beth_{\alpha}}$, and for limit ξ, $\beth_{\xi} = \Sigma_{\alpha<\xi} \beth_{\alpha}$).

<u>Proof</u>. For any cardinal κ, let U_{κ} be the lexicographically ordered set of sequences of 0's and 1's, of length κ. Then $\overline{\overline{U}}_{\kappa} = 2^{\kappa}$, and Sierpiński shows that U_{κ} is a complete linear ordering.

Let H_{κ} be the sub-linear-ordering consisting of those sequences,

S, in U_κ having a last 1, i.e. there is an $\alpha < \kappa$ such that $S_\alpha = 1$ and for all β, $\alpha < \beta < \kappa \Rightarrow S_\beta = 0$. Then Sierpiński shows that H_κ is dense in U_κ, whence it is easily shown that U_κ is the order-completion of H_κ. It remains to determine the cardinality of H_κ. Now, we may imbed any U_α in U_β if $\alpha < \beta$, simply by adjoining a string of 0's to each string of length α. Further, if S is of length κ and has a last 1, this occurs as S_α where $\bar{\bar{\alpha}} < \kappa$, so we conclude that $H_\kappa \subseteq \cup_{\alpha<\kappa} U_\alpha$, whence

$$\bar{\bar{H}}_\kappa \leq \Sigma_{\alpha<\kappa} \bar{\bar{U}}_\alpha = \Sigma_{\alpha<\kappa} 2^\alpha = 2^{\underline{\kappa}}.$$

Hence the theorem is true for any κ such that $\kappa = 2^{\underline{\kappa}}$, and this is clearly the case for $\kappa = \beth_\xi$, where ξ is a limit ordinal.

Taking $U_{\beth_\xi}$ to play the crucial role in our method of proof, we obtain:

THEOREM 4. Let $\mathbb{K}$ be a class of complete partial orderings which contains $U_{\beth_\xi}$ for ξ of arbitrarily high cofinality. Then $\mathbb{K}$ is not an $RPC_\Delta(L_{\kappa\lambda})$ for any κ, λ.

<u>Proof</u>. Suppose $\mathbb{K} \in RPC_\Delta(L_{\kappa\lambda})$ for some κ, λ, $\kappa \geq \lambda$. Let μ, Σ, ϕ have the same meaning as in the preceding proofs. Let ξ be a limit ordinal for which $\beth_\xi \geq \bar{\bar{\mu}}$, $cf(\xi) > \kappa$ and $U_{\beth_\xi} \in \mathbb{K}$. Then $\beth_\xi^{\ \kappa} = \beth_\xi$ and $U_{\beth_\xi} = (\mathfrak{A}' \restriction \phi^{\mathfrak{A}'}) \restriction \{<\}$ for some $\mathfrak{A}' \in Mod_\mu(\Sigma)$.

Since $\overline{\overline{H_{\beth_\xi}}} = \beth_\xi$, by the Downward Löwenheim-Skolem theorem there is $\mathfrak{B}' \prec_{\kappa\lambda} \mathfrak{A}'$ such that $H_{\beth_\xi} \subseteq |\mathfrak{B}'|$ and $\overline{\overline{\mathfrak{B}'}} = \beth_\xi$.

Then $\mathfrak{B} = (\mathfrak{B}' \restriction \phi^{\mathfrak{B}'}) \restriction \{<\}$ is a complete partial ordering. Since $\mathfrak{B}' \prec_{\kappa\lambda} \mathfrak{A}'$, $\mathfrak{B}$ is also a <u>linear</u> ordering and $H_{\beth_\xi} \subseteq \phi^{\mathfrak{B}'} = |\mathfrak{B}|$; so $\mathfrak{B}$ extends $H_{\beth_\xi}$. By completeness, $\mathfrak{B}$ contains an isomorphic copy of $U_{\beth_\xi}$. Therefore $\bar{\bar{\mathfrak{B}}} \geq \beth_{\xi+1}$, contradicting the choice of $\mathfrak{B}'$. This completes the proof.

It seems plausible that this method could be extended to the class of models of any higher-order theory in which a higher-order quantifier ranges over sets of unbounded cardinality, and where no equivalent formulation has quantifiers ranging over sets of bounded cardinality. The problem is to find a class of such models which are in some way generated by means of the power set operation from a set of smaller cardinality. Perhaps this general principle could be formulated as a theorem.

We conclude with a summary of some of the known results on $L_{\kappa\lambda}$-axiomatizability for certain classes:

<u>Classes which are not</u> $RPC_{\Delta}(L_{\kappa\lambda})$ <u>for any</u> κ, λ:

Topological spaces

Compact spaces

T_i-spaces $(i = 0, \ldots, 4)$

Compact and T_i spaces $(i = 0, \ldots, 4)$

Metrizable spaces

Stone spaces,
extremally disconnected Stone spaces

Topological groups,
topological abelian groups,
topological rings,
topological modules, etc.

Complete linear orderings

Complete lattices

Complete distributive lattices

Complete Boolean algebras

Complete uniform spaces (cf. [2])

Classes which are axiomatizable:

Separable spaces $(L_{\omega_1\omega_1})$

Polish spaces $(L_{\omega_1\omega_1})$

Compact metrizable spaces $(L_{\omega_1\omega_1})$

Spaces with a base of cardinality $< \kappa$ $(L_{\kappa\kappa})$

Metric spaces (axiomatizable in $L_{\omega_1\omega}$ in terms of denumerably many predicates)

Well-orderings $(L_{\omega_1\omega_1})$

$(\alpha-\beta)$-distributive Boolean algebras $(L_{(\beta^\alpha)^+(\alpha.\beta)^+}$, but possibly one can do better than this)

It is conjectured that compact topological groups also fail to be an $RPC_\Delta(L_{\kappa\lambda})$. (Our general method should work for the Bohr compactification of a discrete group.) Similarly, the class of Boolean algebras that are completely distributive, i.e., satisfying the $(\alpha-\beta)$-distributive law for all cardinals α, β, should fail to be an $RPC_\Delta(L_{\kappa\lambda})$.

REFERENCES

[1] R. D. Kopperman, Applications of Infinitary Languages to Analysis, in: Applications of Model Theory to Algebra, Analysis and Probability, ed. Luxemburg, 265-273.

[2] R. D. Kopperman, On the axiomatizability of uniform spaces, J. Symb. Logic 32 (1967), 289-294.

[3] W. Sierpiński, Sur un problème concernant les sous-ensembles croissant du continu, Fund. Math. 3 (1922), 109-112.

[4] W. Sierpiński, Sur une propriété des ensembles ordonnés, Fund. Math. 36 (1949), 56-67.

Π_1^1 MODELS AND Π_1^1-CATEGORICITY[1]

Nigel Cutland

Hull, England

Introduction

'Hyperarithmetic model theory' was first investigated by Cleave [2]; in this paper we develop a theory of hyperarithmetic and Π_1^1 models of first order theories, obtaining analogues of results of classical model theory with the analogy:

$$\text{hyperarithmetic} \longleftrightarrow \text{countable}$$
$$\Pi_1^1 \setminus \Sigma_1^1 \longleftrightarrow \text{of cardinality } \aleph_1.$$

For a set S of sets, the analogue of countable is that $S \cap HYP$ is 'bounded' in the hyperarithmetic hierarchy, i.e. there is a hyperarithmetic set H such that $X \in S \cap HYP$ implies X is recursive in H.

We consider Π_1^1 structures $\mathfrak{A}$ whose satisfaction predicate $\models_{\mathfrak{A}}$ is Π_1^1. Then we obtain natural analogues of the downward Löwenheim-Skolem theorem, the notion and construction of saturated structures, the method of model construction due to Ehrenfeucht and Mostowski, and Vaught's two cardinal theorem. §§1-4 of the paper are devoted to the basic definitions and outlines of these analogues.

In §5 we define the notion of Π_1^1-categoricity of a hyperarithmetic

---

[1]This paper contains a summary of the principal results of Part II of the author's Ph.D. thesis [4] submitted to the University of Bristol in August 1970, and written under the supervision of Dr. J. P. Cleave.

theory, showing the existence of such theories and obtaining some of their properties, using the results of the previous sections. Finally, in §6, using the results of Marsh on strongly minimal sets, we show that Π^1_1-categoricity is equivalent to $\aleph_1$-categoricity.

Most proofs given here are somewhat sketchy; for full details, the reader is referred to [4].

Preliminaries

The cardinal of a set X is denoted by card(X).

A finite sequence $(x_0, x_1, \ldots, x_{n-1})$ from ω is represented by the sequence number $\langle x_0, x_1, \ldots, x_{n-1}\rangle = \Pi_{i=0}^{n-1} p_i^{x_i+1}$, where p_i is the ith prime number. If s is a sequence number, lh(s) is the length of the sequence represented by s, and $(s)_i$ is the ith member of the sequence.

If $X \subseteq \omega$, Fs(X) is the set of numbers of finite sequences from X.

The abbreviations r.f. and p.r.f. stand for recursive function and partial recursive function. A Π^1_1 function (of natural numbers) is a function with Π^1_1 graph.

$\mathcal{O}$ is Kleene's set of notations for the recursive ordinals, $<_o$ the partial order on $\mathcal{O}$, and $+_o$ the recursive 'addition' of notations (as e.g. in [12]). The letters α, β, γ, δ will be reserved to denote members of $\mathcal{O}$, and where no ambiguity arises, we omit the subscript $_o$ from $<_o$ and $+_o$. For $\alpha \in \mathcal{O}$, $C(\alpha) = \{\beta : \beta < \alpha\}$.

We work with a fixed Π^1_1 path Z through $\mathcal{O}$ (so Z has length ω_1 = least non-recursive ordinal).

Using a fixed notation $\omega_0 \in \mathcal{O}$ for the ordinal ω, we define an r.f. $\omega_0 \cdot \alpha$ which denotes premultiplication of recursive ordinals by ω;

viz. $\omega_0 \cdot 1 = 1$; $\omega_0 \cdot 2^\alpha = \omega_0 \cdot \alpha +_0 \omega_0$; $\omega_0 \cdot (3.5^d) = 3.5^e$, where $\{e\}(n) = \omega_0 \cdot \{d\}(n)$. This gives a new Π^1_1 path Z_1 through $\mathcal{O}$, where $Z_1 = \bigcup_{\alpha \in Z} C(\omega_0 \cdot \alpha)$.

The sets H_α ($\alpha \in \mathcal{O}$) are defined in the usual way (see e.g. [12]). A set $X \subseteq \omega$ is <u>hyperarithmetic</u> if X is recursive in H_α for some $\alpha \in \mathcal{O}$; or, equivalently, if X is definable in Δ^1_1 form. We abbreviate the term hyperarithmetic by hyp. We write α-recursive, α-p.r.f. for H_α-recursive, H_α-p.r.f. etc. if $\alpha \in \mathcal{O}$.

HYP denotes the set of hyp. subsets of ω; if $\alpha \in \mathcal{O}$, HYP_α denotes those which are α-recursive.

Hyp. sets are (non-uniquely) indexed by natural numbers as follows (after Cleave [3]): if $\alpha \in \mathcal{O}$, $\langle e,\alpha\rangle$ is an index for the set $\{x : \{e\}^{H_\alpha}(x) = 0\}$, which is denoted by $H(\langle e,\alpha\rangle)$. Throughout the paper, an <u>index</u> is an index of this kind.

We work with an arbitrary but fixed countable recursive finitary first order language L (with equality). L has variables $(v_n)_{n<\omega}$, and logical connectives $\sim, \wedge, \vee, \rightarrow, \exists, \forall$. We assume that the sets of formulae, terms, etc. of L are recursive subsets of ω, and that all the usual syntactic operations are recursive. FL_n denotes the set of formulae of L whose free variables are amongst $\{v_0, v_1, \ldots, v_{n-1}\}$. $FL = \bigcup_{n<\omega} FL_n$.

If Γ is a theory in L, $B_n(\Gamma)$ is the nth Lindenbaum algebra of Γ, and $S_n(\Gamma)$ is the set of n-element types of Γ; an element type is a 1-element type.

We will work with an arbitrary but fixed complete hyperarithmetic theory T in L, having only infinite models.

The symbols $\mathfrak{A}$, $\mathfrak{B}$, $\mathfrak{C}$, $\mathfrak{D}$ will be reserved for structures; the universe of a structure is invariably denoted by the corresponding

Latin capital: e.g. the universe of $\mathfrak{A}$ is A etc. Unless otherwise specified we assume that the universe of a structure is a subset of ω. We indicate an exception to this rule by the phrase 'an arbitrary structure'.

A structure $\mathfrak{A}$ is coded by a set $Cd(\mathfrak{A}) \subseteq \omega$, so that $Cd(\mathfrak{A})$ contains all information about $\mathfrak{A}$. (Details of one method of doing this are given in [2] or [4].) Then $\mathfrak{A}$ is hyperarithmetic (Π_1^1, etc.) iff $Cd(\mathfrak{A})$ is hyp. (Π_1^1, etc.), and by an index for $\mathfrak{A}$ we mean one for $Cd(\mathfrak{A})$.

The structure obtained from $\mathfrak{A}$ by naming as individual constants the elements of a set $X \subseteq A$ is denoted by $(\mathfrak{A}, (x)_{x \in X})$.

Structures $\mathfrak{A}$, $\mathfrak{B}$ are Π_1^1-isomorphic if there is a Π_1^1 isomorphism $i : \mathfrak{A} \cong \mathfrak{B}$.

If $\mathfrak{A}$ is a structure, $a_0, a_1, \ldots, a_{n-1} \in A$ and $\phi \in FL_n$, the relation $\mathfrak{A} \models \phi[a_0, a_1, \ldots, a_{n-1}]$ is defined as usual (i.e. with a_0, a_1, ..., a_{n-1} assigned to the free variables $v_0, v_1, \ldots, v_{n-1}$ in ϕ). We define $\mathfrak{A} \models \phi(s)$ iff

$s \in Fs(A)$ & $\phi \in FL_{lh(s)}$ & $\mathfrak{A} \models \phi[(s)_0, (s)_1, \ldots, (s)_{lh(s)-1}]$.

Regarded as a predicate of the two number variables ϕ, s, we call this predicate $\models_{\mathfrak{A}}$. It is clear that if $\mathfrak{A}$ is hyp., so is $\models_{\mathfrak{A}}$, and in [2] it is shown how an index for $\models_{\mathfrak{A}}$ is obtained recursively from one for $\mathfrak{A}$.

If $\phi \in FL_1$, then $\phi^{\mathfrak{A}} = \{a \in A : \mathfrak{A} \models \phi[a]\}$.

$Th(\mathfrak{A})$ denotes the first order theory of a structure $\mathfrak{A}$, and $L(\mathfrak{A})$ denotes the first order language corresponding to $\mathfrak{A}$.

§1. Covered Π_1^1 structures and the Löwenheim-Skolem theorem

1.1 DEFINITION.

(a) $\mathfrak{A}$ is a covered Π_1^1 structure (written $\mathfrak{A} \in M(\Pi_1^1)$) if the satisfaction predicate $\models_{\mathfrak{A}}$ is Π^1_1.

(b) $\mathfrak{A}$ is a <u>properly covered</u> Π^1_1 <u>structure</u> ($\mathfrak{A} \in M'(\Pi^1_1)$) if, in addition, $\mathfrak{A}$ is not hyperarithmetic.

Clearly $\mathfrak{A} \in M(\Pi^1_1)$ implies that $\mathfrak{A}$ is a Π^1_1 structure.

EXAMPLES

(a) If $\mathfrak{A}$ is hyp., then $\mathfrak{A} \in M(\Pi^1_1)$.

(b) Let $\mathfrak{A}_\alpha = \mathfrak{Q}\times\{\alpha\}$ for $\alpha \in Z$, where $\mathfrak{Q}$ is a copy of the rationals with their order. Let $\oplus$ denote the ordinal sum of two ordered sets. Then the structure $\mathfrak{A} = \oplus_{\alpha\in Z}\, \mathfrak{A}_\alpha$ is in $M'(\Pi^1_1)$.

(c) $(\omega, \mathcal{O})$ is a Π^1_1 structure which is <u>not</u> in $M(\Pi^1_1)$.

1.2 LEMMA. Let $\mathfrak{A} \in M(\Pi^1_1)$. The following are equivalent:

(a) $\mathfrak{A} \notin M'(\Pi^1_1)$ (i.e. $\mathfrak{A}$ is hyp.)

(b) A is hyp.

(c) $\models_{\mathfrak{A}}$ is hyp.

<u>Proof</u>. Clearly (c) implies (a), and (a) implies (b). To show that (b) implies (c) suppose that A is hyp.

We know that $\models_{\mathfrak{A}}$ is Π^1_1; and we have $\models_{\mathfrak{A}}(\phi,s) \iff s \in Fs(A)\ \&\ \phi \in FL_{lh(s)}\ \&\ \neg(\models_{\mathfrak{A}}(\sim\phi,s))$ which is a Σ^1_1 definition of $\models_{\mathfrak{A}}$. Hence $\models_{\mathfrak{A}}$ is $\Pi^1_1 \cap \Sigma^1_1$, i.e. hyp.

Our first main result for covered Π^1_1 structures is the following analogue of the downward Löwenheim-Skolem theorem.

1.3 THEOREM. Let $\mathfrak{A} \in M(\Pi^1_1)$. There is an r.f. p such that if w is an index for a hyp. set $X \subseteq A$, then there is a hyp. structure $\mathfrak{B} \prec \mathfrak{A}$ such that $X \subseteq B$, and $\models_{\mathfrak{B}}$ has index p(w).

<u>Proof</u>. The proof is analogous to that of the classical downward Löwenheim-Skolem theorem. First show that there is a Π^1_1 uniform Skolem

function $f : FL \times Fs(A) \to A$ such that

$\mathfrak{A} \models \exists v_0 \phi[\cdot, a_1, a_2, \ldots, a_n] \Rightarrow \mathfrak{A} \models \phi[f(\phi, \langle a_1, \ldots, a_n \rangle), a_1, \ldots, a_n]$.

If $X \subseteq A$ is hyp., then $f(FL, X)$ is hyp., since f is a function. Thus we can obtain an r.f. h such that if $X \subseteq A$ has index w, then h(w) is an index for $cl_f(X)$, the closure of X under $f(FL, \cdot)$.

Let $B = cl_f(X)$; clearly B is the universe of a structure $\mathfrak{B} \prec \mathfrak{A}$ with $X \subseteq B$. Finally observe that

$$\mathfrak{B} \models \phi(s) \Leftrightarrow s \in Fs(B) \mathbin{\&} \phi \in FL_{lh(s)} \mathbin{\&} \mathfrak{A} \models \phi(s)$$
$$\Leftrightarrow s \in Fs(B) \mathbin{\&} \phi \in FL_{lh(s)} \mathbin{\&} \neg \mathfrak{A} \models \sim\phi(s).$$

These are Π^1_1 and Σ^1_1 definitions respectively for $\models_{\mathfrak{B}}$, hence $\models_{\mathfrak{B}}$ is hyp.; using these definitions we obtain from the index h(w) for B an index p(w) for $\models_{\mathfrak{B}}$, with p recursive.

The function p is a <u>cover function</u> for $\mathfrak{A}$; the structures with indices p(w) for w with $H(w) \subseteq A$ constitute a <u>cover</u> of $\mathfrak{A}$ by hyp. elementary substructures. This explains the origin of the term <u>covered Π^1_1 structure</u>; in [4] $M(\Pi^1_1)$ is defined to be the set of Π^1_1 structures having an elementary cover as in the theorem. The equivalence of that definition with (1.1) above follows from (1.3) and the next result which is a converse to (1.3).

1.4 THEOREM. Let $\mathfrak{A}$ be a Π^1_1 structure and p a Π^1_1 function such that whenever w is an index for a hyp. set $X \subseteq A$, then there is a hyp. $\mathfrak{B} \prec \mathfrak{A}$ with $X \subseteq B$, and such that $\models_{\mathfrak{B}}$ has index p(w). Then $\mathfrak{A} \in M(\Pi^1_1)$.

<u>Proof</u>. Take an r.f. f such that if s is a sequence number, then f(s) is an index for $\{(s)_0, \ldots, (s)_{lh(s)-1}\}$.

Then if $s \in Fs(A)$, p(f(s)) is an index for $\models_{\mathfrak{B}}$, where $\mathfrak{B} \prec \mathfrak{A}$ and $s \in Fs(B)$. Hence

$$\mathfrak{A} \models \phi(s) \Longleftrightarrow s \in Fs(A) \mathbin{\&} \models_{\mathfrak{L}} (\phi, s)$$
$$\Longleftrightarrow s \in Fs(A) \mathbin{\&} \langle \phi, s \rangle \in H(p(f(s))),$$

which is a Π_1^1 definition for $\models_{\mathfrak{A}}$.

<u>Note</u>. This, together with (1.3) shows that a Π_1^1 structure having a Π_1^1 cover function in fact has a recursive cover function.

We outline in the next theorem the method of constructing $M(\Pi_1^1)$ structures as unions of elementary chains, which is used throughout this paper.

1.5 THEOREM. Let $(\mathfrak{A}_\alpha)_{\alpha \in Z}$ be a sequence of hyp. structures, and w an r.f. such that

(a) $\mathfrak{A}_\alpha \prec \mathfrak{A}_\beta$ if $\alpha < \beta$

(b) $\models_{\mathfrak{A}_\alpha}$ has index $w(\alpha)$.

Then $\mathfrak{A} = \bigcup_{\alpha \in Z} \mathfrak{A}_\alpha$ is in $M(\Pi_1^1)$. Moreover, if $\mathfrak{A}_\alpha \neq \mathfrak{A}$ for each α, then $\mathfrak{A} \in M'(\Pi_1^1)$.

<u>Proof</u>. We have the following Π_1^1 definition of $\models_{\mathfrak{A}}$:

$$\mathfrak{A} \models \phi(s) \Longleftrightarrow (E\alpha)[\alpha \in Z \mathbin{\&} \mathfrak{A}_\alpha \models \phi(s)]$$
$$\Longleftrightarrow (E\alpha)[\alpha \in Z \mathbin{\&} \langle \phi, s \rangle \in H(w(\alpha))].$$

Hence $\mathfrak{A} \in M(\Pi_1^1)$.

For the last part, suppose $X \subseteq A$ is hyp; then we have $(x)(E\alpha)[x \in X \Rightarrow \alpha \in Z$ and $x \in A_\alpha]$.

By lemma 1 of Kreisel [7], since this is of the form $(x)(Ey)R(x,y)$ with R Π_1^1, there is a hyp. function g such that $(x)[x \in X \Rightarrow g(x) \in Z$ and $x \in A_{g(x)}]$.

The set $\{g(x) : x \in X\}$ is a hyp. subset of Z, so there is γ with

$g(x) < y$ for all $x \in X$. Thus $X \subseteq A_y$.

If now $\mathfrak{A} \notin M'(\Pi_1^1)$, then A is hyp., so taking X = A we have that $\mathfrak{A} = \mathfrak{A}_y$.

<u>Note</u>. It is shown in [4] how the structures $(\mathfrak{A}_\alpha)_{\alpha\in Z}$ may be taken as the elementary cover for $\mathfrak{A}$.

We end this section by describing the method generally used to define the function w in (1.5) above - <u>effective transfinite induction</u> (ETFI).

Suppose we are aiming to construct $\mathfrak{A} \in M(\Pi_1^1)$ having certain properties, as a union of an elementary chain. To this end, we start by defining $\mathfrak{A}_1$ in some prescribed way. Then obtain an r.f. g such that if $\mathfrak{A}_\alpha$ has been defined correctly for the construction, and $\models_{\mathfrak{A}_\alpha}$ has index u, then g(u) is an index of $\models_{\mathfrak{A}_{\alpha+1}}$ for a structure $\mathfrak{A}_{\alpha+1} \succ \mathfrak{A}_\alpha$ which is correct for the construction. In most cases it is sufficient to put $\mathfrak{A}_\delta = \bigcup_{\alpha<\delta} \mathfrak{A}_\alpha$ for limits δ. Now using the recursion theorem as in the Recursion Lemma on p. 398 of [12] it is possible to obtain an r.f. w such that

(a) $w(1)$ is an index for $\models_{\mathfrak{A}_1}$

(b) $w(\alpha +_0 1_0) = g(w(\alpha))$

(c) $w(\delta)$ is an index for $\bigcup_{n<\omega} \models_{\mathfrak{A}_{\{d\}(n)}}$ if $\delta = 3.5^d$; i.e. an index for $\bigcup_{n<\omega} H(w(\{d\}(n)))$.

Then $\mathfrak{A}_\alpha$ is <u>by definition</u> the structure whose satisfaction predicate $\models_{\mathfrak{A}_\alpha}$ has index $w(\alpha)$; and now (1.5) is applied to obtain $\mathfrak{A} \in M(\Pi_1^1)$ as required, where $\mathfrak{A} = \bigcup_{\alpha\in Z} \mathfrak{A}_\alpha$.

In the following three sections we outline the way in which ETFI is used to construct $M(\Pi_1^1)$ models of T having various special properties.

§2. Π_1^1-saturated structures

2.1 DEFINITION. A structure $\mathfrak{A} \in M(\Pi_1^1)$ is Π_1^1-saturated if, for any hyp. $X \subseteq A$, all hyp. element types of $Th(\mathfrak{A}, (x)_{x\in X})$ are realised in $(\mathfrak{A}, (x)_{x\in X})$.

The existence of Π_1^1-saturated models of any hyp. theory having infinite models is given by the following.

2.2 THEOREM. T has a Π_1^1-saturated model.

Proof. To construct a Π_1^1-saturated model, start with any hyp. model $\mathfrak{L}$ of T; then using ETFI we obtain a properly increasing recursive cofinal sequence of limit notations $(\beta_\alpha)_{\alpha\in Z}$ from $\mathcal{O}$, a sequence $(\mathfrak{A}_\alpha)_{\alpha\in Z}$ of models of T, and an r.f. w such that for $\alpha \in Z$:

(a) $\mathfrak{A}_1 \cong \mathfrak{L}$,

(b) $\mathfrak{A}_\alpha \prec \mathfrak{A}_\beta$ if $\alpha < \beta$,

(c) $\mathfrak{A}_\delta = \bigcup_{\alpha<\delta} \mathfrak{A}_\alpha$ if δ is a limit,

(d) $\models_{\mathfrak{A}_\alpha}$ has index $<w(\alpha),\beta_\alpha>$,

(e) Every element type of $Th(\mathfrak{A}_\alpha, (x)_{x\in A_\alpha})$ which is γ-recursive for some $\gamma < \beta_{\alpha+1}$ is realised in $(\mathfrak{A}_{\alpha+1}, (x)_{x\in A_\alpha})$.

Using Henkin's construction for the completeness theorem an r.f. g is obtained so that if $\models_{\mathfrak{A}_\alpha}$ has index $<u,\beta_\alpha>$, then $<g(u),\beta_{\alpha+1}>$ is an index for $\models_{\mathfrak{A}_{\alpha+1}}$ with $\mathfrak{A}_{\alpha+1}$ satisfying (d) and (e) above. w is then obtained as described in §1, by ETFI.

We claim that $\mathfrak{A} = \bigcup_{\alpha\in Z} \mathfrak{A}_\alpha$ is Π_1^1-saturated. By (1.5), $\mathfrak{A} \in M(\Pi_1^1)$. Suppose that $X \subseteq A$, and P is an element type of $Th(\mathfrak{A}, (x)_{x\in X})$, with X, P hyp.

Choose $\alpha \in Z$ so that $X \subseteq A_\alpha$ (from the proof of (1.5) we know that such α exists) <u>and</u> X, P are both β_α-recursive. Then P can be 'translated' into a set P' which can be extended to a hyp. element type P″ of $Th(\mathfrak{A}_\alpha, (x)_{x\in A_\alpha})$, with P″ $\beta_\alpha +_o 2_o$-recursive. By construction, P″ is realised in $(\mathfrak{A}_{\alpha+1}, (x)_{x\in A_\alpha})$, so P is realised in $(\mathfrak{A}_{\alpha+1}, (x)_{x\in X})$ by the same point.

As is remarked in [4], Π_1^1-saturated structures can also be obtained by means of an ultraproduct construction.

The importance of Π_1^1-saturated structures for us is in the following isomorphism theorem.

2.3 THEOREM. If $\mathfrak{A}$, $\mathfrak{B}$ are elementarily equivalent Π_1^1-saturated structures, then they are Π_1^1-isomorphic.

<u>Proof</u>. As for the classical counterpart of this theorem, a back-and-forth argument is used, which is complicated by the requirement that the isomorphism be Π_1^1.

It is necessary first to show that $\mathfrak{A}$, $\mathfrak{B}$ have <u>Π_1^1-saturation functions</u>, i.e. (for $\mathfrak{A}$) a Π_1^1 function $q_{\mathfrak{A}}$ such that if w_1 is an index of $X \subseteq A$, and w_2 is an index of an element type P of $Th(\mathfrak{A}, (x)_{x\in X})$, then P is realised at $q_{\mathfrak{A}}(w_1,w_2)$ in A; and similarly $q_{\mathfrak{B}}$ for $\mathfrak{B}$.

Using these functions, we can obtain an increasing sequence $(X_\alpha)_{\alpha\in Z}$ of hyp. subsets of A, with $A = \cup_{\alpha\in Z} X_\alpha$, and hyp. elementary monomorphisms $f_\alpha : X_\alpha \to B$ such that f_β extends f_α if $\beta > \alpha$, and $B = \cup_{\alpha\in Z} f_\alpha(X_\alpha)$. By ETFI we obtain an r.f. g such that the graph I_α of f_α has index $g(\alpha)$.

Then $I = \cup_{\alpha\in Z} I_\alpha = \cup_{\alpha\in Z} H(g(\alpha))$ is the Π_1^1 graph of an isomorphism $f : \mathfrak{A} \cong \mathfrak{B}$, as required.

In [4] notions of <u>Π_1^1-universal</u> and <u>Π_1^1-homogeneous</u> structures are defined; it is shown there that a structure is Π_1^1-saturated iff it is

Π_1^1-homogeneous and Π_1^1-universal.

We note here for use in a later section that $\mathfrak{A}$ is Π_1^1-<u>universal</u>, if any elementarily equivalent $M(\Pi_1^1)$ structure $\mathfrak{B}$ is Π_1^1-elementarily embeddable in $\mathfrak{A}$; so Π_1^1-saturated structures have this property.

§3. The Construction of Ehrenfeucht and Mostowski

3.1 DEFINITION. Let $\mathfrak{A}$ be a structure and $(X,<)$ an ordered set with $X \subseteq A$. $(X,<)$ is <u>transportable</u> in $\mathfrak{A}$ if, whenever $s_1, s_2 \in Fs(X)$ are order isomorphic sequences of length n, and $\phi \in FL_n$, then

$$\mathfrak{A} \models \phi(s_1) \Leftrightarrow \mathfrak{A} \models \phi(s_2).$$

The method of Ehrenfeucht and Mostowski [6] can be used to obtain:

3.2 THEOREM. There is a model $\mathfrak{A} \in M(\Pi_1^1)$ of T and a Π_1^1 injection $g : Z_1 \to A$ such that $g(Z_1,<_0)$ is transportable for $\mathfrak{A}$.

<u>Proof</u>. We construct an increasing elementary chain of models $(\mathfrak{A}_\alpha)_{\alpha \in Z}$ and injections $g_\alpha : C(\omega_0 \cdot \alpha) \to A_\alpha$ such that

(a) $g_\alpha(C(\omega_0 \cdot \alpha), <_0)$ is transportable for $\mathfrak{A}_\alpha$

(b) $g_\beta | C(\omega_0 \cdot \alpha) = g_\alpha$ if $\alpha < \beta$.

By ETFI obtain r.f.s. w_1, w_2 so that $\models_{\mathfrak{A}_\alpha}$ and the graph of g_α have indices $w_1(\alpha)$, $w_2(\alpha)$ respectively. Now put $\mathfrak{A} = \cup_{\alpha \in Z} \mathfrak{A}_\alpha$ and $g = \cup_{\alpha \in Z} g_\alpha$, and these are both as required.

The following is an analogue of the application by Ehrenfeucht [5] and Morley [9] of the original construction.

3.3 THEOREM. There is a model $\mathfrak{B}$ of T and an r.f. w, with $\mathfrak{B} \in M'(\Pi_1^1)$ and $\mathfrak{B} = \cup_{\alpha \in Z} \mathfrak{B}_\alpha$, where $\mathfrak{B}_\alpha$ is an increasing chain of hyp.

elementary substructures, having the following properties.

(a) $w(\alpha)$ is an index for $\models \mathfrak{B}_\alpha$

(b) if P is an element type of $Th(\mathfrak{B}, (x)_{x \in B_\alpha})$ which is realised in $\mathfrak{B}$, then P is realised in $\mathfrak{B}_{\alpha+1}$.

Proof. Let L^* be the language obtained from L by adding symbols for Skolem functions; let T^* be any complete extension of T + axioms for the Skolem functions. We now apply (3.2) to T^*. Models of T^* are denoted by $\mathfrak{A}^*$, $\mathfrak{B}^*$ etc., and their L-reducts by $\mathfrak{A}$, $\mathfrak{B}$, etc.

Let $\mathfrak{A}^* = \bigcup_{\alpha \in Z} \mathfrak{A}^*_\alpha$ be the model of T^* given by (3.2), and w_1, w_2, g_α, g as in (3.2) also.

Let $g_\alpha(C(\omega_0 \cdot \alpha), <_0) = (X_\alpha, <_\alpha)$ and let $(X, <_1) = g(Z_1, <_0) = \bigcup_{\alpha \in Z} (X_\alpha, <_\alpha)$.

Clearly $X_\alpha \subseteq A_\alpha$ for each α. Now, for $\alpha \neq 1$, let B_α be the closure of X_α under the functions of $\mathfrak{A}^*_\alpha$; then B_α is the universe of a structure $\mathfrak{B}_\alpha \prec \mathfrak{A}_\alpha$. From w_1, w_2 obtain an r.f. w such that $\models \mathfrak{B}_\alpha$ has index $w(\alpha)$. Putting $\mathfrak{B} = \bigcup_{\alpha \in Z} \mathfrak{B}_\alpha$, we have by (1.5) that $\mathfrak{B} \in M(\Pi^1_1)$.

If $\alpha \neq 1$, the fact that $(X_\alpha, <_\alpha)$ is transportable and has no greatest member ensures that $g(\omega_0 \cdot \alpha) \notin B_\alpha$. Hence $B_{\alpha+1} \neq B_\alpha$, and by (1.5) $\mathfrak{B} \in M'(\Pi^1_1)$.

For (b) of the theorem, let $b \in B$ and $\alpha \in Z$. There is a term $t(v_1, \ldots, v_n)$ of L^* and $x_1, \ldots, x_n \in X$ such that $b = t(x_1, \ldots, x_n)$; we may assume without loss of generality that $x_1 <_1 x_2 <_1 \ldots <_1 x_n$.

There is $m \leq n$ such that $x_1, x_2, \ldots, x_m \in X_\alpha$ and $x_{m+1}, \ldots, x_n \notin X_\alpha$. Now take $x'_{m+1}, \ldots, x'_n \in X_{\alpha+1} \setminus X_\alpha$ with $x'_{m+1} <_{\alpha+1} x'_{m+2} <_{\alpha+1} \ldots <_{\alpha+1} x'_n$ and let

$$b' = t(x_1, x_2, \ldots, x_m, x'_{m+1}, \ldots, x'_n) \in B_{\alpha+1}.$$

Then, using the fact that $(X,<_1)$ is transportable for $\mathfrak{A}^*$ and that the elements of B_α are the values of terms in $(\mathfrak{A}^*, (x)_{x\in X_\alpha})$ it is easily seen that b and b' realise the same element type of $Th(\mathfrak{B}, (x)_{x\in B_\alpha})$.

3.4 COROLLARY. There is a model $\mathfrak{B} \in M'(\Pi^1_1)$ of T such that if $X \subseteq B$ is hyp., there is $y \in Z$ such that those hyp. element types of $Th(\mathfrak{B}, (x)_{x\in X})$ which are realised in $\mathfrak{B}$ are all y-recursive.

Proof. Let $\mathfrak{B}$ be the model of (3.3) and take α such that $X \subseteq B_\alpha$. Let $y \in Z$ such that $\models_{(\mathfrak{B}_{\alpha+1},(x)_{x\in X})}$ is y-recursive.

Suppose P is an element type of $Th(\mathfrak{B}, (x)_{x\in X})$ which is realised in $\mathfrak{B}$. Then by (3.3) P is realised in $\mathfrak{B}_{\alpha+1}$, and is hence y-recursive, so y is as required.

§4. Vaught's two cardinal theorem

In this section we give an analogue of the two cardinal theorem of Vaught [11].

4.1 THEOREM. Suppose $\phi \in FL_1$ and there are arbitrary models $\mathfrak{A}$, $\mathfrak{B}$, of T with $\mathfrak{B} \succ \mathfrak{A}$, $\mathfrak{B} \neq \mathfrak{A}$ and $\phi^{\mathfrak{A}} = \phi^{\mathfrak{B}}$, infinite (this is the case, in particular, if T has a model $\mathfrak{B}$ with $\aleph_0 \leq card(\phi^{\mathfrak{B}}) < card(\mathfrak{B})$). Then T has a model $\mathfrak{C} \in M'(\Pi^1_1)$ with $\phi^{\mathfrak{C}}$ hyperarithmetic.

Proof. The proof is similar to that of Vaught's Theorem. Let L' be the language L with a unary predicate U added. Let Σ be T together with L' sentences expressing that in any model $\mathfrak{C}$:

(a) U is the universe of a proper elementary L-structure of the L-reduct of $\mathfrak{C}$,

(b) $\phi^{\mathfrak{C}} \subseteq U$.

Since Σ has $(\mathfrak{B}, A)$ as a model, there is a hyp. model $(\mathfrak{B}_1, A_1)$ of Σ which is weakly α-saturated, for some limit $\alpha \in \mathcal{O}$, in the sense of Cleave [2] (see also [4]). It follows that $\mathfrak{A}_1$ (the structure with universe A_1) and $\mathfrak{B}_1$ are also weakly α-saturated, which means that there is an α-recursive isomorphism $f : \mathfrak{A}_1 \cong \mathfrak{B}_1$ (see [2] or [4]). We have also, by construction, that $\mathfrak{A}_1 \prec \mathfrak{B}_1$, $\mathfrak{A}_1 \neq \mathfrak{B}_1$ and $\phi^{\mathfrak{A}_1} = \phi^{\mathfrak{B}_1}$.

Now we construct a chain $(\mathfrak{C}_\alpha)_{\alpha \in Z}$ of hyp. structures so that for $\alpha, \delta \in \mathcal{O}$:

(a) $\mathfrak{C}_\alpha \cong \mathfrak{A}_1$

(b) $\mathfrak{C}_{\alpha+1} \succ \mathfrak{C}_\alpha$, and $\mathfrak{C}_{\alpha+1}$ bears the same relation to $\mathfrak{C}_\alpha$ as $\mathfrak{B}_1$ does to $\mathfrak{A}_1$ (hence $\phi^{\mathfrak{C}_{\alpha+1}} = \phi^{\mathfrak{C}_\alpha}$)

(c) $\mathfrak{C}_\delta = \cup_{\alpha<\delta} \mathfrak{C}_\alpha$ if δ is a limit.

To show that $\mathfrak{C}_\delta \cong \mathfrak{A}_1$ for limits δ, we use the fact that weakly α-saturated structures are homogeneous (see [2] or [4]) and proceed exactly as in the classical case of Vaught's theorem.

Finally $\mathfrak{C} = \cup_{\alpha \in Z} \mathfrak{C}_\alpha$ is as required; for $\phi^{\mathfrak{C}} = \cup_{\alpha \in Z} \phi^{\mathfrak{C}_\alpha} = \phi^{\mathfrak{C}_1}$. An r.f. w is obtained so that $\mathfrak{C}_\alpha$ has index $w(\alpha)$, and hence $\mathfrak{C} \in M'(\Pi^1_1)$, by (1.5).

§5. Π^1_1-categoricity

In this section we define the notion of Π^1_1-categoricity, and establish some properties of Π^1_1-categorical theories.

5.1 DEFINITION. T is Π^1_1-categorical iff any two models of T in $M'(\Pi^1_1)$ are Π^1_1-isomorphic.

The existence of Π^1_1-categorical theories will follow from the next result.

5.2 THEOREM. (Cf. Theorem 5.5 of [9]). T is Π_1^1-categorical iff all $M'(\Pi_1^1)$ models of T are Π_1^1-saturated.

Proof. If T is Π_1^1-categorical, take a Π_1^1-saturated model $\mathfrak{A}$ of T (by (2.2)). If $\mathfrak{B} \in M'(\Pi_1^1)$ is another model of T, then $\mathfrak{B}$ is Π_1^1-isomorphic to $\mathfrak{A}$, and it is easily seen that $\mathfrak{B}$ is Π_1^1-saturated.

The converse follows from the isomorphism theorem (2.3).

5.3 COROLLARY. An $\aleph_1$-categorical (hyp.) theory is Π_1^1-categorical.

Proof. Let $\mathfrak{A} \in M'(\Pi_1^1)$ be a model of an $\aleph_1$-categorical theory. By (5.2) it is sufficient to show that $\mathfrak{A}$ is Π_1^1-saturated.

Let $X \subseteq A$ be hyp. Using the elementary cover of $\mathfrak{A}$ given by (1.3), there is a properly increasing chain $\mathfrak{A}_0 \prec \mathfrak{A}_1 \prec \mathfrak{A}_2 \prec \ldots \prec \mathfrak{A}$ of hyp. elementary submodels of $\mathfrak{A}$, with $X \subseteq A_0$. The proof of Theorem 4 of [9] shows that all element types of $Th(\mathfrak{A}, (x)_{x \in A_0})$ are realised in $\mathfrak{A}$. Hence $\mathfrak{A}$ is Π_1^1-saturated.

Thus we have as examples of Π_1^1-categorical theories the standard $\aleph_1$-categorical theories - algebraically closed fields of given characteristic, torsion free divisible abelian groups, etc.

In the remainder of this section we describe some properties of Π_1^1-categorical theories which are analogues of those possessed by $\aleph_1$-categorical theories, and which we will use to obtain the converse to (5.3) in establishing the equivalence theorem (6.7) of the final section.

First we discuss a property analogous to total transcendence [9] or stability of a theory.

5.4 DEFINITION. T is H-stable iff for every hyp. model $\mathfrak{A}$ and hyp. subset $X \subseteq A$, there is $y \in \mathcal{O}$ such that

$$S_1(\mathrm{Th}(\mathfrak{A},(x)_{x\in X})) \cap \mathrm{HYP} \subseteq \mathrm{HYP}_\gamma .$$

5.5 THEOREM. (Cf. Theorem 3.8 of [9].) A Π_1^1-categorical theory is H-stable.

Proof. Suppose T is Π_1^1-categorical, and let $\mathfrak{B}$ be the model of T given by (3.4). By (5.2), $\mathfrak{B}$ is Π_1^1-saturated, so if $\mathfrak{A}$ is any hyp. model of T, there is a hyp. elementary embedding $e : \mathfrak{A} \to \mathfrak{B}$.

Let $X \subseteq A$ be hyp.; then $Y = e(X)$ is hyp., and since $\mathfrak{B}$ is Π_1^1-saturated, all hyp. members of $S_1(\mathrm{Th}(\mathfrak{B}, (y)_{y\in Y}))$ are realised in $\mathfrak{B}$. So by (3.4) there is $\gamma \in Z$ such that all hyp. members of $S_1(\mathrm{Th}(\mathfrak{A}, (x)_{x\in X})) = S_1(\mathrm{Th}(\mathfrak{B}, (y)_{y\in Y}))$ are γ-recursive, i.e.

$$S_1(\mathrm{Th}(\mathfrak{A},(x)_{x\in X})) \cap \mathrm{HYP} \subseteq \mathrm{HYP}_\gamma$$

as required.

H-stable theories have several properties analogous to those of totally transcendental theories, among them the following:

5.6 THEOREM. (Cf. Theorem 4.3 of [9].) Let $\mathfrak{A}$ be a hyp. model of an H-stable theory T and $X \subseteq A$ be hyp. Then $\mathrm{Th}(\mathfrak{A}, (x)_{x\in X})$ has a hyp. prime model (i.e. a model of T prime over X).

Proof. It can be shown (see [4]) that if $\mathfrak{B}$ is a hyp. Boolean algebra and $\gamma \in \mathcal{O}$ such that $S(\mathfrak{B}) \cap \mathrm{HYP} \subseteq \mathrm{HYP}_\gamma$ ($S(\mathfrak{B})$ is the Stone space of $\mathfrak{B}$), then $\mathfrak{B}$ is atomistic (just as $S(\mathfrak{B})$ countable implies $\mathfrak{B}$ atomistic).

If now T is H-stable, and $\mathfrak{A}$, X are as in the theorem, we can show (by induction on n) that for each n there is $\gamma_n \in \mathcal{O}$ such that

$$S_n(T') \cap \mathrm{HYP} \subseteq \mathrm{HYP}_{\gamma_n}$$

where $T' = Th(\mathfrak{A}, (x)_{x \in X})$.

From the above, we deduce that $B_n(T')$ is atomistic for each n, and hence T' has a prime model. In [4] it is shown how this may be taken to be hyp.

We follow with an application of (4.1) to Π_1^1-categorical theories.

5.7 THEOREM. (Cf. the proof of Theorem 1 of [10].) If T is Π_1^1-categorical and $\phi \in FL_1$ defines an infinite set in all models of T, and $\mathfrak{A} \prec \mathfrak{B}$ are arbitrary models of T with $\phi^{\mathfrak{A}} = \phi^{\mathfrak{B}}$, then $\mathfrak{A} = \mathfrak{B}$.

<u>Proof</u>. If the conclusion does not hold, by (4.1) there is a model $\mathfrak{C} \in M'(\Pi_1^1)$ with $\phi^{\mathfrak{C}}$ hyp. Then $\{v_0 \neq x : x \in \phi^{\mathfrak{C}}\} \cup \{\phi\}$ is a consistent hyp. set of sentences which is not satisfiable in $\mathfrak{C}$, but by (5.2) $\mathfrak{C}$ is Π_1^1-saturated, which is a contradiction.

§6. <u>Strongly minimal sets and Π_1^1-categorical theories</u>

In this section we indicate how the work of Marsh on strongly minimal sets can be applied to Π_1^1-categorical theories, using the results of the previous section. We are then able to show that for hyp. theories the notions of Π_1^1-categoricity and $\aleph_1$-categoricity are equivalent.

An exposition of Marsh's work [8] appears in [1]; the definitions and results which we need are given below.

6.1 DEFINITION. Let $\mathfrak{A}$ be a structure, $X, Y \subseteq A$ and let $\mathfrak{A}' = (\mathfrak{A}, (x)_{x \in X})$.

(a) the <u>algebraic closure</u> of X (in $\mathfrak{A}$), cl(X), is defined by $cl(X) = \cup\{\phi^{\mathfrak{A}'} : \phi \in FL_1(\mathfrak{A}') \text{ and } \phi^{\mathfrak{A}'} \text{ is finite}\}$.

(b) X <u>spans</u> Y if $Y \subseteq cl(X)$.

(c) X is <u>independent</u> if $x \notin cl(X \setminus \{x\})$ for each $x \in X$.

(d) X is a <u>basis</u> for Y if $X \subseteq Y$, X is independent, and spans Y.

6.2 DEFINITION. Let $\mathfrak{A}$ be a structure, $\phi \in FL_1$.

(a) the set $\phi^{\mathfrak{A}}$ is <u>minimal</u> (and the formula ϕ is minimal for $Th(\mathfrak{A})$) if $\phi^{\mathfrak{A}}$ is infinite, and for every $\psi \in FL_1$, either $(\phi \wedge \psi)^{\mathfrak{A}}$ or $(\phi \wedge \sim\psi)^{\mathfrak{A}}$ is finite.

(b) ϕ is <u>strongly minimal</u> (and ϕ is strongly minimal for $Th(\mathfrak{A})$) if ϕ is minimal for any consistent extension of $Th(\mathfrak{A})$ by constants.

6.3 LEMMA. (Marsh) Let $\phi^{\mathfrak{A}}$ be strongly minimal in $\mathfrak{A}$, and $X \subseteq \phi^{\mathfrak{A}}$. If $Z \subseteq X$ is independent, then Z can be extended to a basis for X. Further, any two bases for X have the same cardinality.

This allows us to define the <u>dimension</u>, dim(X), of a subset X of a strongly minimal set as the cardinality of any basis for X.

6.4 LEMMA. (Marsh) Let $\mathfrak{A}$, $\mathfrak{B}$ be models of a complete theory T having a strongly minimal formula ϕ. If $X \subseteq \phi^{\mathfrak{A}}$, $Y \subseteq \phi^{\mathfrak{B}}$, X and Y are independent, and $f : X \longrightarrow Y$ is an injection, then f is an elementary monomorphism.

We now have the following result which allows us to apply these results of Marsh (cf. Lemma 9 of [1]).

6.5 THEOREM. If T is Π_1^1-categorical, then there is a principal hyp. extension T' of T by a finite number of constants, such that T' has a strongly minimal formula.

<u>Proof</u>. This is proved by adapting the argument used in the proof of the corresponding result for $\aleph_1$-categorical theories (Lemma 9 of [1]). Essential use is made of the fact that T is H-stable (5.5), and (5.7).

This enables us to prove the following.

6.6 THEOREM. (Cf. Theorem 1 of [10].) If T is Π_1^1-categorical, then every hyp. model of T has a hyp. prime elementary extension.

Proof. Let $\mathfrak{A}$ be a hyp. model of T, and let T' be the extension of T guaranteed by (6.5), with strongly minimal formula ϕ. Then T' is Π_1^1-categorical, and there are elements $a_0, \ldots, a_n \in A$ such that $\mathfrak{A}' = (\mathfrak{A}, a_0, a_1, \ldots, a_n)$ is a model of T'.

Take a hyp. structure $\mathfrak{B}' \succ \mathfrak{A}'$, with $\mathfrak{B}' \neq \mathfrak{A}'$. By (5.7) there is $b \in \phi^{\mathfrak{B}'} \setminus \phi^{\mathfrak{A}'}$; by (5.5) and (5.6), $Th(\mathfrak{B}', (x)_{x \in A}, b)$ has a prime model $(\mathfrak{C}', (x)_{x \in A}, b)$ say, with $\mathfrak{A}' \prec \mathfrak{C}' \prec \mathfrak{B}'$; we claim that the L-reduct $\mathfrak{C}$ of $\mathfrak{C}'$ is the required prime elementary extension.

Let $\mathfrak{D} \succ \mathfrak{A}$, with $\mathfrak{D} \neq \mathfrak{A}$. Let $\mathfrak{D}' = (\mathfrak{D}, a_0, \ldots, a_n)$; then $\mathfrak{D}' \succ \mathfrak{A}'$ and so there is $d \in \phi^{\mathfrak{D}'} \setminus \phi^{\mathfrak{A}'}$; so by (6.4)

$$(\mathfrak{D}', (x)_{x \in A}, d) \equiv (\mathfrak{C}', (x)_{x \in A}, b)$$

since $\{d\}$, $\{b\}$ are independent subsets. Now since $(\mathfrak{C}', (x)_{x \in A}, b)$ is prime, there is an elementary embedding e of this structure into $(\mathfrak{D}', (x)_{x \in A}, d)$. Then e is a fortiori an elementary embedding of $\mathfrak{C}$ into $\mathfrak{D}$ which is the identity on A, as required.

We can now prove the converse of (5.3) to obtain:

6.7 THEOREM. T is $\aleph_1$-categorical iff T is Π_1^1-categorical.

Proof. If T is $\aleph_1$-categorical, the result is given by (5.3).

For the converse, we employ the following characterisation of $\aleph_1$-categorical theories due to Morley [10]:

T is $\aleph_1$-categorical iff every countable model of T has a prime elementary extension.

First, suppose that T is Π_1^1-categorical and has a strongly minimal formula ϕ; let $\mathfrak{A}$ be an arbitrary countable model of T, and $(a_i)_{i<k}$ be a basis for $\phi^{\mathfrak{A}}$, where $\dim(\phi^{\mathfrak{A}}) = k \leq \omega$. We show that $\mathfrak{A}$ has a prime elementary extension.

We can obtain a hyp. model $\mathfrak{B}$ of T with $\dim(\phi^{\mathfrak{B}}) = k$, having a basis $(b_i)_{i<k}$ and such that $(\mathfrak{B}, (b_i)_{i<k})$ is prime.

By (6.4), $(\mathfrak{B}, (b_i)_{i<k}) \equiv (\mathfrak{A}, (a_i)_{i<k})$, so there is $\mathfrak{A}_1 \prec \mathfrak{A}$ with

$$(\mathfrak{B},(b_i)_{i<k}) \cong (\mathfrak{A}_1,(a_i)_{i<k}) \prec (\mathfrak{A},(a_i)_{i<k}).$$

Now $(a_i)_{i<k}$ is a basis for $\phi^{\mathfrak{A}}$, and so

$$\phi^{\mathfrak{A}} \subseteq \mathrm{cl}((a_i)_{i<k}) \subseteq A_1.$$

Hence, $\phi^{\mathfrak{A}} = \phi^{\mathfrak{A}_1}$, and by (5.7), $\mathfrak{A}_1 = \mathfrak{A}$, so $\mathfrak{B} \cong \mathfrak{A}$.

By (6.6), $\mathfrak{B}$ has a prime elementary extension, hence so does $\mathfrak{A}$. Thus T is $\aleph_1$-categorical, by Morley's criterion.

If T has no strongly minimal formula, extend T to a principal hyp. extension T' of T by constants, having a strongly minimal formula (6.5). T' is Π^1_1-categorical, and so by the above, T' is $\aleph_1$-categorical.

Now any model of T can be enlarged to a model of T'; hence T is $\aleph_1$-categorical also.

It is known that there is no logical relationship between $\aleph_0$-categoricity and $\aleph_1$-categoricity. It follows from the above results that for hyp. theories there is no connection between $\aleph_0$-categoricity and Π^1_1-categoricity, although both notions involve isomorphisms of countable models, e.g. the theory of dense ordered sets without end points (the rationals with their order) is $\aleph_0$- but not Π^1_1-categorical.

We conclude with one open problem from [4], concerning an analogue of Morley's characterisation of $\aleph_1$-categorical theories quoted above.

In [4] we indicated that an index for a prime elementary extension of a hyp. model $\mathfrak{A}$ of a Π_1^1-categorical theory can be obtained effectively from an index for $\mathfrak{A}$. If, conversely, we have a hyp. theory T such that each hyp. model has such an 'effective' prime elementary extension, does it follow that T is Π_1^1-categorical (or, equivalently, $\aleph_1$-categorical)? If so, is effectiveness essential?

REFERENCES

[1] J. T. Baldwin and A. H. Lachlan, On strongly minimal sets, Journ. Symbolic Logic 36 (1971), 79-96.

[2] J. P. Cleave, Hyperarithmetic model theory, unpublished.

[3] J. P. Cleave, Hyperarithmetic ultrafilters, Proc. Summer School in Logic, Lecture Notes in Mathematics, Vol. 70, Springer, Berlin (1968), 223-240.

[4] N. J. Cutland, The theory of hyperarithmetic and Π_1^1-models, Ph.D. thesis, University of Bristol (1970).

[5] A. Ehrenfeucht, Theories having at least continuum many non-isomorphic models in each power, Abstract 550-23, Notices Amer. Math. Soc., 5 (1958), 680-681.

[6] A. Ehrenfeucht and A. Mostowski, Models of axiomatic theories admitting automorphisms, Fund. Math., 43 (1956), 50-68.

[7] G. Kreisel, The axiom of choice and the class of hyperarithmetic functions, Indag. Math., 24 (1962), 307-319.

[8] W. Marsh, On ω_1-categorical but not ω-categorical theories, Doctoral dissertation, Dartmouth College (1966).

[9] M. Morley, Categoricity in power, Trans. Amer. Math. Soc., 114 (1965), 514-538.

[10] M. Morley, Countable models of $\aleph_1$-categorical theories, Israel Jour. of Math., 5 (1967), 65-72.

[11] M. Morley and R. L. Vaught, Homogeneous universal models, Math. Scand., 11 (1962), 37-57.

[12] H. Rogers Jr., *Theory of Recursive Functions and Effective Computability*, McGraw-Hill, New York (1967).

INFINITARY PROPERTIES, LOCAL FUNCTORS, AND SYSTEMS OF ORDINAL FUNCTIONS

Solomon Feferman

Stanford University

1. Introduction

The principal results of this paper take the following form:

(1) whenever $\vec{\mathfrak{A}} = \langle \mathfrak{A}_i \rangle$, $\vec{\mathfrak{B}} = \langle \mathfrak{B}_i \rangle$ are in $\Pi_{i\in I}\,\mathcal{C}_i$ and for each i, $\mathfrak{A}_i \equiv_L \mathfrak{B}_i$, then $F(\vec{\mathfrak{A}}) \equiv_L F(\vec{\mathfrak{B}})$.

(1) is established for a variety of operations $F : \Pi_{i\in I}\,\mathcal{C}_i \longrightarrow \mathfrak{D}$ where each $\mathcal{C}_i$, $\mathfrak{D}$ is a collection of structures and L is one of the infinitary languages $L_{\infty,\kappa}$ or its fragments $L^{\alpha}_{\infty,\kappa}$. We say that F preserves $\equiv_L$ when (1) holds. The results are also accompanied by ones in which F preserves $\preccurlyeq_L$ (a semantical substructure relation).

§2 is taken up with preliminaries on structures, the languages L and some simple but useful lemmas on preservation. It concludes with a slight reformulation of a characterization of $\equiv$ in $L^{\alpha}_{\infty,\kappa}$, the so-called back-and-forth criterion. This is the main tool for the further general results.

The notion of κ-local functor is introduced in §3. Roughly speaking, for the case $F : \mathcal{C} \rightarrow \mathfrak{D}$ (i.e. where I is a singleton), F is κ-local if it preserves $\subseteq$ and if, whenever $\mathfrak{A}$ is in $\mathcal{C}$, each set $Z \subseteq |F(\mathfrak{A})|$ with $\mathrm{Card}(Z) < \kappa$ depends only on some $X \subseteq |\mathfrak{A}|$ with $\mathrm{Card}(X) < \kappa$. For $\kappa = \omega$ this is equivalent to the condition that F preserves direct limits.

The general result obtained in §3 is that

(2) any κ-local functor preserves $\equiv_L$ and $\preccurlyeq_L$ for $L = L_{\infty,\kappa}, L^{\alpha}_{\infty,\kappa}$.

The proof of this is straightforward, once matters are arranged properly.

A number of applications of an algebraic character are made in §4. Among these are to the operations (i) $F(\mathfrak{A})$ = ring of polynomials over (a ring) $\mathfrak{A}$, (ii) $F(\mathfrak{A})$ = ring of formal power series over $\mathfrak{A}$, (iii) $F(\mathfrak{A})$ = free Γ-model on the set $|\mathfrak{A}|$ of generators (Γ a set of equations), (iv) $F(\mathfrak{A}, \mathfrak{B}) = \mathfrak{A} \otimes \mathfrak{B}$ (the tensor product of modules over a ring) and (v) the generalized products $F(\vec{\mathfrak{A}})$ of [13]. F is ω-local in (i) and $\aleph_1$-local in (ii). The other cases are not κ-local as they stand, but are obtained as compositions of some κ-local functor with other operations which trivially preserve $\equiv$ and $\preccurlyeq$.

The applications which are given the most attention are (in §5) to <u>relatively categorical systems</u> $\vec{f}$ <u>of ordinal functions</u>, as defined in [11]. Each such $\vec{f}$ has directly associated with it an ω-local functor $F_{\vec{f}}$ from substructures of the ordinals (OR,$\leqslant$), to $\vec{f}$-closed substructures. For example, when $\mathfrak{A} = (\alpha,\leqslant)$, $F_{+}(\mathfrak{A}) = (\omega^{\alpha},\leqslant,+)$, and $F_{+,\cdot}(\mathfrak{A}) = (\omega^{\omega^{\alpha}},\leqslant,+,\cdot)$; for the function $f_1 = \lambda\xi,\eta.(\omega^{\xi} + \eta)$ which may be used to build up Cantor normal forms, $F_{f_1}(\mathfrak{A}) = (\epsilon_{\alpha},\leqslant,f_1)$. Combining (2) with Ehrenfeucht's result [9] that $(\omega^{\omega},\leqslant) \preccurlyeq_L (\mathrm{OR},\leqslant)$ in $L = L^{\omega}_{\infty,\omega}$, one immediately recovers the further conclusions of [9] for + and for +, $\cdot$, as well as similar results for much more general $\vec{f}$. It is also shown in §5 that each $F_{\vec{f}}$ has a natural extension to the collection $\mathcal{C}$ of arbitrary simply ordered structures $\mathfrak{A}$.

Most of the comparisons with earlier work have been placed at the end of the paper, in §6. In particular, the principal results and methods for operations F which preserve $\equiv$ in the usual finitary language $L_{\omega,\omega}$ are recalled. Some of the F mentioned above, for example tensor product, do not preserve $\equiv$ in $L_{\omega,\omega}$. The restriction here to infinitary languages is thus a matter of necessity.

Notation. k, ℓ, m, n, p range over the set ω of natural numbers. $\alpha, \beta, \gamma, \xi, \eta, \zeta$ range over the ordinals OR; each ordinal is identified with its set of predecessors. Cardinals are initial ordinals; κ is any cardinal. Card(X) or $\overline{\overline{X}}$ is the cardinality of X. $\mathcal{P}(A) = \{X \mid X \subseteq A\}$, $\mathcal{P}_\kappa(A) = \{X \mid X \subseteq A, \overline{\overline{X}} < \kappa\}$. h is a partial function from A to B if it is a function $h \subseteq A \times B$; we write $h : \underset{\smile}{A} \longrightarrow B$ in this case. $\mathcal{D}h$, $\mathcal{R}h$ are the domain and range (set of values) resp. of h. B^A is the set of all $h : \underset{\smile}{A} \longrightarrow B$ with $\mathcal{D}h = A$; we write $h : A \longrightarrow B$ for $h \in B^A$. In contrast to the usage in category theory, B is not determined by h; the notation for morphisms in categories will be discussed below. For any A, $id_A : A \longrightarrow A$ is the identity map on A. $\vec{a} = \langle a_i \rangle_{i \in I}$, (or $\langle a_i \rangle_i$ when I is fixed) is written for a function $\vec{a}$ with $\mathcal{D}\vec{a} = I$, $\vec{a}(i) = a_i$ for each $i \in I$. One also says that $\vec{a}$ is the map $i \longmapsto a_i$. The λ-symbolism is used for functions of ordinals, i.e. $\lambda\xi.a_\xi$ instead of $\langle a_\xi \rangle_{\xi \in OR}$. The use of $i, j \longmapsto a_{i,j}$ and $\lambda\xi,\eta.a_{\xi,\eta}$ is explained similarly.

2. Semantic relations preserved by operations on structures

2.1. Types of structures. A signature is a triple $\sigma = (\langle M_\nu \rangle_{\nu \in \sigma_1}, \langle N_\nu \rangle_{\nu \in \sigma_2}, \sigma_3)$, where the σ_κ, M_ν, N_ν are all sets and each M_ν, N_ν is non-empty.[1] $\mathfrak{A}$ is a structure of type σ if it has the form

$$(1) \qquad \mathfrak{A} = (A, \langle R_\nu \rangle_{\nu \in \sigma_1}, \langle f_\nu \rangle_{\nu \in \sigma_2}, \langle c_\nu \rangle_{\nu \in \sigma_3})$$

where each $R_\nu \subseteq A^{M_\nu}$, $f_\nu : A^{N_\nu} \longrightarrow A$, and $c_\nu \in A$. A is called the under-

[1] Note that there is otherwise no restriction on the cardinality of these sets.

lying set of $\mathfrak{A}$, also written $A = |\mathfrak{A}|$.[2] The collection of all $\mathfrak{A}$ of type σ is denoted by $\mathcal{S}(\sigma)$. σ is called relational when $\sigma_2 \cup \sigma_3 = 0$.

A more informal notation than (1) is used in applications, e.g. $\mathfrak{A} = (A, R_1,\ldots,R_j, f_1,\ldots,f_\ell, c_1,\ldots,c_p)$ when the σ_k are finite. Also, if $\mathfrak{A} = (A,\ldots)$ and $S \subseteq A^p$, the structure $(\mathfrak{A},S) = (A,S,\ldots)$ obtained by adjoining S to $\mathfrak{A}$ is of type σ', obtained from σ by adjoining an index for a new p-ary relation.

Let $\mathfrak{A} = (A,\ldots)$, $\mathfrak{B} = (B,\ldots)$ be of the same type σ. We write $\mathfrak{A} \subseteq \mathfrak{B}$ if $\mathfrak{A}$ is a substructure of $\mathfrak{B}$, i.e. $A \subseteq B$, the designated individuals c_ν are the same, and the relations and operations of $\mathfrak{A}$ are the restrictions to A of the corresponding relations and operations in $\mathfrak{B}$ (under which A must be closed). For any $X \subseteq A$, $\mathrm{Gen}_{\mathfrak{A}}(X)$ is the substructure of $\mathfrak{A}$ generated by X; this may be empty if σ is relational and $X = 0$. The empty structure is excluded only when considering models of sentences.

If $h : A \xrightarrow[\text{onto}]{1-1} B$ establishes an isomorphism between $\mathfrak{A}$ and $\mathfrak{B}$ we write $h : \mathfrak{A} \cong \mathfrak{B}$. Given $h : A \rightharpoonup B$, we put $h \in J^0(\mathfrak{A},\mathfrak{B})$ if for some $\bar{h}$,

(2) $$h \subseteq \bar{h} \text{ and } \bar{h} : \mathrm{Gen}_{\mathfrak{A}}(\mathfrak{D}h) \cong \mathrm{Gen}_{\mathfrak{B}}(\mathfrak{R}h).$$

There is at most one such $\bar{h}$, called the partial isomorphism induced by h or the closure of h. h is said to be closed if $h = \bar{h}$, i.e. if $h \in J^0(\mathfrak{A},\mathfrak{B})$ and $\mathfrak{D}h = |\mathrm{Gen}_{\mathfrak{A}}(\mathfrak{D}h)|$. If $h \in J^0(\mathfrak{A},\mathfrak{B})$ with $\mathfrak{D}h = A$ (i.e. if h establishes an isomorphism of $\mathfrak{A}$ into $\mathfrak{B}$) we write $h : \mathfrak{A} \rightarrowtail \mathfrak{B}$.

---

[2] The results of §§2,3 extend to many-sorted structures $\mathfrak{A} = (\langle A_\nu\rangle_{\nu\in\sigma_0},\ldots)$ by the usual device of unifying the domains A_ν. Single-sorted domains are treated for simplicity, and suffice for most of the examples to be discussed.

2.2. <u>The languages</u> $L_{\infty,\kappa}$ ($\kappa > 1$). Fix any signature σ. Let V be an unlimited stock of <u>variables</u>; in some cases below we take it that with each object e is associated $v_e \in V$ (in a 1-1 manner). In the following, v ranges over subsets of V. The class of <u>terms</u> of the language for σ-structures is defined inductively by: (i) each variable and constant $\underset{\sim}{c}_\nu$ ($\nu \in \sigma_3$) is a term; (ii) if $\nu \in \sigma_2$ and $\vec{t} = \langle t_\mu \rangle_{\mu \in N_\nu}$ where each t_μ is a term then $\underset{\sim}{f}_\nu(\vec{t})$ is a term. The <u>atomic formulas</u> are all those of the form $(t_1 = t_2)$ where t_1, t_2 are terms, and $\underset{\sim}{R}_\nu(\vec{t})$ where $\nu \in \sigma_1$ and $\vec{t} = \langle t_\mu \rangle_{\mu \in N_\nu}$ is a sequence of terms. The class of <u>formulas</u> is inductively defined by: (i) each atomic formula is a formula; (ii) if ϕ is a formula so also is its <u>negation</u> $\sim\phi$; (iii) ('∞') if Γ is any non-empty set of formulas, then the <u>disjunction</u> $\bigvee\!\!\!\bigvee_{\phi \in \Gamma} \phi$ is a formula; (iv) ('κ') if ϕ is a formula and $v \in \mathcal{P}_\kappa(V)$ with $v \neq 0$, then the result $(\exists v)\phi$ of <u>existential quantification</u> w.r. to v is a formula. The operations of <u>conjunction</u> $\bigwedge\!\!\!\bigwedge$ and <u>universal quantification</u> $(\forall v)$ $(0 < \bar{\bar{v}} < \kappa)$ are defined as usual from these. fr is the function which assigns to each term or formula its <u>set of free variables</u>. ϕ is a <u>sentence</u> if $fr(\phi) = 0$. If ψ is a subformula of a sentence then $card(fr(\psi)) < max(\omega, \kappa)$. However, no general restriction is made on $fr(\phi)$. We also write $L_{\infty,\omega}$ for the class of all formulas of this language.

It should be noted for comparison with the literature (e.g. [3], [15]) that $L_{\infty,2}$ is often denoted $L_{\infty,\omega}$. While these have the same expressive power, there is a useful technical distinction to be considered next.

The <u>quantifier-rank</u> $qr(\phi)$ is an ordinal defined recursively for ϕ in $L_{\infty,\kappa}$ by:

(1) (i) $qr(\phi) = 0$ <u>for</u> ϕ <u>atomic</u>,

(ii) $qr(\sim\phi) = qr(\phi)$,

(iii) $qr(\bigvee\!\!\!\bigvee_{\phi\in\Gamma}\phi) = \sup\{qr(\phi) \mid \phi \in \Gamma\}$,

(iv) $qr((\exists v)\phi) = qr(\phi) + 1$.

We write $L^{\alpha}_{\infty,\kappa}$ for the class of formulas ϕ of $L_{\infty,\kappa}$ with $qr(\phi) \leq \alpha$.

Note, again for comparison with the literature, that $qr(\phi) = 1$ for ϕ of the form $\bigvee\!\!\!\bigvee_{n<\omega} (\exists\{v_0,\ldots,v_n\})\phi_n$ when each ϕ_n is atomic. In [15] this formula would be construed as $\bigvee\!\!\!\bigvee_{n<\omega} (\exists v_0)\ldots(\exists v_n)\phi_n$ in $L_{\infty,2}$ and would be given rank ω.

2.3. Semantical notions. Given $\mathfrak{A} = (A,\ldots)$ with $A \neq 0$, and $v \subseteq V$, an assignment a to v in $\mathfrak{A}$ (or A) is a map $a : v \longrightarrow A$. When a is an assignment to fr(t) in A, the value of t under a in $\mathfrak{A}$ is denoted by $t_{\mathfrak{A}}[a]$ or $t[a]$. We write $\mathfrak{A} \models \phi[a]$ if a is an assignment to fr(ϕ) in $\mathfrak{A}$ and a satisfies ϕ in $\mathfrak{A}$; $\mathfrak{A} \models \phi$ is written if ϕ is a sentence true in $\mathfrak{A}$.

Suppose $\mathfrak{A} = (A,\ldots)$, $\mathfrak{B} = (B,\ldots)$ are of the same type, $h : A \rightharpoonup B$ and L is any set of formulas. We define $\mathfrak{A} \equiv \mathfrak{B}$ at h in L to hold if

(1) for any ϕ in L and any assignment a to fr(ϕ) in $\mathcal{D}$ h and for b = ha we have: $\mathfrak{A} \models \phi[a] \iff \mathfrak{B} \models \phi[b]$.

Let L be $L_{\infty,\kappa}$ or some $L^{\alpha}_{\infty,\kappa}$. The usual relations $\equiv_L$ and $\preccurlyeq_L$ of elementary equivalence in L, resp. elementary substructure in L, are obtained as special cases of this notion:

(2)
(i) $\mathfrak{A} \equiv_L \mathfrak{B} \iff \mathfrak{A} \equiv \mathfrak{B}$ at 0 in L,
(ii) $\mathfrak{A} \preccurlyeq_L \mathfrak{B} \iff \mathfrak{A} \subseteq \mathfrak{B}$ and $\mathfrak{A} \equiv \mathfrak{B}$ at id_X in L for every $X \subseteq A$ with $card(X) < \max(\omega,\kappa)$.

2.4. Operations on structures. Suppose given $I \neq 0$ and similarity types σ_i ($i \in I$) and τ. Let $\mathcal{C}_i \subseteq \mathcal{S}(\sigma_i)$ and $\mathfrak{D} \subseteq \mathcal{S}(\tau)$. We deal in general with operations

$$(1) \qquad F : \Pi_{i \in I} \mathcal{C}_i \longrightarrow \mathfrak{D},$$

i.e. where $F(\vec{\mathfrak{A}})$ is in $\mathfrak{D}$ for any $\vec{\mathfrak{A}} = \langle \mathfrak{A}_i \rangle_{i \in I}$ with each $\mathfrak{A}_i \in \mathcal{C}_i$. In all the applications to be made, $|F(\vec{\mathfrak{A}})|$ is also a function of $\vec{A} = \langle A_i \rangle_{i \in I}$ where $A_i = |\mathfrak{A}_i|$; this need not be assumed for the general results. The sequence notation is dropped in the case of unary and binary operations, e.g. $F(\mathfrak{A})$ is written when $F : \mathcal{C} \rightarrow \mathfrak{D}$ and $\mathfrak{A} \in \mathcal{C}$.

Suppose $\mathcal{E}$ is a relation between structures of the same, but arbitrary, type. We say F <u>preserves</u> $\mathcal{E}$ (when F is of the form (1)) if

$$(2) \qquad \begin{array}{l} \text{whenever } \vec{\mathfrak{A}} = \langle \mathfrak{A}_i \rangle_i, \ \vec{\mathfrak{B}} = \langle \mathfrak{B}_i \rangle_i \text{ are in } \Pi_{i \in I} \mathcal{C}_i, \\ \text{and } \mathfrak{A}_i \,\mathcal{E}\, \mathfrak{B}_i \text{ for all } i \in I \text{ then } F(\vec{\mathfrak{A}}) \,\mathcal{E}\, F(\vec{\mathfrak{B}}). \end{array}$$

In particular, we are concerned with F which preserve $\equiv_L$ and $\preccurlyeq_L$ for L as in §2.3. In some cases we make use of intermediate results for a pair of relations $\mathcal{E}$, $\mathcal{E}_1$ where the conclusion in (2) is replaced by $F(\vec{\mathfrak{A}}) \,\mathcal{E}_1 F(\vec{\mathfrak{B}})$.

2.5. Two lemmas on preservation. These are both quite trivial, but are helpful to separate the work for the applications in §§4, 5.

Let $\mathcal{C} = \mathcal{S}(\sigma)$ and suppose θ is in $L_{\infty,\kappa}$ (for σ). Let $fr(\theta) = u$. Associated with each $\mathfrak{A} = (A,\ldots)$ in $\mathcal{C}$ is the relation

$$(1) \qquad S^{\theta}_{\mathfrak{A}} \subseteq A^u; \quad a \in S^{\theta}_{\mathfrak{A}} \Longleftrightarrow \mathfrak{A} \models \theta[a].$$

Let τ be the type of structures $(\mathfrak{A},S) = (A,S,\ldots)$ where $\mathfrak{A}$ is in $\mathcal{C}$

and $S \subseteq A^u$; take $\mathfrak{D} = \mathcal{S}(\tau)$. Then $D_\theta : \mathcal{C} \to \mathfrak{D}$ where

(2) $$\text{for each } \mathfrak{A} = (A,\ldots),\ D_\theta(\mathfrak{A}) = (\mathfrak{A}, S^\theta_{\mathfrak{A}}).$$

In other words, $D_\theta(\mathfrak{A})$ simply expands $\mathfrak{A}$ by the relation explicitly defined by θ in $\mathfrak{A}$.

LEMMA 1. Suppose θ is in $L^\beta_{\infty,\kappa}$. If $\mathfrak{A} \equiv \mathfrak{B}$ at h in $L^{\beta+\alpha}_{\infty,\kappa}$ then $D_\theta(\mathfrak{A}) \equiv D_\theta(\mathfrak{B})$ at h in $L^\alpha_{\infty,\kappa}$.

COROLLARY 1(a). If θ is in $L^\beta_{\infty,\kappa}$ and $\beta + \alpha = \alpha$ and $L = L^\alpha_{\infty,\kappa}$ then D_θ preserves $\equiv_L$ and $\preccurlyeq_L$.

COROLLARY 1(b). D_θ preserves $\equiv_L$ and $\preccurlyeq_L$ when $L = L_{\infty,\kappa}$ and θ is in L.

One obtains similar results for expansions $D_{\theta_1,\ldots,\theta_n}$.

The second lemma is a standard one for congruence relations. Let σ be a type for structures $\mathfrak{A} = (A,E,\ldots)$ with a designated binary relation E. Let $\mathcal{C} \subseteq \mathcal{S}(\sigma)$ consist of all $\mathfrak{A}$ in which E is an equivalence relation with respect to which all the relations and functions of $\mathfrak{A}$ are preserved. Let $\mathfrak{D} = \mathcal{S}(\sigma)$. Take $M : \mathcal{C} \to \mathfrak{D}$ with

(3) $$M(\mathfrak{A}) = \mathfrak{A}/E.$$

LEMMA 2. For any $\mathfrak{A}$ in $\mathcal{C}$, formula ϕ and assignment a to $fr(\phi)$ in $\mathfrak{A}$, if ϕ does not contain the equality symbol then

$$\mathfrak{A} \models \phi[a] \iff \mathfrak{A}/E \models \phi[a/E].$$

When $\mathfrak{A} = (A,E,\ldots)$, $\mathfrak{B} = (B,E',\ldots)$ in $\mathcal{C}$ and $h \in J^o(\mathfrak{A},\mathfrak{B})$, determine $M(h) : A/E \to B/E'$ by $M(h)(x/E) = h(x)/E'$.

COROLLARY 2(a). If $\mathfrak{A}, \mathfrak{B} \in \mathcal{C}$ and $\mathfrak{A} \equiv \mathfrak{B}$ at h in $L^{\alpha}_{\infty,\kappa}$ then $M(\mathfrak{A}) \equiv M(\mathfrak{B})$ at $M(h)$ in $L^{\alpha}_{\infty,\kappa}$.

COROLLARY 2(b). M preserves $\equiv_L$ and $\preccurlyeq_L$ for any $L = L^{\alpha}_{\infty,\kappa}$ or $L_{\infty,\kappa}$.

2.6. The back-and-forth criterion. As stated in Theorem 3 and its corollaries below, this is the main tool in establishing the general preservation result of §3. It is formulated essentially as in [5]; for background and remarks on the proof, cf. §6.1 below. Fix any σ and $\kappa > 1$.

For each $\mathfrak{A} = (A,\dots)$, $\mathfrak{B} = (B,\dots)$ and α, a set $J^{\alpha}_{\kappa}(\mathfrak{A},\mathfrak{B})$ or $J^{\alpha}(\mathfrak{A},\mathfrak{B})$ (when κ is fixed) is defined by the following recursion:

(1) (i) $h \in J^0(\mathfrak{A},\mathfrak{B}) \Leftrightarrow h$ induces a partial isomorphism between $\mathfrak{A}$ and $\mathfrak{B}$ (2.1(2));

(ii) $h \in J^{\alpha+1}(\mathfrak{A},\mathfrak{B}) \Leftrightarrow$ (a) for each $X \subseteq A$ with $\mathrm{card}(X) < \kappa$ there exists $h' \in J^{\alpha}(\mathfrak{A},\mathfrak{B})$ with $h \subseteq h'$ and $X \subseteq \mathfrak{D}h'$, and (b) for each $Y \subseteq B$ with $\mathrm{card}(Y) < \kappa$ there exists $h' \in J^{\alpha}(\mathfrak{A},\mathfrak{B})$ with $h \subseteq h'$ and $Y \subseteq \mathfrak{R}h'$.

(iii) $J^{\alpha}(\mathfrak{A},\mathfrak{B}) = \bigcap_{\beta<\alpha} J^{\beta}(\mathfrak{A},\mathfrak{B})$ for limit α.

The following points should be noted. First, $J^0(\mathfrak{A},\mathfrak{B}) \supseteq J^1(\mathfrak{A},\mathfrak{B}) \supseteq \dots \supseteq J^{\alpha}(\mathfrak{A},\mathfrak{B}) \supseteq \dots$. Further, if $h \in J^{\alpha}(\mathfrak{A},\mathfrak{B})$ and $h_1 \subseteq h$ then $h_1 \in J^{\alpha}(\mathfrak{A},\mathfrak{B})$. Finally, if $h \in J^{\alpha}(\mathfrak{A},\mathfrak{B})$ then its closure $\bar{h} \in J^{\alpha}(\mathfrak{A},\mathfrak{B})$.

THEOREM 3. $\mathfrak{A} \equiv \mathfrak{B}$ at h in $L^{\alpha}_{\infty,\kappa} \Leftrightarrow h \in J^{\alpha}(\mathfrak{A},\mathfrak{B})$.

COROLLARY 3(a). For $L = L^{\alpha}_{\infty,\kappa}$, $\mathfrak{A} \equiv_L \mathfrak{B} \Leftrightarrow 0 \in J^{\alpha}(\mathfrak{A},\mathfrak{B}) \Leftrightarrow J^{\alpha}(\mathfrak{A},\mathfrak{B}) \neq 0$.

COROLLARY 3(b). For $L = L^{\alpha}_{\infty,\kappa}$, $\mathfrak{A} \preccurlyeq_L \mathfrak{B} \Leftrightarrow \mathfrak{A} \subseteq \mathfrak{B}$ and $id_X \in J^{\alpha}(\mathfrak{A},\mathfrak{B})$ for all $X \subseteq A$ with $\mathrm{card}(X) < \max(\omega,\kappa)$.

3. Semantic relations preserved by local functors

3.1. Categories and functors for structures.[3] For each σ, $\mathcal{S}(\sigma)$ is considered as a category with morphisms $h : \mathfrak{A} \rightarrow \mathfrak{B}$, i.e. triples $(h, \mathfrak{A}, \mathfrak{B})$ where $h : A \longrightarrow B$ establishes an isomorphism of $\mathfrak{A}$ into $\mathfrak{B}$. For $\mathfrak{A} \subseteq \mathfrak{B}$, $h = id_A$ determines the inclusion morphism, written $h : \mathfrak{A} \subseteq \mathfrak{B}$. Any $\mathcal{C} \subseteq \mathcal{S}(\sigma)$ can be considered as a subcategory, using all $h : \mathfrak{A} \rightarrow \mathfrak{B}$ for $\mathfrak{A}, \mathfrak{B}$ in $\mathcal{C}$.

Suppose given an operation $F : \mathcal{C} \longrightarrow \mathcal{D}$ where $\mathcal{C} \subseteq \mathcal{S}(\sigma)$, $\mathcal{D} = \mathcal{S}(\tau)$. For F to be regarded as a functor we must also have an associated map (also denoted by F) on morphisms in $\mathcal{C}$ to morphisms in $\mathcal{D}$, which satisfies:

(1) if $h : \mathfrak{A} \rightarrow \mathfrak{B}$ in $\mathcal{C}$ then $F(h) : F(\mathfrak{A}) \longrightarrow F(\mathfrak{B})$ in $\mathcal{D}$.

In addition, it is required that $F(id_{|\mathfrak{A}|}) = id_{|F(\mathfrak{A})|}$ for $id_{|\mathfrak{A}|} : \mathfrak{A} \longrightarrow \mathfrak{A}$, and that F preserves composition of morphisms. By (1), if $\mathfrak{A} \subseteq \mathfrak{B}$ then $F(\mathfrak{A})$ is isomorphic to a substructure of $F(\mathfrak{B})$. Following the terminology of 2.4, when considered simply as an operation on structures, F preserves $\subseteq$ if $\mathfrak{A} \subseteq \mathfrak{B}$ implies $F(\mathfrak{A}) \subseteq F(\mathfrak{B})$. In addition, when considered as a functor, it is also required that

(2) $h : \mathfrak{A} \subseteq \mathfrak{B}$ implies $F(h) : F(\mathfrak{A}) \subseteq F(\mathfrak{B})$.

Every functor is equivalent to one which preserves $\subseteq$, at least among substructures of any given structure. For simplicity, we shall arrange to deal primarily with such functors.

Suppose $\mathcal{C}$ is closed under substructures and that F is a functor

[3] No knowledge of category theory is presumed beyond acquaintance with the basic notions; [18] is one of several good references.

which preserves $\subseteq$. To see when F preserves direct limits, it is sufficient to consider the cases $\mathcal{B} = \varinjlim \mathcal{H}$ where

(3) (i) $\mathcal{H}$ is a non-empty collection of substructures of $\mathcal{B}$,

(ii) for each $\mathfrak{A}_1, \mathfrak{A}_2 \in \mathcal{H}$ there exists $\mathfrak{A}' \in \mathcal{H}$ with $\mathfrak{A}_i \subseteq \mathfrak{A}'$ ($i = 1, 2$), and

(iii) $|\mathcal{B}| = \cup_{\mathfrak{A} \in \mathcal{H}} |\mathfrak{A}|$.

Then F preserves $\varinjlim \mathcal{H}$ simply if $|F(\mathcal{B})| = \cup_{\mathfrak{A} \in \mathcal{H}} |F(\mathfrak{A})|$.

Now consider an operation $F : \Pi_{i \in I} \mathcal{C}_i \to \mathcal{D}$ where $\mathcal{C}_i \subseteq \mathcal{S}(\sigma_i)$, $\mathcal{D} = \mathcal{S}(\tau)$. $\Pi_{i \in I} \mathcal{S}(\sigma_i)$ forms a category with morphisms $\vec{h} : \vec{\mathfrak{A}} \to \vec{\mathcal{B}}$ where $\vec{\mathfrak{A}} = \langle \mathfrak{A}_i \rangle_i$, $\vec{\mathcal{B}} = \langle \mathcal{B}_i \rangle_i$, $\vec{h} = \langle h_i \rangle_i$ and each $h_i : \mathfrak{A}_i \to \mathcal{B}_i$. The restriction of this to $\Pi_{i \in I} \mathcal{C}_i$ determines the latter as a subcategory. For F to be regarded as a functor we must again have an associated map F on morphisms such that

(4) if $h_i : \mathfrak{A}_i \longrightarrow \mathcal{B}_i$ for each $i \in I$ and $\vec{h} = \langle h_i \rangle_i$ then $F(\vec{h}) : F(\vec{\mathfrak{A}}) \longrightarrow F(\vec{\mathcal{B}})$,

and such that F preserves identity morphisms and composition. F is said to <u>preserve $\subseteq$ as a functor</u> if whenever $h_i : \mathfrak{A}_i \subseteq \mathcal{B}_i$ for each $i \in I$ then $F(\vec{h}) : F(\vec{\mathfrak{A}}) \subseteq F(\vec{\mathcal{B}})$.

3.2. <u>κ-local functors (unary case)</u>. Suppose $\kappa > 1$. A unary operation $F : \mathcal{C} \to \mathcal{D}$ is called a κ-<u>local functor</u> if the following conditions hold:

(1) (i) the domain $\mathcal{C}$ of F is closed under substructures;

(ii) F is a functor which preserves $\subseteq$; and

(iii) whenever $\mathfrak{A} \in \mathcal{C}$ and $Z \in \mathcal{P}_\kappa(|F(\mathfrak{A})|)$ then for some $X \in \mathcal{P}_\kappa(|A|)$ we have $Z \subseteq |F(\mathrm{Gen}_{\mathfrak{A}}(X))|$.

The general notion of κ-local functor seems not to have been considered in category theory. The following shows that it is equivalent to a categorical notion when $\kappa = \omega$.

LEMMA 4. (i) If F is a κ-local functor and $\kappa_1 \geq \max(\omega,\kappa)$, then F is κ_1-local.

(ii) If F satisfies 1(i),(ii) above, then F is ω-local if and only if F preserves direct limits.

Proof. (i) Suppose $\kappa_1 > \kappa$, κ_1 infinite. Given $Z \in \mathcal{P}_{\kappa_1}(|F(\mathfrak{A})|)$ choose $X_z \in \mathcal{P}_{\kappa}(|\mathfrak{A}|)$ for each $z \in Z$, so that $z \in F(|\mathrm{Gen}_{\mathfrak{A}}(X_z)|)$. Then $X = \bigcup_{z \in Z} X_z$ serves for Z since F preserves $\subseteq$; further $\mathrm{card}(X) < \kappa_1$.

(ii) Suppose F is ω-local and $\mathfrak{B} = \varinjlim \mathcal{H}$, where $\mathcal{H}$ satisfies (3)(i)-(iii) of the preceding section. It is sufficient to show $|F(\mathfrak{B})| \subseteq \bigcup_{\mathfrak{A} \in \mathcal{H}} |F(\mathfrak{A})|$. Given $z \in |F(\mathfrak{B})|$, there is $X = \{x_1,\ldots,x_n\}$ with $X \subseteq |\mathfrak{B}|$ and $z \in |F(\mathrm{Gen}_{\mathfrak{B}}(X))|$. By 3(ii), for some $\mathfrak{A} \in \mathcal{H}$, $X \subseteq |\mathfrak{A}|$; since F preserves $\subseteq$, also $z \in |F(\mathfrak{A})|$.

Suppose conversely that F preserves direct limits. Each $\mathfrak{A} = \varinjlim \{\mathrm{Gen}_{\mathfrak{A}}(X) \mid X \in \mathcal{P}_{\omega}(|\mathfrak{A}|)\}$. Hence each element z of $|F(\mathfrak{A})|$ belongs to $|F(\mathrm{Gen}_{\mathfrak{A}}(X))|$ for some $X \subseteq |\mathfrak{A}|$ with X finite. By the same argument as in (i), F is ω-local.

Assume in the following that F is a κ-local functor, and that $\mathfrak{A}, \mathfrak{B} \in \mathcal{C}$. $J^{\alpha}(\mathfrak{A},\mathfrak{B})$ is defined as in §2.6(1) relative to κ. If $h \in J^{0}(\mathfrak{A},\mathfrak{B})$ and h is closed then $h : \mathrm{Gen}_{\mathfrak{A}}(\mathcal{D} h) \longrightarrow \mathrm{Gen}_{\mathfrak{B}}(\mathcal{R} h)$. The domain and range structures of h are substructures of $\mathfrak{A}, \mathfrak{B}$ resp., hence both belong to $\mathcal{C}$ by (1)(i). Then by the functorial requirement, $F(h) : F(\mathrm{Gen}_{\mathfrak{A}}(\mathcal{D} h)) \longrightarrow F(\mathrm{Gen}_{\mathfrak{B}}(\mathcal{R} h))$ in $\mathfrak{D}$, i.e. $F(h) \in J^{0}(F(\mathfrak{A}),F(\mathfrak{B}))$.

LEMMA 5. Suppose $h \in J^{\alpha}(\mathfrak{A},\mathfrak{B})$ and that h is closed; then $F(h) \in J^{\alpha}(F(\mathfrak{A}),F(\mathfrak{B}))$.

Proof. By induction on α. For $\alpha = 0$ it is true by the preceding.

It is trivial by definition for limit α, assuming it is true for all smaller ordinals.

Suppose the statement true for α; it will be shown for $\alpha + 1$. Let h be closed, $h \in J^{\alpha+1}(\mathfrak{A},\mathfrak{B})$. To show $F(h) \in J^{\alpha+1}(F(\mathfrak{A}),F(\mathfrak{B}))$, first consider any $Z \subseteq |F(\mathfrak{A})|$ with card(Z) $< \kappa$. By (1)(iii) there exists $X \in \mathcal{P}_\kappa(A)$ with $Z \subseteq |F(\mathrm{Gen}_{\mathfrak{A}}(X))|$. By definition of $J^{\alpha+1}$, there exists $h' \in J^{\alpha}(\mathfrak{A},\mathfrak{B})$ with $h \subseteq h'$ and $X \subseteq \mathfrak{D}h'$. We can assume further that h' is closed. $F(h) \subseteq F(h')$, since F preserves $\subseteq$. For the same reason, $F(\mathrm{Gen}_{\mathfrak{A}}(X)) \subseteq F(\mathrm{Gen}_{\mathfrak{A}}(\mathfrak{D}h'))$; hence $Z \subseteq |F(\mathrm{Gen}_{\mathfrak{A}}(\mathfrak{D}h'))| = \mathfrak{D}F(h')$. Finally, $F(h') \in J^{\alpha}(F(\mathfrak{A}),F(\mathfrak{B}))$ by induction hypothesis. By symmetry, for each $W \subseteq |F(\mathfrak{B})|$ with card(W) $< \kappa$ there exists h' satisfying $F(h') \in J^{\alpha}(F(\mathfrak{A}),F(\mathfrak{B}))$, $F(h) \subseteq F(h')$ and $W \subseteq \mathfrak{R}F(h')$.

Main preservation theorem (unary case)

THEOREM 6. Suppose F is a κ-local functor and $L = L^{\alpha}_{\infty,\kappa}$. Then F preserves $\equiv_L$ and $\preccurlyeq_L$.

Proof. This is now a direct consequence of Lemma 5 and the back-and-forth criterion. If $\mathfrak{A} \equiv \mathfrak{B}$ in $L^{\alpha}_{\infty,\kappa}$ then $J^{\alpha}(\mathfrak{A},\mathfrak{B}) \neq 0$, hence $J^{\alpha}(F(\mathfrak{A}),F(\mathfrak{B})) \neq 0$. Suppose $\mathfrak{A} \preccurlyeq \mathfrak{B}$ in $L^{\alpha}_{\infty,\kappa}$. Since κ-local implies $\max(\omega,\kappa)$-local, for each $Z \subseteq |F(\mathfrak{A})|$ with card(Z) $< \max(\omega,\kappa)$ there exists $X \subseteq A$ with $Z \subseteq |F(\mathrm{Gen}_{\mathfrak{A}}(X))|$ and card(X) $< \max(\omega,\kappa)$. $\mathfrak{A} \subseteq \mathfrak{B}$ and $h = \mathrm{id}_X \in J^{\alpha}(\mathfrak{A},\mathfrak{B})$. Then $\bar{h} = \mathrm{id}_{|\mathrm{Gen}_{\mathfrak{A}}(X)|}$ and $\bar{h} : \mathrm{Gen}_{\mathfrak{A}}(X) \subseteq \mathfrak{B}$, so $F(\bar{h}) : F(\mathrm{Gen}_{\mathfrak{A}}(X)) \subseteq F(\mathfrak{B})$. It follows by Lemma 5 that $F(\bar{h}) \in J^{\alpha}(F(\mathfrak{A}),F(\mathfrak{B}))$; but $\mathrm{id}_Z \subseteq F(\bar{h})$ so also $\mathrm{id}_Z \in J^{\alpha}(F(\mathfrak{A}),F(\mathfrak{B}))$.

3.3. κ-local functors (general case). An operation $F : \Pi_{i\in I}\, \mathcal{C}_i \longrightarrow \mathfrak{D}$ is called a κ-local functor if the following conditions hold:

(1) (i) Each $\mathcal{C}_i$ is closed under substructures;

(ii) F is a functor which preserves $\subseteq$; and

(iii) whenever $\vec{\mathfrak{A}} = \langle \mathfrak{A}_i \rangle_i \in \Pi_{i \in I} \mathcal{C}_i$ and $Z \in \mathcal{P}_\kappa(|F(\vec{\mathfrak{A}})|)$ then for some $\vec{X} = \langle X_i \rangle_i$ we have $X_i \in \mathcal{P}_\kappa(A_i)$ for each i and $Z \subseteq |F(\langle Gen_{\mathfrak{A}_i}(X_i) \rangle_i)|$.

When $\vec{h} = \langle h_i \rangle_i$ with each $h_i \in J^o(\mathfrak{A}_i, \mathfrak{B}_i)$, we call $\vec{h}$ closed if each h_i is closed; in this case $F(\vec{h})$ is defined, and $F(\vec{h}) \in J^o(F(\vec{\mathfrak{A}}), F(\vec{\mathfrak{B}}))$ by (1)(i),(ii). The methods of proof in §3.2 extend directly to yield the following.

LEMMA 7. Suppose $\vec{h} = \langle h_i \rangle_i$ is closed, where each $h_i \in J^\alpha(\mathfrak{A}_i, \mathfrak{B}_i)$; then $F(\vec{h}) \in J^\alpha(F(\vec{\mathfrak{A}}), F(\vec{\mathfrak{B}}))$.

Main preservation theorem (general case)

THEOREM 8. Suppose F is a κ-local functor and $L = L^\alpha_{\infty,\kappa}$. Then F preserves $\equiv_L$ and $\preccurlyeq_L$.

4. Applications to some universal constructions and product operations

Each of the (more or less familiar) operations F considered in 4.1-4.7, and some of those falling under 4.8, has a categorical characterization in the sense that for each $\vec{\mathfrak{A}}$ of a certain kind, $F(\vec{\mathfrak{A}})$ is a universal solution to a certain mapping problem. To obtain preservation results for these by the present methods we must return instead to explicit descriptions of $F(\vec{\mathfrak{A}})$. In any case, only a few of the generalized product operations treated in 4.8 have known algebraic characterizations.

In general, $|F(\vec{\mathfrak{A}})|$ is given as a set of finite or infinite "words" which may be identified under some equivalence relation E. The preservation results for these F follow from the main theorems of §3 in combination with the lemmas of §2.5. In most cases, the conclusions follow directly from the description of $F(\vec{\mathfrak{A}})$; a few require more detailed discussion. The examples were primarily chosen to illustrate different features of the general results and for instructive comparisons with known situations in finitary languages (§6).

We now write L_κ for any $L^\alpha_{\infty,\kappa}$ or $L_{\infty,\kappa}$.

4.1. Polynomials over a ring. Let $\mathcal{C}$ be the class of rings $\mathfrak{A} = (A, +, \cdot, 0)$. For each $\mathfrak{A}$ in $\mathcal{C}$, let $\mathfrak{A}[x]$ be the ring of polynomials $p = \Sigma_{i=0}^{n} p_i x^i$ in one indeterminate x, over $\mathfrak{A}$. $\mathfrak{A}[x]$ is isomorphic to $F(\mathfrak{A}) = (A_0^\omega, \oplus, \circ, \vec{0})$, where A_0^ω consists of all sequences $\vec{p} = \langle p_n \rangle_{n<\omega}$ of elements of A which are eventually 0, $\vec{p} \oplus \vec{q} = \langle p_n + q_n \rangle_n$, $\vec{p} \circ \vec{q} = \langle \Sigma_{i=0}^{n} p_i q_{n-i} \rangle_n$ and $\vec{0} = \langle 0 \rangle_n$. F can be regarded as an ω-local functor from $\mathcal{C}$ to $\mathcal{C}$. Hence, the operation $\mathfrak{A} \mapsto \mathfrak{A}[x]$ preserves $\equiv_{L_\kappa}$ and $\preccurlyeq_{L_\kappa}$ for any $\kappa \geqslant \omega$.

4.2. Formal power series over a ring. For $\mathcal{C}$ as in 4.1 and $\mathfrak{A}$ in $\mathcal{C}$, $\mathfrak{A}[[x]]$ is the ring of formal power series $p = \Sigma_{i=0}^{\infty} p_i x^i$ in x, over $\mathfrak{A}$. Here $\mathfrak{A}[[x]] \cong F(\mathfrak{A}) = (A^\omega, \oplus, \circ, \vec{0})$ with $\oplus$, $\circ$, $\vec{0}$ as before. In this case, F can be regarded as an $\aleph_1$-local functor from $\mathcal{C}$ to $\mathcal{C}$, so that the operation $\mathfrak{A} \mapsto \mathfrak{A}[[x]]$ preserves $\equiv_{L_\kappa}$ and $\preccurlyeq_{L_\kappa}$ for any $\kappa \geqslant \aleph_1$.

4.3. Fields of quotients. Let $\mathcal{C}$ be the class of integral domains $\mathfrak{A} = (A, +, \cdot, 0, 1)$. Let $Q(\mathfrak{A})$ be the field of quotients a/b for $a, b \in A$, $b \neq 0$. $Q(\mathfrak{A}) \cong F(\mathfrak{A})/E$ where

$$F(\mathfrak{A}) = (A \times (A - \{0\}), E, \oplus, \circ, \bar{0}, \bar{1}),$$

$(a,b)E(a',b') \Leftrightarrow ab' = a'b$, $(a,b)\oplus(a',b') = (ab' + a'b, bb')$, $(a,b)\circ(a',b') = (aa', bb')$, $\bar{0} = (0,1)$, and $\bar{1} = (1,1)$. F can be considered as an ω-local functor from $\mathcal{C}$ to $\mathcal{C}$, so that the operation $\mathfrak{A} \longmapsto Q(\mathfrak{A})$ preserves $\equiv_{L_\kappa}$ and $\preccurlyeq_{L_\kappa}$ for any $\kappa > \omega$.

4.4. Free Γ-models. Let σ be the signature for sets, i.e. structures $\mathfrak{A} = (A)$; take $\mathcal{C} = \mathcal{S}(\sigma)$. Let $\tau = (0, \langle N_\nu\rangle_{\nu\in\tau_2}, \tau_3)$ be a signature for algebraic structures (i.e. with no relations), $\mathfrak{B} = (B, \langle f_\nu\rangle_{\nu\in\tau_2}, \langle c_\nu\rangle_{\nu\in\tau_3})$; take $\mathfrak{D} = \mathcal{S}(\tau)$. For any set Γ of equations in the language of $\mathfrak{D}$ and for any A we have a structure $G_\Gamma(A)$ which is a free Γ-model on the set A of generators; this may be constructed explicitly as follows. Start with the set Tm of all terms in the language for $\mathfrak{D}$. Let Tm(A) consist of all terms t such that $fr(t) \subseteq \{v_a \mid a \in A\}$, and let $\Gamma(A)$ consist of all equations $(t_1' = t_2')$ which result from equations $(t_1 = t_2)$ in $\Gamma \cup \{v_0 = v_0\}$ by substituting members of Tm(A) throughout for the variables of $(t_1 = t_2)$. Finally, take E_A to be the set of pairs (t_1, t_2) for which the equation $t_1 = t_2$ is derivable from $\Gamma(A)$ by means of the following rules of inference:

$$\frac{t_1 = t_2}{t_2 = t_1} \qquad \frac{t_1 = t_2,\ t_2 = t_3}{t_1 = t_3} \qquad \frac{t_\mu = t'_\mu \ (\mu \in N_\nu)}{f_\nu(\langle t_\mu\rangle_\mu) = f_\nu(\langle t'_\mu\rangle_\mu)}.$$

Put $F(\mathfrak{A}) = (Tm(A), E_A, \langle f_\nu\rangle_{\nu\in\tau_2}, \langle c_\nu\rangle_{\nu\in\tau_3})$. Then E_A is a congruence relation for $F(\mathfrak{A})$ and $F(\mathfrak{A})/E_A$ is a free Γ-model on the set A of generators (w.r. to the embedding $a \longmapsto v_a/E_A$). F preserves $\subseteq$, for if $A \subseteq B$ and $t_1, t_2 \in Tm(A)$ and $\Gamma(B) \vdash (t_1 = t_2)$ then we can find a derivation of $(t_1 = t_2)$ from $\Gamma(A)$. Let κ be a regular cardinal with $card(N_\nu) < \kappa$ for each $\nu \in \tau_1$. If $t \in Tm(A)$ then $t \in Tm(A_1)$ for some $A_1 \in \mathcal{P}_\kappa(A)$. Hence F is a κ-local functor. It follows that G_Γ preserves $\equiv_{L_\kappa}$ and $\preccurlyeq_{L_\kappa}$.

This can be extended more generally to the formation of free structures for sets Γ of conditional equations (e.g. to obtain preservation results for free Boolean algebras with countable meets and joins.)

4.5. Completions of metric spaces. Let $\mathcal{C}$ be the collection of two-sorted structures $\mathfrak{A} = (A, \mathbb{R}^+, m)$ where $\mathbb{R}^+$ is fixed throughout as the set of non-negative real numbers and $m : A^2 \longrightarrow \mathbb{R}^+$ is a metric on A. Let $C_{\mathfrak{A}}$ be the collection of Cauchy sequences $\vec{x} = \langle x_n \rangle_{n<\omega}$ in A and E the usual equivalence relation between such sequences. Extend m to M on $C_{\mathfrak{A}}$ by $M(\vec{x},\vec{y}) = \lim_{n\to\infty} m(x_n,y_n)$. Then $F(\mathfrak{A}) = (C_{\mathfrak{A}}, \mathbb{R}^+, E, M)$ determines F as an $\aleph_1$-local functor. $F(\mathfrak{A})/E$ is isomorphic to the completion $\overline{\mathfrak{A}}$ of $\mathfrak{A}$. Hence $\mathfrak{A} \mapsto \overline{\mathfrak{A}}$ preserves $\equiv_{L_\kappa}$ and $\preccurlyeq_{L_\kappa}$ for any $\kappa \geq \aleph_1$.

4.6. Group rings.* Let $\mathcal{C}_0$ be the collection of rings $\mathfrak{A} = (A, +, \cdot, 0)$ and $\mathcal{C}_1$ the collection of groups $\mathfrak{B} = (B, *)$. The group ring $\mathfrak{A}\mathfrak{B}$ is isomorphic to

$$F(\mathfrak{A}, \mathfrak{B}) = (A_0^B, \oplus, \circ, \vec{0}),$$

where A_0^B consists of all $\vec{a} = \langle a_g \rangle_{g\in B}$ in A^B such that $a_g \neq 0$ for only finitely many g, $\vec{a} \oplus \vec{b} = \langle a_g + b_g \rangle_{g\in B}$, $\vec{a} \circ \vec{b} = \langle \Sigma_{h_1 * h_2 = g}\, a_{h_1} \cdot b_{h_2} \rangle_{g\in B}$, and $\vec{0} = \langle 0 \rangle_{g\in B}$. F preserves $\subseteq$ in the first argument, but not as it stands in the second argument. This is corrected by taking A_0^B to consist instead of all $\vec{a} : \underset{\smile}{B} \longrightarrow A - \{0\}$ with $\mathfrak{D}\vec{a}$ finite, modifying the operations accordingly. Then F is an ω-local functor, so that it and also the operation $\mathfrak{A}, \mathfrak{B} \mapsto \mathfrak{A}\mathfrak{B}$ preserves $\equiv_{L_\kappa}$ and $\preccurlyeq_{L_\kappa}$ for any $\kappa \geq \omega$. (As pointed out to me by G. Sabbagh, this extends directly to the collection $\mathcal{C}_1$ of monoids, thereby including 4.1 as a special case.)

4.7. Tensor products of modules.* Fix any commutative ring $\mathcal{R} = (R, +_R, \cdot_R)$ and let $\mathcal{C}$ be the class of modules $\mathfrak{A} = (A, +, \langle m_r \rangle_{r\in R})$

---

*Cf. [17] for the categorical and explicit descriptions of this example.

over $\mathcal{R}$, where $m_r(x) = rx$. For $\mathcal{A}, \mathcal{B} \in \mathcal{C}$, the tensor product $\mathcal{A} \otimes \mathcal{B}$ is again in $\mathcal{C}$. Described informally, its underlying set $A \otimes B$ is generated under addition from elements denoted by $a \otimes b$, for all $a \in A$, $b \in B$. Sums of these are identified only as required by the laws for modules and by the conditions: $(a_1 + a_2) \otimes b = (a_1 \otimes b) + (a_2 \otimes b)$, $a \otimes (b_1 + b_2) = (a \otimes b_1) + (a \otimes b_2)$, and $r(a \otimes b) = (ra) \otimes b = a \otimes (rb)$. In contrast to the preceding examples, the operation $G(\mathcal{A}, \mathcal{B}) = \mathcal{A} \otimes \mathcal{B}$ does not preserve $\subseteq$.[5]

The following is one way of spelling out this description of the tensor product. First define

$$F(\mathcal{A}, \mathcal{B}) = ((A \times B)^{(\omega)}, S_1, S_2, \frown, +_1, +_2, \langle m_r^1 \rangle_{r \in R}, \langle m_r^2 \rangle_{r \in R}),$$

where (i) $(A \times B)^{(\omega)}$ is the set of all finite non-empty sequences $t = ((a_1, b_1), \ldots, (a_n, b_n))$ of members of $A \times B$, (ii) S_1, S_2 are binary relations where for $t = ((a_1, b_1), \ldots, (a_n, b_n))$, $t' = ((a_1', b_1'), \ldots, (a_m', b_m'))$ we have $(t, t') \in S_1 \Leftrightarrow n = m = 1$ and $a_1 = a_1'$ and $(t, t') \in S_2 \Leftrightarrow n = m = 1$ and $b_1 = b_1'$, (iii) $t \frown t' = ((a_1, b_1), \ldots, (a_n, b_n), (a_1', b_1'), \ldots, (a_m', b_m'))$, (iv) $t +_1 t' = (a_1 + a_1', b_1)$ and $t +_2 t' = (a_1, b_1 + b_1')$, (v) $m_r^1(t) = ((ra_1, b_1), \ldots, (ra_n, b_n))$, and $m_r^2(t) = ((a_1, rb_1), \ldots, (a_n, rb_n))$. The intention is that $((a_1, b_1), \ldots, (a_n, b_n))$ will correspond to $\Sigma_{i=1}^{n} a_i \otimes b_i$. Then $\frown$ corresponds to addition and both m_r^1, m_r^2 correspond to multiplication by r. S_1, S_2 are the relations which hold only between singletons $t = ((a_1, b_1))$, $t' = ((a_1', b_1'))$, having the same 1st term, resp. same 2nd term. In these cases $t +_1 t'$ corresponds to $(a_1 + a_1') \otimes b_1$ and $t +_2 t'$ to $a_1 \otimes (b_1 + b_1')$. F preserves $\subseteq$ and is an ω-local functor.

---

[5] For an example (related to another one in §6), consider modules over the integers Z. Let Q be the rationals; then $Q \otimes Q/Z$ is trivial while $Z \otimes Q/Z$ is not.

Consider the following "rules" ρ_i $(i = 1,\ldots,7)$ for "deriving" pairs (t_1,t_2).

$$\rho_1.\quad \frac{(t_1,t_2)}{(t_2,t_1)} \qquad\qquad \rho_2.\quad \frac{(t_1,t_2)\ (t_2,t_3)}{(t_1,t_3)}$$

$$\rho_3.\quad \frac{(t_1,t_2)}{(m_r^1(t_1),m_r^1(t_2))} \qquad\qquad \rho_4.\quad \frac{(t_1,t_2)}{(m_r^2(t_1),m_r^2(t_2))}$$

$$\rho_5.\quad \frac{(t_1,t_2)\ (t_1',t_2')}{(t_1 \frown t_1',\ t_2 \frown t_2')}$$

$$\rho_6.\quad \frac{(t_1,t_2)\ (t_1',t_2')}{(t_1 +_1 t_1',\ t_2 +_1 t_2')} \qquad \rho_7.\quad \frac{(t_1,t_2)\ (t_1',t_2')}{(t_1 +_2 t_1',\ t_2 +_2 t_2')}$$

Let the "axioms" α_i $(i = 1,\ldots,5)$ be all pairs (t,t') of the following forms:

$\alpha_1.\quad (t,t) \qquad \alpha_2.\quad (m_r^1(t),\ m_r^2(t)) \qquad \alpha_3.\quad (t_1 \frown t_2,\ t_2 \frown t_1)$

$\alpha_4.\quad (t_1 \frown t_2,\ t_1 +_1 t_2)$ for each $(t_1,t_2) \in S_2$

$\alpha_5.\quad (t_1 \frown t_2,\ t_1 +_2 t_2)$ for each $(t_1,t_2) \in S_1$.

By a "derivation" we mean a finite sequence of pairs (t_j,t_j') each term of which is one of the axioms α_i or is obtained from one or two of the preceding pairs by means of one of the rules ρ_i. Let $(t,t') \in E_{F(\mathfrak{A},\mathfrak{B})} \Longleftrightarrow$ there is a derivation in which (t,t') is the final term. Then $E = E_{F(\mathfrak{A},\mathfrak{B})}$ is a congruence relation in $F(\mathfrak{A},\mathfrak{B})$ and

$$\mathfrak{A}\otimes\mathfrak{B} \cong ((A \times B)^{(\omega)}/E,\ \frown/E,\ \langle m_r^1/E\rangle_{r\in R}).$$

Furthermore, we have a formula θ in $L^2_{\infty,\omega}$ (with two free variables) such that

$$F(\mathfrak{A},\mathfrak{B}) \models \theta[t,t'] \Leftrightarrow (t,t') \in E_{F(\mathfrak{A},\mathfrak{B})}.$$

θ simply expresses that there is a derivation which puts (t,t') in E.

Suppose $\mathfrak{A} \equiv_L \mathfrak{A}'$, $\mathfrak{B} \equiv_L \mathfrak{B}'$ in $L = L^{\alpha}_{\infty,\kappa}$ with $\omega \leq \kappa$. By Theorem 8, $F(\mathfrak{A},\mathfrak{B}) \equiv_L F(\mathfrak{A}',\mathfrak{B}')$. Then by Cor. 1(a), $D_\theta(F(\mathfrak{A},\mathfrak{B})) \equiv_L D_\theta(F(\mathfrak{A}',\mathfrak{B}'))$, provided also $\alpha \geq \omega$. Hence by Cor. 2(b) $F(\mathfrak{A},\mathfrak{B})/E_{F(\mathfrak{A},\mathfrak{B})} \equiv_L F(\mathfrak{A}',\mathfrak{B}')/E_{F(\mathfrak{A}',\mathfrak{B}')}$, i.e. $\mathfrak{A}\otimes\mathfrak{B} \equiv_L \mathfrak{A}'\otimes\mathfrak{B}'$. By the same argument $\otimes$ preserves $\preccurlyeq_L$ in $L = L^{\alpha}_{\infty,\kappa}$ for $\alpha \geq \omega$, and $\otimes$ preserves $\equiv_{L_{\infty,\kappa}}$ and $\preccurlyeq_{L_{\infty,\kappa}}$ for any $\kappa \geq \omega$. (Added in proof: G. Sabbagh has found a direct proof of this for $\otimes$ and other functors using only their categorical properties, by means of an interesting strengthening of Theorem 6.)

4.8. <u>Generalized products</u>. Let σ be arbitrary, $\mathcal{C} = \mathcal{S}(\sigma)$, $I \neq 0$. These products are operations of the form $G : \mathcal{C}^I \to \mathcal{D}$ where $\mathcal{D} = \mathcal{S}(\tau)$. For simplicity we illustrate with τ the type of structures (P,R) where $R \subseteq P^2$. Consider any $\theta_1, \ldots, \theta_m$ with two free variables v_0, v_1 and any condition $\Phi(X_1,\ldots,X_m)$ on subsets X_j of I. Given $\vec{\mathfrak{A}} = \langle\mathfrak{A}_i\rangle_i$ in $\mathcal{C}^I$, let $P = \Pi_{i\in I}A_i$. The relation $R = R_\Phi^{\theta_1,\ldots,\theta_m} \subseteq P^2$ is defined for $\vec{x} = \langle x_i\rangle_i$, $\vec{y} = \langle y_i\rangle_i$ in P by:

$$(\vec{x},\vec{y}) \in R \Leftrightarrow \Phi(K^{\theta_1}_{\vec{\mathfrak{A}}}(\vec{x},\vec{y}),\ldots,K^{\theta_m}_{\vec{\mathfrak{A}}}(\vec{x},\vec{y})) \text{ holds,}$$

where $K^{\theta}_{\vec{\mathfrak{A}}}(\vec{x},\vec{y}) = \{i \in I \mid \mathfrak{A}_i \models \theta[x_i,y_i]\} = \{i \in I \mid (x_i,y_i) \in S^{\theta}_{\vec{\mathfrak{A}}_i}\}$. We take $G(\vec{\mathfrak{A}}) = (P,R)$.

G need not preserve $\subseteq$ unless each θ_j is quantifier-free. To get around this, first apply the operation $D : \mathcal{C} \to \mathcal{C}_1$ on members $\mathfrak{A}$ of $\mathcal{C}$ to structures $(\mathfrak{A},S^{(1)},\ldots,S^{(m)})$ with m additional binary relations, given by

$$D(\mathfrak{A}) = (\mathfrak{A},S^{\theta_1}_{\mathfrak{A}},\ldots,S^{\theta_m}_{\mathfrak{A}}).$$

Let $G_1 : \mathcal{C}_1^I \to \mathcal{D}$ be the generalized product operation for Φ and the formulas $\mathfrak{S}^{(1)}(v_0,v_1), \ldots, \mathfrak{S}^{(m)}(v_0,v_1)$. Then $G(\vec{\mathfrak{A}}) = G_1(\langle D(\mathfrak{A}_i)\rangle_i)$. Furthermore, G_1 is a 1-local functor. Let $\beta = \max_j(qr(\theta_j))$, $\beta + \alpha = \alpha$, $\kappa \geq 2$ and $L = L^{\alpha}_{\infty,\kappa}$ or $L_{\infty,\kappa}$. By Cor. 1(a),(b) and the results of §3, G preserves $\equiv_L$ and $\preccurlyeq_L$.

Note, for comparison with [13], that Φ can be any condition on subsets of I (not necessarily given by a formula for a structure on $\mathcal{P}(I)$) and the θ_j need not be finitary formulas. As it happens, all of the familiar instances of generalized products (direct products, direct sums, reduced products, various ordered sums and products) use quantifier-free finitary θ_j.

(Added in proof: another, quite interesting, algebraic application of Theorem 6 has been given by Eklof [24], providing the strongest and mathematically most natural formulation yet of <u>Lefschetz' Principle</u>.)

5. <u>Applications to systems of ordinal functions</u> [6]

5.1. <u>Relatively categorical systems of functions</u>. In this section we briefly review the relevant notions and results from [11]; cf. also [12]. A (finitary) system of ordinal functions is of the form $\vec{f} = \langle f_1,\ldots,f_n\rangle$ where $f_i : OR^{m_i} \longrightarrow OR$. Let

(1) $$\Omega_{\vec{f}} = (OR, \leq, \vec{f}, 0).$$

Then for any set X of ordinals, $Cl_{\vec{f}}(X) = |Gen_{\Omega_{\vec{f}}}(X)|$ is the closure of X under $\vec{f}$. The class $In_{\vec{f}} = \{\xi \mid Cl_{\vec{f}}(\xi) = \xi\}$ of $\vec{f}$-<u>inaccessibles</u> is closed and unbounded; it is enumerated by a unique normal function $\vec{f}'$, called the <u>critical function</u> of $\vec{f}$. (One also refers to the critical function g' of a normal function $g : OR \longrightarrow OR$, where g' enumerates $\{\xi \mid g(\xi) = \xi\}$; the connection is that $g' = \vec{f}'$, when g is cofinal with $\vec{f}$.)

[6] The new results of this section were presented in a talk entitled 'Systems of ordinal functions and functionals' given at the meeting of the American Mathematical Society at Claremont, Nov. 22, 1969. It was only realized somewhat later that the main ideas for the proofs could be formulated in much more general terms, as given above.

$\vec{f}$ is called <u>replete</u> if for each α, $Cl_{\vec{f}}(\alpha) \in OR$, i.e. if the closure of each initial segment is again an initial segment. Note that

(2) if $\vec{f}$ is replete then for each α, $\vec{f}'(\alpha) = Cl_{\vec{f}}(\mathcal{R}(\vec{f}'\restriction\alpha))$

(and conversely). We also write $In_{\vec{f}}\restriction\alpha$ for $\mathcal{R}(\vec{f}'\restriction\alpha)$, i.e. for $\{\vec{f}'(\beta) \mid \beta < \alpha\}$.

Let τ be the type of $\Omega_{\vec{f}}$ and take $Tm_{\vec{f}}$ to be the set of terms in the language for $\mathcal{S}(\tau)$. $\vec{f}$ is called <u>relatively categorical</u> if there is a relation R such that

$$(3)\qquad \begin{array}{c} \text{for all } t_1, t_2 \in Tm_{\vec{f}} \text{ and } a : fr(t_1) \cup fr(t_2) \longrightarrow In_{\vec{f}}, \\ t_1[a] \leqslant t_2[a] \Longleftrightarrow R(t_1, t_2, Dg_{\leqslant}(a)). \end{array}$$

Here $t[a]$ is the value of t in $\Omega_{\vec{f}}$ under the assignment $a\restriction fr(t)$; $Dg_{\leqslant}(a) = \{(k,l) \mid v_k, v_l \in \mathcal{D}a \text{ and } a(v_k) \leqslant a(v_l)\}$.[7] In other words, $\vec{f}$ is rel. categorical if the orderings of given terms built up from $\vec{f}$-inaccessibles depends only on the ordering of these inaccessibles.

Call $\vec{f}$ <u>increasing</u> if each of its members f_i is increasing, i.e. for any $j = 1, \ldots, m_i$ and $\alpha_1, \ldots, \alpha_{m_i}$ we have $\alpha_j \leqslant f_i(\alpha_1, \ldots, \alpha_{m_i})$. We shall use for each normal $g : OR \longrightarrow OR$ with $g(0) > 0$ a certain "hierarchy" $\langle g^{(\alpha)} \rangle_{\alpha \in OR}$ of normal functions derived from g by iteration of the critical process: $g^{(0)} = g$ and for $\alpha > 0$, $g^{(\alpha)}$ enumerates $\{\xi \mid g^{(\gamma)}(\xi) = \xi \text{ for all } \gamma < \alpha\}$. Then $\lambda\alpha,\beta.g^{(\alpha)}(\beta)$ is increasing and $\lambda\alpha.g^{(\alpha)}(0)$ is normal. For example if $g(\alpha) = 1 + \alpha$, we have $g^{(\alpha)}(\beta) = \omega^\alpha + \beta$. We write $\chi^{(\alpha)}(\beta)$ when the initial function $g(\alpha) = \chi^{(0)}(\alpha) = \omega^\alpha$; thus $\chi^{(1)}(\alpha) = \epsilon_\alpha$. The αth fixed point of $\lambda\xi.\chi^{(\xi)}(0)$ is

[7] There is an implicit restriction here, but not in the next section, to terms with variables v_k, $k \in \omega$. This permits an <u>effective</u> version of the notion of relative categoricity, when R can be chosen to be recursive; cf. [11] for the formulation of the results with this notion.

frequently denoted by Γ_α.

The following was proved in [11]:

(4) Suppose $\vec{f}$ is rel. categorical (resp., replete) and $g = \vec{f}'$. Then the same holds of

(i) $\vec{f} ^\frown \langle g \rangle$, and

(ii) $\vec{f} ^\frown \langle \lambda\xi,\eta . g^{(\xi)}(\eta) \rangle$.

More general schemes for iterating the critical process on systems of functions have been considered in [12], but these go far beyond the familiar systems considered for illustration here.

(5) The following systems $\vec{f}$ are relatively categorical, replete and increasing:

(i) $\vec{f} = \langle \lambda\xi.(1+\xi) \rangle$, $\vec{f}'(\alpha) = \omega + \alpha$;

(ii) $\vec{f} = \langle \lambda\xi,\eta.(\omega^\xi + \eta) \rangle$, $\vec{f}'(\alpha) = \epsilon_\alpha$;

(iii) $\vec{f} = \langle \lambda\xi,\eta.(\omega^\xi + \eta), \lambda\xi.\epsilon_\xi \rangle$, $\vec{f}'(\alpha) = \chi_\alpha^{(1)}$;

(iv) $\vec{f} = \langle \lambda\xi,\eta.(\omega^\xi + \eta), \lambda\xi,\eta.\chi^{(\xi)}(\eta) \rangle$, $\vec{f}'(\alpha) = \Gamma_\alpha$;

(v) $\vec{f} = \langle \lambda\xi,\eta.(\xi+\eta) \rangle$, $\vec{f}'(\alpha) = \omega^\alpha$;

(vi) $\vec{f} = \langle \lambda\xi,\eta.(\xi+\eta), \lambda\xi,\eta.(\xi\cdot\eta) \rangle$, $\vec{f}'(0) = 1$, $f'(1+\alpha) = \omega^{\omega^\alpha}$.

(i) is trivial. (ii)-(iv) follow alternately by (3)(ii), (3)(i). The statements for (v), (vi) do not fall out as special cases of (3), but it is easy to establish them directly.

5.2. Functors associated with relatively categorical systems. Let σ be the type of structures $\mathfrak{A} = (A,S)$ with $S \subseteq A^2$. $\mathfrak{A}$ is said to be pre-ordered if S is reflexive in A, transitive, and connected; we write $x \leqslant_{\mathfrak{A}} y$ for $(x,y) \in S$ in this case. Take $(x,y) \in E_{\mathfrak{A}} \iff x \leqslant_{\mathfrak{A}} y$ and $y \leqslant_{\mathfrak{A}} x$. $\mathfrak{A}$ is linearly ordered if $E_{\mathfrak{A}}$ is the identity relation. $\mathcal{C}_p$, $\mathcal{C}_l$, $\mathcal{C}_w$ denote respectively the categories of pre-ordered, linearly ordered, and (linearly) well-ordered structures. $M : \mathcal{C}_p \longrightarrow \mathcal{C}_l$ is the functor given by $M(\mathfrak{A}) = \mathfrak{A}/E_{\mathfrak{A}}$. We write $\mathfrak{A} = (A, \leq)$ for $\mathfrak{A} \subseteq (OR, \leq)$.

For $\mathfrak{A}, \mathfrak{B}$ in $\mathcal{C}_l$, $\mathfrak{A}$ is called an <u>initial segment</u> of $\mathfrak{B}$ if $\mathfrak{A} \subseteq \mathfrak{B}$ and if $|\mathfrak{A}|$ is an initial segment of $|\mathfrak{B}|$ in $\leq_{\mathfrak{B}}$; we write $\mathfrak{A} \lessdot \mathfrak{B}$ in this case. It is then clear what is meant by a functor which preserves initial segments. Each $\mathfrak{B}$ in $\mathcal{C}_l$ has a maximal well-ordered $\mathfrak{A} \lessdot \mathfrak{B}$, denoted by $W(\mathfrak{B})$; W is not a functor.

With τ as in §5.1, $\mathcal{S}(\tau)$ is the type of structures $\mathfrak{A} = (A,S,\vec{g},c)$. The <u>retract</u> operation $\mathfrak{A} \longmapsto (A,S)$ is a functor from $\mathcal{S}(\tau)$ to $\mathcal{S}(\sigma)$. $\mathcal{D}_p$, $\mathcal{D}_l$, $\mathcal{D}_w$ denote respectively the category of $\mathfrak{A}$ in $\mathcal{S}(\tau)$ for which the retract of $\mathfrak{A}$ is correspondingly in $\mathcal{C}_p$, $\mathcal{C}_l$, $\mathcal{C}_w$. One also speaks of $F : \mathcal{C}_l \to \mathcal{D}_l$ preserving initial segments by composing retraction with F. When $\mathfrak{A}$ is in $\mathcal{C}_p$ and $E_{\mathfrak{A}}$ is a congruence relation in $\mathfrak{A}$, we also write $M(\mathfrak{A})$ for $\mathfrak{A}/E_{\mathfrak{A}}$.

Each $\mathfrak{A} = (A, \leq_{\mathfrak{A}})$ in $\mathcal{C}_w$ has an <u>order-type</u> $\|\mathfrak{A}\|$ which is an ordinal or is equal to OR. $e : (\|\mathfrak{A}\|, \leq) \cong \mathfrak{A}$ for a unique isomorphism e, called <u>the enumeration of</u> $\mathfrak{A}$. One also speaks more generally of order-types of linearly ordered $\mathfrak{A}$, i.e. representatives $\|\mathfrak{A}\|$ of the isomorphism types.

Suppose $\vec{f}$ is relatively categorical. Let

$$(1) \qquad G(\mathfrak{A}) = Gen_{\Omega_{\vec{f}}}(A) = (Cl_{\vec{f}}(A), \leq, \vec{f}, 0), \quad \text{for } \mathfrak{A} = (A, \leq) \subseteq (In(\vec{f}), \leq).$$

By definition of rel. categoricity, 5.1(3), G can be considered as a functor from substructures of $(In(\vec{f}), \leq)$ to substructures of $\Omega_{\vec{f}}$. This leads naturally to a functor F_1 on $\mathcal{C}_w$; given well-ordered $\mathfrak{A} = (A, \leq_{\mathfrak{A}})$

first find $e^{-1}(\mathfrak{A}) = (\alpha, \leq)$ and then form $G(In_{\vec{f}} \restriction \alpha, \leq)$. In particular, $F(\alpha, \leq) = Gen_{\Omega_{\vec{f}}}(In_{\vec{f}} \restriction \alpha)$. The following gives an extension of F_1 to a functor F on $\mathcal{C}_1$.

THEOREM 9. Suppose $\vec{f}$ is relatively categorical. Then we have a functor $F : \mathcal{C}_1 \longrightarrow \mathcal{D}_1$ satisfying the following conditions:

(i) F preserves direct limits,

(ii) for each α, $F(\alpha, \leq) \cong (Cl_{\vec{f}}(In_{\vec{f}} \restriction \alpha), \leq, \vec{f}, 0)$, and

(iii) $F \restriction \mathcal{C}_w : \mathcal{C}_w \longrightarrow \mathcal{D}_w$.

In addition,

(iv) if $\vec{f}$ is increasing then F preserves initial segments, and

(v) $W(F(\mathfrak{A})) = F(W(\mathfrak{A}))$.

Finally,

(vi) if $\vec{f}$ is replete then $\|F(\alpha, \leq)\| = \vec{f}'(\alpha)$ for each α.[8]

<u>Proof</u>. This makes use of an intermediate functor $F^* : \mathcal{C}_1 \rightarrow \mathcal{S}(\tau)$ defined as follows. Consider any linearly ordered $\mathfrak{A} = (A, \leq_{\mathfrak{A}})$. Let $Tm_{\vec{\mathfrak{f}}}(A)$ consist of all $t \in Tm_{\vec{\mathfrak{f}}}$ with $fr(t) \subseteq \{v_x \mid x \in A\}$. For $t_1, t_2 \in Tm_{\vec{\mathfrak{f}}}(A)$, take

$$(2) \quad t_1 \leq^*_{\mathfrak{A}} t_2 \Leftrightarrow \text{for every assignment } a : fr(t_1) \cup fr(t_2) \longrightarrow In_{\vec{f}} \text{ such that for each } x, y \in \mathcal{D}a,\ x \leq_{\mathfrak{A}} y \Leftrightarrow a(v_x) \leq a(v_y), \text{ we have } t_1[a] \leq t_2[a].$$

---

[8] This theorem brings out the <u>categorical content</u> of Theorem 2.10 of [11], but is more general in that $F(\mathfrak{A})$ was determined there only for $\mathfrak{A} = (A, \leq_{\mathfrak{A}})$ with $A \subseteq \omega$. But one loses its <u>recursive content</u>, by which if f is effectively rel. categorical (ftn. 7) and $\mathfrak{A}$ is an ordering of natural numbers then $\leq_{F(\mathfrak{A})}$ is recursive in $\leq_{\mathfrak{A}}$. However, the idea of the proof is essentially the same. {Correction to the statement of [11] 2.10: j is partial recursive <u>in P</u>.}

Observe that by relative categoricity, this equivalence holds equally well if 'for every assignment' is replaced by 'for some assignment.' Take

$$(3) \qquad F^*(\mathfrak{A}) = (Tm_{\vec{f}}(A), \leq^*_{\mathfrak{A}}, \vec{f}, Q).$$

To determine F^* as a functor, given $h : \mathfrak{A} \to \mathfrak{B}$ and $t \in Tm_{\vec{f}}(A)$ let $(F^*(h))(t) = \mathrm{Subst}(v_{h(x)}^{v_x} \mid v_x \in fr(t))t$. Let $E^*_{\mathfrak{A}}$ be the intersection of $\leq^*_{\mathfrak{A}}$ and its converse. The following is straightforward from these definitions and the observation about (2).

LEMMA 10. F^* is an ω-local functor from $\mathcal{C}_1$ to $\mathcal{D}_p$; for each $\mathfrak{A}$ in $\mathcal{C}_1$, $E^*_{\mathfrak{A}}$ is a congruence relation in $F^*_{\mathfrak{A}}$.

We can now define F by:

$$(4) \qquad F(\mathfrak{A}) = F^*(\mathfrak{A})/E^*_{\mathfrak{A}},$$

and, of course, $F(h) = M(F^*(h))$. Then (i) is immediate. (F need not preserve $\subseteq$ since equivalence classes can change in extensions.) For (ii), use the map $a : \{v_\xi \mid \xi < \alpha\} \to In_{\vec{f}}$, $a(v_\xi) = \vec{f}'(\xi)$; this induces the map $t \mapsto t[a]$ of $Tm_{\vec{f}}(\alpha)$ onto $Cl_{\vec{f}}(In_{\vec{f}} \restriction \alpha)$. (iii) follows directly from (ii), since every $\mathfrak{A}$ in $\mathcal{C}_w$ is isomorphic to $(\|\mathfrak{A}\|, \leq)$.

Now suppose $\vec{f}$ is increasing. To prove (iv), (v) we return to $F^*(\mathfrak{A})$ and first show:

$$(5) \qquad v_y \leq^*_{\mathfrak{A}} t \Leftrightarrow \text{for some } v_x \in fr(t),\ y \leq_{\mathfrak{A}} x.$$

The hypothesis on $\vec{f}$ is used to obtain $v_x \leq^*_{\mathfrak{A}} t$ for each $v_x \in fr(t)$, giving the $\Leftarrow$ direction. For the converse, suppose $x <_{\mathfrak{A}} y$ for all $v_x \in fr(t)$. Consider any $a : fr(t) \cup \{v_y\} \to In(\vec{f})$ with each $a(v_x) < a(v_y)$. Then $t[a] = t[a \restriction fr(t)] < a(v_y)$, since $a(v_y)$ is inaccessible.

To show (iv), consider $\mathfrak{A} \leqslant \mathfrak{B}$ and elements $t_1 \in Tm_{\vec{f}}(B)$, $t_2 \in Tm_{\vec{f}}(A)$ with $t_1 \leqslant^*_{\mathfrak{B}} t_2$. Let $fr(t_1) = \{v_{y_1}, \ldots, v_{y_n}\}$, $fr(t_2) = \{v_{x_1}, \ldots, v_{x_n}\}$. By (5) each $v_{y_i} \leqslant^*_{\mathfrak{B}} t_2$, and then there exists j with $v_{y_i} \leqslant^*_{\mathfrak{B}} v_{x_j}$; hence $y_i \leqslant_{\mathfrak{B}} x_j$ and so $y_i \in A$. In other words, t_1 must be in $Tm_{\vec{f}}(A)$ in this case. It follows that $t_2/E^*_{\mathfrak{B}} = t_2/E^*_{\mathfrak{A}}$, and further that $F(\mathfrak{A}) \leqslant F(\mathfrak{B})$.

$W(\mathfrak{A}) \leqslant \mathfrak{A}$ so $F(W(\mathfrak{A})) \leqslant F(\mathfrak{A})$ by (iv); further $F(W(\mathfrak{A}))$ is well-ordered by (iii). It can then be seen using (5) that $|F(\mathfrak{A})| - |F(W(\mathfrak{A}))|$ has no least element, hence that (v) holds.

(vi) is just a restatement of §5.1(2), using (ii). This completes the proof of the theorem.

F is determined up to equivalence by (i) and (ii), even with (ii) just for finite α, since each $\mathfrak{A} = \underrightarrow{\lim}\{(X, \leqslant_{\mathfrak{A}}) \mid X \subseteq A,\ X \text{ finite}\}$.[9] We shall denote this functor by $F_{\vec{f}}$, and call it an <u>ordinal system functor</u>. In case $\vec{f}$ is also replete, it is reasonable by (v) to denote the order-type $\|F_{\vec{f}}(\mathfrak{A})\|$ by $\vec{f}'(\|\mathfrak{A}\|)$ for arbitrary linearly ordered $\mathfrak{A}$. This gives, for example, a meaning to $\omega^{\|\mathfrak{A}\|}$, using $\vec{f} = \langle\lambda\xi,\eta.(\xi+\eta)\rangle$, a meaning to $\epsilon_{\|\mathfrak{A}\|}$, using $\vec{f} = \langle\lambda\xi,\eta.(\omega^{\xi}+\eta)\rangle$, etc.

For relationships of this theorem with other work on functorial aspects of well-ordered structures, cf. §6.3 below.

5.3. <u>Properties preserved by ordinal system functors</u>. The following are now immediate consequences of Theorem 9, Lemma 10, the main preservation theorem for (unary) ω-local functors and Cor. 1(a), Cor. 2(b).

---

[9] One can also characterize F in terms of a rather special <u>adjoint functor</u> situation, but which is not particularly illuminating.

THEOREM 11. Suppose $\vec{f}$ is relatively categorical, $F = F_{\vec{f}}$, and $L = L^{\alpha}_{\infty,\kappa}$ with $\kappa \geq \omega$. Then:

(i) F preserves $\equiv_L$ and $\preccurlyeq_L$, and

(ii) $(\beta, \leq) \preccurlyeq_L (\gamma, \leq)$ implies $(\vec{f}'(\beta), \leq, \vec{f}) \preccurlyeq_L (\vec{f}'(\gamma), \leq, \vec{f})$, provided that $\vec{f}$ is also replete.

Applications of this are found by combining examples of rel. categorical $\vec{f}$, such as those listed in §5.1(5), with interesting examples of linearly or well-ordered $\mathfrak{A}, \mathfrak{B}$ with $\mathfrak{A} \equiv_L \mathfrak{B}$ or $\mathfrak{A} \preccurlyeq_L \mathfrak{B}$ for various L. We mention three from the literature for illustration:

(1) (Ehrenfeucht [9]). $(\omega^{\omega}, \leq) \preccurlyeq_{L^{\omega}_{\infty,\omega}} (OR, \leq)$.

(2) (Chang [7]). Also for $\kappa > \omega$, $(\kappa^{\kappa}, \leq) \preccurlyeq_{L^{\kappa}_{\infty,\kappa}} (OR, \leq)$.

(3) (Karp [15]). For any α there are well-ordered $\mathfrak{A}$, non-well-ordered $\mathfrak{B}$ with $\mathfrak{A} \equiv_{L^{\alpha}_{\infty,\omega}} \mathfrak{B}$.

By (1), (2) and Theorem 11(ii) we have:

COROLLARY 12. If $\vec{f}$ is rel. categorical and replete and $L = L^{\kappa}_{\infty,\kappa}$ with $\kappa \geq \omega$ then $(\vec{f}'(\kappa^{\kappa}), \leq, \vec{f}) \preccurlyeq_L (OR, \leq, \vec{f})$.

In particular:

(4) (i) $(\omega^{\kappa^{\kappa}}, \leq, +) \preccurlyeq_L (OR, \leq, +)$,

(ii) $(\omega^{\omega^{\kappa^{\kappa}}}, \leq, +, .) \preccurlyeq_L (OR, \leq, +, .)$,

(iii) $(\epsilon_{\kappa^{\kappa}}, \leq, \lambda\xi,\eta.(\omega^{\xi}+\eta)) \preccurlyeq_L (OR, \leq, \lambda\xi,\eta.(\omega^{\xi}+\eta))$,

(iv) $(\Gamma_{\kappa^{\kappa}}, \leq, \lambda\xi,\eta.(\omega^{\xi}+\eta), \lambda\xi,\eta.\chi^{(\xi)}(\eta)) \preccurlyeq_L$ $(OR, \leq, \lambda\xi,\eta.(\omega^{\xi}+\eta), \lambda\xi,\eta.\chi^{(\xi)}(\eta))$

and so on. It is easily checked that for any $\kappa > \omega$, $\omega^{\kappa^{\kappa}} = \kappa^{\kappa^{\kappa}}$ and $\omega^{\omega^{\kappa^{\kappa}}} = \kappa^{\kappa^{\kappa^{\kappa}}}$, so that one recaptures from (4)(i)(ii) both the original results of [9] for $\kappa = \omega$ and of [7] for $\kappa > \omega$ for the systems with + and +, ., resp.

(3) shows that the class of well-ordered $\mathfrak{A} = (A, \leq_{\mathfrak{A}})$ cannot be characterized by any single $L_{\infty,\omega}$ sentence. This lifts to the following.

COROLLARY 13. If $\vec{f}$ is rel. categorical, replete and increasing, $F = F_{\vec{f}}$, then for any α there are $\mathfrak{A}, \mathfrak{B}$ with $F(\mathfrak{A})$ well-ordered, $F(\mathfrak{B})$ not well-ordered and $F(\mathfrak{A}) \equiv_{L^{\alpha}_{\infty,\omega}} F(\mathfrak{B})$.

Proof. Choose $\mathfrak{A}, \mathfrak{B}$ satisfying (iii). By Theorem 9(v), $W(F(\mathfrak{B})) = F(W(\mathfrak{B}))$. Since $W(\mathfrak{B})$ is a proper initial segment of $\mathfrak{B}$, $W(F(\mathfrak{B}))$ cannot be all of $F(\mathfrak{B})$. Hence $F(\mathfrak{B})$ is not well-ordered. The conclusion is by Theorem 11(i).

Thus the class of well-ordered $F_{\vec{f}}(\mathfrak{A})$ cannot be characterized by a single $L_{\infty,\omega}$ sentence. Of course, the class of well-ordered $\mathfrak{A}$ is characterized in $L^{1}_{\infty,\aleph_1}$ from which one obtains a characterization of the class of well-ordered $F_{\vec{f}}(\mathfrak{A})$ by using the definition of relative categoricity as given in §5.1(3).

6. Comparisons with previous work

Instead of the more familiar notation $L_{\omega,\omega}$, we use here $L_{\omega,2}$ to denote the finitary 1st-order language. It can be identified with the set of formulas in $L^{\omega}_{\infty,2}$ in which all disjunctions are finite. $L^{*}_{\omega,\omega}$ is written for the weak 2nd-order language, i.e. where one has (in addition) variables ranging over all finite subsets of the domain. Every sentence of $L^{*}_{\omega,\omega}$ is equivalent to a sentence of $L^{\omega}_{\infty,\omega}$.

6.1. The back-and-forth criterion. This originates with Fraissé's [14] characterization of $\equiv$ in $L_{\omega,2}$, which would be put here as follows: if σ is a finite relational signature, then $\mathfrak{A} \equiv \mathfrak{B}$ in $L_{\omega,2} \Leftrightarrow 0 \in J^{\omega}_{2}(\mathfrak{A},\mathfrak{B})$. (Neither additional hypothesis on σ can be dropped.) Ehrenfeucht [9] essentially made use of: $0 \in J^{\omega}_{\omega}(\mathfrak{A}, \mathfrak{B}) \Rightarrow \mathfrak{A} \equiv \mathfrak{B}$ in $L^{*}_{\omega,\omega}$.[10]

Karp found the criterion for the $L^{\alpha}_{\infty,2}$ and $L_{\infty,2}$.[11] Benda [5] and, independently, Calais [6] extended this to the $L^{\alpha}_{\infty,\kappa}$.[12]

There are several minor differences in the formulation of the criterion here from those just mentioned. One is that signatures for structures with relations and functions of unrestricted "arity" are also covered. Further, the result of [5] takes the form: $\mathfrak{A} \equiv \mathfrak{B}$ in $L^{\alpha}_{\infty,\kappa} \Leftrightarrow$ there exists a sequence $\langle \mathcal{F}^{\beta} \rangle_{\beta \leq \alpha}$ of partial isomorphisms between $\mathfrak{A}$ and $\mathfrak{B}$ satisfying a certain condition $C_{\kappa}(\langle \mathcal{F}^{\beta} \rangle_{\beta \leq \alpha})$. In fact, it is established that if $\mathfrak{A} \equiv \mathfrak{B}$ in $L^{\alpha}_{\infty,\kappa}$ then $C_{\kappa}(\langle J^{\beta}(\mathfrak{A}, \mathfrak{B}) \rangle_{\beta \leq \alpha})$. The converse direction is no weaker in the present statement (Cor. 3(a)) since $C_{\kappa}(\langle \mathcal{F}^{\beta} \rangle_{\beta \leq \alpha})$ implies $\mathcal{F}^{\beta} \subseteq J^{\beta}(\mathfrak{A}, \mathfrak{B})$ for each $\beta \leq \alpha$, and then also $C_{\kappa}(\langle J^{\beta}(\mathfrak{A}, \mathfrak{B}) \rangle_{\beta \leq \alpha})$.

6.2. Some preservation results and counter-examples

(i) Rings of polynomials. Scott [22] gives an example of (algebraically closed) fields $\mathfrak{A}, \mathfrak{B}$ such that $\mathfrak{A} \equiv \mathfrak{B}$ in $L_{\omega,2}$ but $\mathfrak{A}[x] \not\equiv \mathfrak{B}[x]$ in $L_{\omega,2}$.

(ii) Rings of formal power series. Ax and Kochen [2] proved (by quite special methods) that if $\mathfrak{A}, \mathfrak{B}$ are fields of characteristic 0 and $\mathfrak{A} \equiv \mathfrak{B}$ in $L_{\omega,2}$ then $\mathfrak{A}[[x]] \equiv \mathfrak{B}[[x]]$ in $L_{\omega,2}$. The general status of preservation of $\equiv$ both in $L_{\omega,2}$ and $L_{\infty,\omega}$ is unknown for the

---

[10] Cf. also [10] for a quick review of [14], [9] and especially of Ehrenfeucht's applications to ordinal structures.

[11] Barwise [3] reviews a good deal of work using this characterization. Barwise and Eklof [4] also contains an interesting extension of the criterion to certain subclasses of these languages, for example the classes of existential and of positive formulas.

[12] These papers also contain more delicate statements for the $L^{\alpha}_{\lambda,\kappa}$, i.e. where the size of disjunctions in formulas is bounded by the cardinal λ.

operation $\mathfrak{A} \mapsto \mathfrak{A}[[x]]$ applied to arbitrary rings $\mathfrak{A}$.

(iii) Free Γ-models. Tarski and Vaught [23] showed that the operation G_Γ of 4.4 for finitary τ preserves $\equiv$ and $\preccurlyeq$ in $L_{\omega,2}$. The proof uses automorphisms of $G_\Gamma(A)$ induced by permutations of A; this method can be extended to $L_{\infty,2}$.

(iv) Tensor products. The question whether $\otimes$ preserves $\equiv$ in $L_{\omega,2}$ was raised in [13]. The following (unpublished) counter-example was shown to me by Yu. L. Eršov. Take the additive groups of the rationals Q and the reals $\mathbb{R}$ as Z-modules. Then $Q/Z \equiv \mathbb{R}/Z$ but $Q \otimes Q/Z$ is trivial while $Q \otimes \mathbb{R}/Z$ is not. Here we are dealing with modules as structures in the sense of §4.7.

One may also consider modules $\mathfrak{A}$ over a ring $\mathcal{R}$ as two-sorted structures (A,R; ...). A counter-example due to Eklof and Olin for this understanding is given in [19]. It makes use of the fact that one may have isomorphisms of modules over given $\mathcal{R}$, (A,R; ...) $\cong$ (B,R; ...) which are non-trivial on R. The same counter-example also works for $L_{\infty,\omega}$. Of course, there is no conflict with §4.7.

(v) Generalized and special products. The preservation results relative to $L_{\omega,2}$ for generalized products [13] were obtained by an elimination-of-quantifiers type of argument. This method also gives much other information of interest. However, it does not seem to extend to any of the infinitary languages considered.

Calais [6] obtained preservation results for $\Pi_{i \in I}$ and $\Sigma_{i \in I}$ in the $L^{\alpha}_{\infty,\kappa}$.[13] I understand from P. Eklof that Mr. William Brown extended this to generalized products relative to quantifier-free formulas $\theta_1, \ldots, \theta_n$. Ollmann [21] treats them, without such a restriction, by

[13] One should also mention Olin's [19], [20] giving counter-examples for these operations applied to modules as two-sorted structures with one sort fixed, yet positive results for direct powers and multiples.

the back-and-forth method; explicit statements and details are given only for $L_{\omega,2}$, with extensions to infinitary languages loosely indicated.

(vi) Relatively categorical systems of functions. The case $\kappa = \omega$ of Cor. 12 in §5.3 was first established by H. Friedman. His proof (unpublished) made use of the isomorphism of $(\omega^{\omega}, \leq)$ and $(OR, \leq)$ in non-standard models of certain systems of set theory.

(vii) Systems of ordinal functions. Doner [8] obtains results of the form

$$(\vec{O}'_{\alpha}(\omega^{\omega}.\delta), \leq, \vec{O}_{\alpha}) \preccurlyeq (OR, \leq, \vec{O}_{\alpha}) \text{ in } L^{*}_{\omega,\omega},$$

where $\delta \neq 0$ and $\vec{O}_{\alpha} = \langle O_{\beta} \rangle_{\beta \leq \alpha}$ is a sequence of binary ordinal operations defined by Tarski: $\xi O_0 \eta = \xi + \eta$, $\xi O_1 \eta = \xi \cdot \eta$, and the further O_{β} are obtained by a certain iteration of the recursive schemes of definition of O_0, O_1. The method of proof is along the lines of [9], using special normal forms developed by Doner.

The O_{α} grow at roughly the same rate as the $\chi^{(\alpha)}$. This suggests that the facts worked out for the $\vec{O}_{\alpha}$ could be used to place them under the general theory of §5,[14] and that one could then obtain correspondingly stronger results analogous to Cor. 12.

6.3. Categorical and functorial aspects of well-ordered systems

The notation of §5 is used here in the description of previous work.

(i) Normal functors. This notion was introduced and studied by Aczel [1].[15] A functor $G : \mathcal{C}_1 \rightarrow \mathcal{C}_1$ is called normal if it preserves

[14] Strictly speaking, under a suitable extension of the theory to systems with infinitely long sequences $\vec{f}$.

[15] The following is also based on mimeographed notes by Aczel detailing his work.

direct limits and initial segments. With each normal G is associated a certain normal G' such that $G'(\mathfrak{A}) \cong G(G'(\mathfrak{A}))$ for all linearly ordered $\mathfrak{A}$ (by a natural isomorphism). In addition, $\mathcal{C}_W$ is closed under G' if it is closed under G; then the function of ordinals associated with G' is the critical function of that associated with G. However, no categorical characterization is given of $G'(\mathfrak{A})$ in general.

Suppose $\vec{f}$ is replete and increasing; let $G_{\vec{f}}(\mathfrak{A})$ be the retract of $F_{\vec{f}}(\mathfrak{A})$. By Theorem 9, $G_{\vec{f}}$ is a normal functor under which $\mathcal{C}_W$ is closed. It can also be seen that if $\vec{f}$ is replete then $G_{\vec{f}\frown\langle\vec{f}'\rangle} = (G_{\vec{f}})'$ (as defined by Aczel).

Theorem 9 provides systematic means for generating normal functors, relative to means for generating rel. categorical, replete and increasing systems $\vec{f}$ of ordinal functions. Given $\vec{f}$, $G_{\vec{f}}$ can be characterized categorically. It would be preferable to have a self-contained treatment of these functors.

QUESTION: Is there a categorical characterization of the class of functors $G_{\vec{f}}$?

(ii) Natural well-orderings. Proof theory makes use of what are called natural well-orderings $\leq_{\mathfrak{A}}$ obtained from natural systems of ordinal representation $\mathfrak{A} = (A, \leq_{\mathfrak{A}}, \vec{f}, 0)$ with $A = Cl_{\vec{f}}(0)$. The first and paradigm example is Gentzen's use of an ordering of type ϵ_0 obtained from the representation of ordinals $< \epsilon_0$ in Cantor normal form. Kreisel [16] has stressed the canonical nature of the familiar systems $\mathfrak{A} = (A, \leq_{\mathfrak{A}}, \vec{f}, 0)$, given by suitable characterizations up to isomorphism. However, one has no general explanation of the notion of being natural in this context, even as applied to systems with build-up functions. It is tempting to look for such an explanation in the framework of category theory, given its success in telling what is natural in many mathematical contexts. This seems to me the eventual point of work such as that in §5.2 and just discussed.

REFERENCES

[1] P. Aczel, Normal functors on linear orderings (abstract), J. Symbolic Logic 32 (1967), 430.

[2] J. Ax and S. Kochen, Diophantine problems over local fields, III. Decidable fields, Ann. of Math. 83 (1966), 437-456.

[3] J. Barwise, Back and forth thru infinitary logic (to appear).

[4] J. Barwise and P. Eklof, Infinitary properties of Abelian torsion groups, Ann. of Math. Logic, 2 (1970), 25-68.

[5] M. Benda, Reduced products and non-standard logics, J. Symbolic Logic 34 (1969), 424-436.

[6] J. P. Calais, La méthode de Fraissé dans les langages infinis, C. R. Acad. Sci. Paris 268 (1969), 785-788.

[7] C. C. Chang, Infinitary properties of models generated by indiscernibles, Logic, Methodology and Philos. of Sci. III (v. Rootselaar and Staal, eds.), Amsterdam (1968), 9-21.

[8] J. E. Doner, An extended arithmetic of ordinal numbers and its metamathematics, Dissertation, Berkeley (1969).

[9] A. Ehrenfeucht, An application of games to the completeness problem for formalized theories, Fund. Math. 49 (1961) 129-141.

[10] S. Feferman, Some recent work of Ehrenfeucht and Fraissé, Summary of Talks at A.M.S. Summer Inst. in Logic at Cornell, 1957, 201-209.

[11] S. Feferman, Systems of predicative analysis, II: representations of ordinals, J. Symbolic Logic, 33 (1968) 193-220.

[12] S. Feferman, Hereditarily replete functionals over the ordinals, Intuitionism and Proof Theory (Myhill, Kino, Vesley, eds.) Amsterdam (1970) 289-302.

[13] S. Feferman and R. L. Vaught, The first-order properties of products of algebraic systems, Fund. Math. 47 (1959) 57-103.

[14] R. Fraissé, Sur quelques classifications des relations, basées sur des isomorphismes restreints, Publ. Sci. Univ. d'Alger Sér. A, vol. II (1955), Part I, 15-60, Part II, 273-295.

[15] C. Karp, Finite-quantifier equivalence, The Theory of Models (Addison, Henkin, Tarski, eds.), Amsterdam (1965), 407-412.

[16] G. Kreisel, A survey of proof theory, (Part I) J. Symbolic Logic 33 (1968), 321-388, (Part II to appear).

[17] J. Lambek, Lectures on rings and modules, Waltham (1966), viii + 183 pp.

[18] B. Mitchell, Theory of categories, New York (1965), xi + 273 pp.

[19] P. Olin, Direct multiples and powers of modules (to appear in Fund. Math.).

[20] P. Olin, Products of two-sorted structures (to appear in J. Symbolic Logic).

[21] L. T. Ollmann, Operators on models, Dissertation, Cornell (1970).

[22] D. Scott, Definability in polynomial rings (abstract), Notices A.M.S. 5 (1958), 221-222.

[23] A. Tarski and R. L. Vaught, Arithmetical extensions of relational systems, Compos. Math. 13 (1957), 81-102.

[24] P. Eklof, Lefschetz' principle and local functors (to appear).

LOGICS CONTAINING S4 WITHOUT THE FINITE MODEL PROPERTY

Kit Fine

St. John's College, Oxford

In [1], Harrop asked whether there were logics containing the intuitionistic logic IL which lack the finite model property. Jankov gave examples of such logics, but they were not finitely axiomatizable. By the Tarski-McKinsey translation, Harrop's problem relates to the question of whether there exist extensions of the modal logic S4 without fmp. Makinson [2] showed that there are extensions of the modal logic M without fmp, but he could not extend his results to S4. In this paper, I shall exhibit logics containing both IL and S4 which lack fmp, but are finitely axiomatized and decidable.

We begin with S4. Let X be the following formula:

$$[s \wedge \Box(s \rightarrow \Diamond(-s \wedge \Diamond s)) \wedge \Diamond p \wedge \Diamond q \wedge \Diamond r \wedge \Box(p \rightarrow -\Diamond q \wedge -\Diamond r) \wedge$$
$$\Box(q \rightarrow -\Diamond p \wedge -\Diamond r) \wedge \Box(r \rightarrow -\Diamond p \wedge -\Diamond q)] \rightarrow \Diamond(\Diamond p \wedge \Diamond q \wedge -\Diamond r).$$

Let L be the (normal) logic obtained by adding X as an axiom to S4.

THEOREM. L lacks fmp.

<u>Proof</u>. We must show that there is a formula Y such that:

(1) Y is consistent in L

(2) Any model for Y that verifies L is infinite.

To show (1) it suffices to show that there is a structure $\mathfrak{A} = (W,R,\phi)$

such that

(1') $\mathfrak{A}$ is a model for Y,

(1") $\mathcal{F} = (W,R)$ is a frame for L.

We let Y be the antecedent of X and let $\mathfrak{A}$ be the structure with the following diagram:

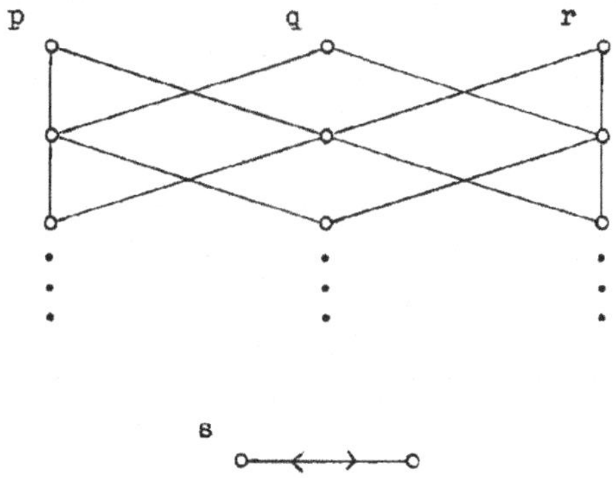

To be precise, let

$$W_0 = \{1,2,3\},$$
$$W_{n+1} = \{\{x,y\} : x \neq y \text{ and } x, y \in W_n\}, \quad \text{and}$$
$$W = \cup W_n \cup \{0,-1\}.$$

For x, y $\in$ W, we say: xRy iff $y \in TC(\{x\})$ or $x = 0$ or $x = -1$, (TC is transitive closure). Finally, $\phi(s) = \{0\}$, $\phi(p) = \{1\}$, $\phi(q) = \{2\}$, and $\phi(r) = \{3\}$.

First, we show that $\mathfrak{A}$ is a model for Y. s is true at 0 alone. 0R-1 and -1R0. Therefore $s \wedge \Box(s \rightarrow \Diamond(-s \wedge \Diamond s))$ is true at 0. p (q, r) are true at 1 (2, 3) alone. 0R1, $1 \not R 2$ and $1 \not R 3$. Therefore $\Diamond p \wedge \Box(p \rightarrow -\Diamond q \wedge -\Diamond r)$ is true at 0. Similarly, for q and r. Hence Y is true at 0 and $\mathfrak{A}$ is a model for Y.

Now we show that $\mathfrak{F}$ is a frame for L, i.e. that X is true at x in $\mathfrak{L} = (W, R, \psi)$ for any element x of W and any valuation ψ. Suppose Y is true at x. Then $s \wedge \Box(s \rightarrow \Diamond(-s \wedge \Diamond s))$ is true at x and there is an infinite sequence $x = x_0, x_1, x_2, \ldots$ such that $x_i R x_{i+1}$ and $x_i \neq x_{i+1}$ for $i = 0, 1, 2, \ldots$. Therefore $x = 0$ or $x = -1$ and xRz for each z in W. Also $\Diamond p \wedge \Diamond q \wedge \Diamond r$ is true at x. So there are x_1, x_2 and x_3 such that p, q and r are true at x_1, x_2 and x_3 respectively. But $\Box(p \rightarrow -\Diamond q \wedge -\Diamond r)$ is true at x. So $x_1 \not R x_2$ and $x_1 \not R x_3$. Similarly, $x_2 \not R x_1$, $x_2 \not R x_3$, $x_3 \not R x_1$ and $x_3 \not R x_2$. It is then clear that $x_1, x_2, x_3 \in W_n$ for some $n \geq 0$.

Now consider $y = \{x_1, x_2\}$. yRx_1 and yRx_2. So $\Diamond p \wedge \Diamond q$ is true at y. Suppose $\Diamond r$ is true at y. Then r is true at y or $\Diamond r$ is true at x_1 or x_2, contrary to the fact that $\Box(r \rightarrow -\Diamond p \wedge -\Diamond q)$, $\Box(p \rightarrow -\Diamond r)$ and $\Box(q \rightarrow -\Diamond r)$ are true at x. So $-\Diamond r$ is true at y, and $\Diamond(\Diamond p \wedge \Diamond q \wedge -\Diamond r)$ is true at x.

Finally, we must show that any L-model for Y is infinite. Suppose Y is true at x in some L-model $\mathfrak{C}$. Let

$$A_0 = p, \quad B_0 = q, \quad C_0 = r,$$
$$A_{i+1} = \Diamond A_i \wedge \Diamond B_i \wedge -\Diamond C_i,$$
$$B_{i+1} = \Diamond A_i \wedge \Diamond C_i \wedge -\Diamond B_i, \quad \text{and}$$
$$C_{i+1} = \Diamond B_i \wedge \Diamond C_i \wedge -\Diamond A_i.$$

First we show that $\Diamond A_i$, $\Diamond B_i$ and $\Diamond C_i$ are true at x for each $i > 0$. Suppose $i = 1$. Now Y is true at x. Also, axiom $X = Y \rightarrow \Diamond A_1$ is true at x. So $\Diamond A_1$ is true at x. By permuting p, q and r, it follows that $\Diamond B_1$ and $\Diamond C_1$ are true at x. Now suppose that $\Diamond A_i$, $\Diamond B_i$ and $\Diamond C_i$ are true at x for $i > 0$. $A_i \rightarrow -\Diamond C_{i-1}$, $B_i \rightarrow \Diamond C_{i-1} \in S4$, so that $\Box(A_i \rightarrow -\Diamond B_i)$

$\in S4$. Similarly, $\Box(A_i \rightarrow -\Diamond C_i)$, $\Box(B_i \rightarrow -\Diamond A_i \wedge -\Diamond C_i)$, $\Box(C_i \rightarrow -\Diamond A_i \wedge -\Diamond B_i) \in S4$. Therefore, by applying axiom X, $\Diamond(\Diamond A_i \wedge \Diamond B_i \wedge -\Diamond C_i) = \Diamond A_{i+1}$ is true at x. By permuting A_i, B_i and C_i, it follows that $\Diamond B_{i+1}$ and $\Diamond C_{i+1}$ are also true at x.

Second, we must show that for i, j > 0, $A_i \rightarrow -A_{i-j} \in S4$. Suppose j = 1. As before, $A_{i-1} \rightarrow -\Diamond B_{i-1} \in S4$. But $A_i \rightarrow \Diamond B_{i-1} \in S4$. Therefore $A_i \rightarrow -A_{i-1} \in S4$. Now suppose j > 1. $A_i \rightarrow \Diamond B_{i-1}$, $B_{i-1} \rightarrow \Diamond C_{i-2} \in S4$. So $A_i \rightarrow \Diamond C_{i-2} \in S4$. $C_{i-2} \rightarrow \Diamond B_{i-3} \in S4$. So $A_i \rightarrow \Diamond B_{i-3} \in S4$. Proceeding thus, it follows that $A_i \rightarrow \Diamond B_{i-j} \in S4$ or $A_i \rightarrow \Diamond C_{i-j} \in S4$. But $A_{i-j} \rightarrow -\Diamond B_{i-j} \wedge -\Diamond C_{i-j} \in S4$. Therefore $A_i \rightarrow -A_{i-j} \in S4$.

Since $\Diamond A_i$ is true at x, A_i is true at some x_i in $\mathcal{C}$ such that xRx_i, i = 1, 2, ... Since $A_i \rightarrow -A_{i-j} \in S4$, $x_i \neq x_{i-j}$ for i, j > 0. Hence $\mathcal{C}$ is infinite.

This establishes (2) and the proof is complete.

The above argument may be modified to show that there is a logic containing IL without fmp. We add to IL the axiom:

$$[(-s \wedge (p \rightarrow q \vee r)) \rightarrow ((q \rightarrow p \vee r) \vee (r \rightarrow p \vee q))] \rightarrow$$
$$[(-s \wedge p \rightarrow q \vee r) \vee (-s \wedge q \rightarrow p \vee r) \vee (-s \wedge r \rightarrow p \vee q) \vee -s].$$

We then show that the consequent of the axiom is not a theorem in the resulting logic and that any structure which rejects the consequent and verifies the axiom is infinite.

Finally, it should be noted that we can add axioms to the logics described above so as to obtain logics which are decidable, finitely axiomatized, complete for their intended interpretation, and yet without fmp.

REFERENCES

[1] Harrop, R., On the existence of finite models and decision procedures, Proceedings of the Cambridge Philosophical Society, vol. 54 (1958), 1-16.

[2] Makinson, D., A Normal Modal Calculus Between T and S4 Without the Finite Model Property, Journal of Symbolic Logic, vol. 34, Number 1 (1969), 35-38.

AN ϵ-CALCULUS SYSTEM FOR FIRST-ORDER S4[1]

Melvin Fitting

Lehman College (CUNY)

§1. Introduction

We give a formulation of the first-order modal logic S4 patterned after the classical ϵ-calculus of Hilbert (see [3]) and prove (non-constructively) that it is a conservative extension of the usual first-order S4. Other modal logics may be similarly treated.

A natural first attempt at such a formulation would be to add S4 axioms and rules to a classical ϵ-calculus base. This fails, and the reason is not hard to find. If $X(x)$ is a formula with one free variable, x, classically ϵxX is intended to be the name of a constant such that, if $(\exists x)X(x)$ is true, $X(\epsilon xX)$ is true. But in a Kripke S4 model [2], $(\exists x)X(x)$ may be true in two possible worlds but yet there may be no single constant, c, such that $X(c)$ is true in both worlds. Thus ϵxX can not be thought of as the name of a constant in an ϵ-calculus S4. Instead we treat ϵxX as a function defined on the collection of possible worlds and such that, if $(\exists x)X(x)$ is true in some possible world, the value of ϵxX at that world is a constant, c, such that $X(c)$ is true there.

Unfortunately, there is no syntactic machinery in ordinary first-order S4 to handle ϵ-terms. They are neither constants nor variables,

---

[1]This research was supported by City University of New York Faculty Research Program, Grant number 1049.

but rather 'world-dependent' terms, naming different constants in different worlds. Consequently we work with an extension of S4 due to Stalnaker and Thomason [5,6], which they created partly to treat definite descriptions, which are also 'world-dependent' terms. We add an abstraction operator, λ, to the language, so that $(\lambda x\ X)(t)$ is a formula if X is a formula with only x free and t is a 'world-dependent' term. We say $(\lambda x\ X)(t)$ is true in a given possible world if X(c) is true there, where c is the value of t at that world. Thus we can make the necessary distinction between $(\lambda x \Diamond X)(t)$ and $\Diamond(\lambda x\ X)(t)$. A fuller discussion of this point may be found in [5,6].

We give an ϵ-calculus S4 system and a suitable model theory. We mention various results which may be derived in it, and show it is a conservative extension of more customary formulations of first-order S4. The system we give is adequate to prove all theorems of first-order S4 (without parameters), but is not complete in terms of its own model theory. In a later paper we will extend the system, by adding more structural axioms, to produce a complete system.

§2. An ϵ-calculus S4 ($\epsilon S4^{\circ}$)

We take as primitive symbols $\wedge$, $\sim$, $\exists$, $\Diamond$, ϵ, λ,) and (, and use $\vee$, $\supset$, $\equiv$, $\forall$ and $\Box$ as abbreviations as usual. We are informal about parentheses. We assume a countable collection of n-place predicate letters for each natural number n, and countably many variables. (We choose not to have parameters in our basic system, though this is of no real significance.)

Following [3], when we use the words formula or term we mean there are no free variables present. In the more general situation we use quasi-formula or quasi-term. A proper definition of these concepts is straightforward, and contains the following clauses. If X is a quasi-formula, x is a variable and t is a quasi-term, $(\lambda x\ X)(t)$ is a quasi-formula. The free variables of $(\lambda x\ X)(t)$ are those of X except for x, together with those of t. Similarly, $\epsilon x X$ is a quasi-term; its free variables are those of X other than x. We will use x, y, z, ... to stand for variables, and t for a quasi-term. We may use subscripts.

We use $(\lambda x_1 \ldots x_n\ X)(t_1,\ldots,t_n)$ as an abbreviation for $(\lambda x_1(\lambda x_2 \ldots (\lambda x_n\ X)(t_n) \ldots)(t_2))(t_1)$. We denote by $X(x/t)$ the result of substituting t for free x in X. We often use $\underset{\sim}{x}$ and $\underset{\sim}{t}$ for sequences of variables and quasi-terms respectively, and write $(\lambda \underset{\sim}{x}\ X)(\underset{\sim}{t})$ for $(\lambda x_1 \ldots x_n\ X)(t_1,\ldots,t_n)$.

Let X be a quasi-formula whose variables are among $x_1, \ldots, x_n$. Let $t_1, \ldots, t_n$ be quasi-terms. If $(\lambda x_1 \ldots x_n\ X)(t_1,\ldots,t_n)$ is a formula we call it a λ-closure of X. If X has no free variables we consider it to be a λ-closure of itself.

We use the phrase t is free for x in X in the standard way to mean that, on replacing free x by t in X, no free variable of t becomes bound by a quantifier, abstract or ϵ symbol of X.

The axioms and rules of $\epsilon S4^{o}$ are as follows.

RULES:

R1 $$\frac{X \quad X \supset Y}{Y}$$ where X and Y are formulas

R2 $$\frac{X}{\Box X}$$ where X is a formula

AXIOM SCHEMAS: Let X and Y be quasi-formulas. We take as axioms all λ-closures of the following quasi-formulas.

First, structural axioms.

A1 if y is not free in X, but y is free for x in X, $(\lambda x\ X)(t) \equiv [\lambda y\ X(x/y)](t)$

A2 if x is not free in X, $(\lambda x\ X)(t) \equiv X$

A3 if $x_1 \neq x_2$, x_1 is not free in t_2, x_2 is not free in t_1, $(\lambda x_1 x_2\ X)(t_1, t_2) \equiv (\lambda x_2 x_1\ X)(t_2, t_1)$

A4 $[\lambda \underset{\sim}{x}\ (X \wedge Y)](\underset{\sim}{t}) \equiv [(\lambda \underset{\sim}{x}\ X)(\underset{\sim}{t}) \wedge (\lambda \underset{\sim}{x}\ Y)(\underset{\sim}{t})]$

A5 $(\lambda \underset{\sim}{x}\ {\sim}X)(\underset{\sim}{t}) \equiv {\sim}(\lambda \underset{\sim}{x}\ X)(\underset{\sim}{t})$

A6 if y is not free in any quasi-term of $\underset{\sim}{t}$ and y is not in the sequence $\underset{\sim}{x}$, $[\lambda \underset{\sim}{x}\ (\exists y)X](\underset{\sim}{t}) \equiv (\exists y)[(\lambda \underset{\sim}{x}\ X)(\underset{\sim}{t})]$

Next, propositional axioms

A7 X, where X is a tautology

A8 $\Box(X \supset Y) \supset (\Box X \supset \Box Y)$

A9 $\Box X \supset X$

A10 $\Box X \supset \Box\Box X$

Finally, quantification

A11 $(\lambda x\ X)(t) \supset (\lambda x\ X)(\epsilon x X)$

A12 $(\lambda x\ \Diamond X)(t) \supset \Diamond(\lambda x\ X)(\epsilon x X)$

A13 $(\exists x)X \equiv (\lambda x\ X)(\epsilon x X)$

This completes the system $\epsilon S4^{o}$.

§3. $\epsilon S4^{\circ}$ model theory

We give a Kripke type model theory for $\epsilon S4^{\circ}$. It is based on that for first-order S4, as found in [1,2,4], extended along the lines of [5,6].

The system $\epsilon S4^{\circ}$ above has no constant symbols or parameters. For this section only, let us add them, and treat them as terms. We use a, b, c, ... to represent them.

By FS4 we mean ordinary first-order S4, as found in [1] or [4] say. Its language is that part of $\epsilon S4^{\circ}$ with parameters, not containing abstracts or ϵ-symbols. We begin with a model theory for FS4.

By an FS4 model we mean a quadruple, $\langle \mathcal{G}, \mathcal{R}, \models, \mathcal{P} \rangle$ where: $\mathcal{G}$ is a non-empty set; $\mathcal{R}$ is a transitive, reflexive relation on $\mathcal{G}$; $\mathcal{P}$ is a function on $\mathcal{G}$ ranging over non-empty sets of parameters; and $\models$ is a relation between elements of $\mathcal{G}$ and formulas of FS4. These are to satisfy the following, where $\Gamma \in \mathcal{G}$.

1) if $\Delta \in \mathcal{G}$ and $\Gamma \mathcal{R} \Delta$ then $\mathcal{P}(\Gamma) \subseteq \mathcal{P}(\Delta)$.

2) if $\Gamma \models X$, all parameters of X are in $\mathcal{P}(\Gamma)$.

3) if all constants of X and Y are in $\mathcal{P}(\Gamma)$ then

$\Gamma \models (X \wedge Y)$ if and only if $\Gamma \models X$ and $\Gamma \models Y$

$\Gamma \models \sim X$ if and only if not-$\Gamma \models X$.

4) if X is a quasi-formula with at most one free variable, x, and all parameters in $\mathcal{P}(\Gamma)$, then

$\Gamma \models (\exists x)X$ if and only if $\Gamma \models X(x/c)$ for some $c \in \mathcal{P}(\Gamma)$.

5) if all parameters of X are in $\mathcal{P}(\Gamma)$, then

$\Gamma \models \Diamond X$ if and only if for some $\Delta \in \mathcal{G}$ such that $\Gamma \mathcal{R} \Delta$, $\Delta \models X$.

An FS4 formula X is called valid in the FS4 model $<\mathcal{G}, \mathcal{R}, \models, \mathcal{P}>$ if $\Gamma \models X$ for every $\Gamma \in \mathcal{G}$ such that all constants of X belong to $\mathcal{P}(\Gamma)$. Proofs may be found in [1,2,4] that the set of theorems of FS4 coincides with the set of formulas valid in all FS4 models.

By an $\epsilon S4^{o}$ model we mean a quintuple, $<\mathcal{G}, \mathcal{R}, \models, \mathcal{P}, \mathcal{F}>$ where: $<\mathcal{G}, \mathcal{R}, \models, \mathcal{P}>$ is an FS4 model (save that now $\models$ is a relation between elements of $\mathcal{G}$ and formulas of $\epsilon S4^{o}$) and $\mathcal{F}$ is a collection of functions defined on subsets of $\mathcal{G}$. These are to satisfy:

6) if ϵxX is a term, there is an element, $f_{\epsilon xX}$ in $\mathcal{F}$ such that: $f_{\epsilon xX}$ is a function with domain the set of Γ in $\mathcal{G}$ such that $\mathcal{P}(\Gamma)$ contains all parameters of X; if $\Gamma \in$ domain $f_{\epsilon xX}$ then $f_{\epsilon xX}(\Gamma) \in \Gamma$; if $\Gamma \models (\exists x)X$ then $\Gamma \models X(x/f_{\epsilon xX}(\Gamma))$.

[For simplicity in stating the next two items, if c is a parameter, let f_c be the function with domain the set of Γ in $\mathcal{G}$ such that $c \in \mathcal{P}(\Gamma)$, and whose value is given by $f_c(\Gamma) = c$.]

7) if $(\lambda x\, X)(t)$ is a formula,

$$\Gamma \models (\lambda x\, X)(t) \text{ if and only if } \Gamma \models X(x/f_t(\Gamma)).$$

8) if P is an n-place predicate letter and $t_1, \ldots, t_n$ are terms,

$$\Gamma \models P(t_1, \ldots, t_n) \text{ if and only if } \Gamma \models P(f_{t_1}(\Gamma), \ldots, f_{t_n}(\Gamma)).$$

An $\epsilon S4^{o}$ formula X is called valid in the $\epsilon S4^{o}$ model $<\mathcal{G}, \mathcal{R}, \models, \mathcal{P}, \mathcal{F}>$ if $\Gamma \models X$ for every $\Gamma \in \mathcal{G}$ such that all parameters of X are in $\mathcal{P}(\Gamma)$. We leave the reader to verify (by induction on the length of the proof)

THEOREM 3.1. All theorems of $\epsilon S4^{o}$ are valid in all $\epsilon S4^{o}$ models.

Moreover, any FS4 model $< \mathcal{G}, \mathcal{R}, \models, \mathcal{P} >$ can be extended to an $\epsilon S4^o$ model $< \mathcal{G}, \mathcal{R}, \models, \mathcal{P}, \mathcal{F} >$ by extending $\models$ and defining $\mathcal{F}$ by induction on the degree of formulas. Thus we have, using the above and the completeness of FS4,

THEOREM 3.2. Let X be a formula of FS4 with no parameters. If X is a theorem of $\epsilon S4^o$, X is a theorem of FS4.

§4. Development of $\epsilon S4^o$

In this section we merely sketch how $\epsilon S4^o$ can be developed as a practical calculus, and show that it extends the parameter-free part of FS4. We no longer allow parameters in $\epsilon S4^o$ formulas.

We use the notation $\vdash X$ to mean all λ-closures of X are provable. Our axiom schemas are of this form; our rules have analogous generalizations. Thus one may show: if X and Y are quasi-formulas,

1) $$\frac{\vdash X \qquad \vdash X \supset Y}{\vdash Y}$$

2) $$\frac{\vdash X}{\vdash \Box X}.$$

Next one may show a replacement theorem in the following form:

3) Let A, B, X and Y be quasi-formulas. Let Y be the result of replacing, in X, the quasi-formula A at some or all of its occurrences (except within quasi-terms) by B. Then

$$\frac{\vdash A \equiv B}{\vdash X \equiv Y}.$$

This is somewhat different than the usual form, but that follows using

4) (closure theorem) Let us denote by $\forall X$ any universal closure of the quasi-formula X. Then $\vdash X$ if and only if $\forall X$ is provable.

Finally we show that $\epsilon S4^o$ is an extension of the parameter-free part of FS4.

Let $X_1, X_2, \ldots, X_n$ be a proof of X_n in some FS4 axiom system, say that of [1] or [4]. Let $a_1, a_2, \ldots, a_k$ be all the parameters occurring in this proof, and let $x_1, x_2, \ldots, x_k$ be variables not used in any formula of the proof. For each $i = 1, 2, \ldots, n$, let $X_i^* = X_i(\underset{\sim}{a}/\underset{\sim}{x})$. We claim $\vdash X_i^*$. If X_i is an axiom of the FS4 system, this is straightforward. Modus ponens becomes 1) above, the rule of necessitation, 2), and the property corresponding to the rule of universal generalization is easily shown. Thus $\vdash X_n^*$. Now, if X_n has no parameters, $X_n^* = X_n$. Thus we have

THEOREM 4.5. If X has no parameters and is a theorem of FS4, then X is a theorem of $\epsilon S4^o$.

REFERENCES

[1] G. E. Hughes and M. J. Cresswell, An Introduction to Modal Logic, Methuen and Co. Ltd., London (1968).

[2] S. Kripke, Semantical considerations on modal and intuitionistic logic, Acta Philosophica Fennica, Modal and Many Valued Logics, Vol. 16 (1963), 83-94.

[3] A. C. Leisenring, Mathematical Logic and Hilbert's ε-Symbol, MacDonald Technical and Scientific, London (1969).

[4] K. Schütte, Vollständige Systeme Modaler und Intuitionistischer Logik, Springer-Verlag, Berlin (1968).

[5] R. Stalnaker and R. Thomason, Abstraction in first-order modal logic, Theoria, Vol. 34 (1968), 203-207.

[6] R. Thomason and R. Stalnaker, Modality and reference, Nous, 2 (1968), 359-372.

CRAIG'S INTERPOLATION THEOREM FOR MODAL LOGICS

Dov M. Gabbay

The Hebrew University of Jerusalem
Mathematics Institute, Oxford University

§0. Introduction

In this paper we present a uniform method of construction that yields Robinson's joint consistency theorem (and hence Craig's interpolation theorem) for many modal predicate and propositional systems.

Our method of proof is semantical. For some of these systems (those for which a natural deduction formulation is known) a syntactical proof can be given. The semantical method is illuminating, since it applies uniformly to many systems, and also since it shows what can be done with Kripke models.

The predicate systems considered are the following (without the Barcan formula):

(a) C2, D2, E2, E3, $S2^m$, S2, S3, K, T, S4

(b) $S4.1 = S4 + \Diamond\Box\phi \vee \Diamond\Box\sim\phi$

(c) $K + \Box\phi \longrightarrow \Box^{m+1}\phi$

(d) $K + \bigwedge_{n<m} \Box^n(\forall x_1 \ldots x_r)(\Diamond^{m-n}\Box\phi \longrightarrow \Box^{m+1-n}\phi)$

where x_i are all the free variables of ϕ

(e) $K + \bigwedge_{n<m} \Box^n(\forall x_1 \ldots x_r)(\Diamond^{m-n}\phi \longrightarrow \Box^{m-n}\phi)$

(f) $S4.2 = S4 + \Box\Diamond\phi \vee \Box\Diamond\sim\phi$.

We shall present the proof for the case of K. The reader will see immediately that the same proof yields the theorem for the systems in (a), (b) and (c), and further lemmas will establish the result for (d), (e) and (f).

§1. Craig's theorem for the modal system K

In this section we shall prove Robinson's theorem for modal K without the Barcan formula.

First, let us describe the system K: K has in addition to the axioms and rules of the classical predicate calculus the following axioms and rules:

(1) $\Box(\phi \rightarrow \psi) \rightarrow (\Box\phi \rightarrow \Box\psi)$

(2) $\vdash \phi \Rightarrow \vdash \Box\phi$.

We now describe the Kripke semantics for which K is complete: the structures of the form $(S, R, \underline{A}_t, 0)_{t \in S}$, where S is the set of possible worlds, $0 \in S$ is the real world, R is a binary relation on S, and $\underline{A}_t$, for $t \in S$, is a classical structure. Also, for all s, $t \in S$ if tRs then $A_t \subseteq A_s$, where A_t is the domain (universe) of $\underline{\mathbf{A}}_t$.

The truth value of a formula ϕ at a world t for the assignment assigning the members $x_1, \ldots, x_n$ of A_t to its variables is denoted by $[\phi(x_1, \ldots, x_n)]_t$ and is defined as follows:

(3) $[P(x_1, \ldots, x_n)]_t$ = the truth value given by $\underline{A}_t$ to atomic predicate P.

(4) $[\phi \wedge \psi]_t = T$ iff $[\phi]_t = T$ and $[\psi]_t = T$.

(5) $[\exists x\ \phi(x)]_t = T$ iff for some $x \in A_t$ we have $[\phi(x)]_t = T$.

(6) $[\sim\phi]_t = T$ iff $[\phi]_t = F$.

(7) $[\Box\phi]_t = T$ iff for all s such that tRs we have $[\phi]_s = T$.

ϕ is said to hold in the structure $(S,R,\underline{A}_t,0)$ in case $[\phi]_0 = T$.

THEOREM 8. (Completeness) Every consistent theory has a model.

THEOREM 9. (Robinson-type) Let Δ_0 and Θ_0 be two complete and consistent theories in the languages L_0 and M_0 respectively. Let $\Delta_0 \cap \Theta_0$ be a complete theory in the common language $L_0 \cap M_0$, then $\Delta_0 \cup \Theta_0$ is consistent. We assume that L_0 and M_0 have the same individual constants.

Proof. We shall now proceed to construct a model for $\Delta_0 \cup \Theta_0$.

We begin with some notation. Assume our languages L_0 and M_0 are of power λ. Let C_0, C_1, C_2 ... be a sequence of pairwise disjoint sets of individual constants each of power λ. Define for $n \geq 0$:

$$L_{n+1} = L_n + C_n$$
$$M_{n+1} = M_n + C_n$$
$$\text{and} \quad L = \cup_n L_n, \quad M = \cup_n M_n$$

where by L + C we mean the language obtained from L by adding to it all the individual constants of C.

We use δ_n to denote a formula in the language L_n and ϵ_n to denote a formula of the language M_n. ψ_n is reserved for formulas in $L_n \cap M_n$ (i.e. in the common language).

We have $L_n \cap M_n = L_0 \cap M_0 + C_0 + \dots + C_{n-1}$.

LEMMA. Any consistent theory in the language L_n can be extended to a complete and saturated theory Δ in the language L_{n+1}, i.e. the following holds:

(10) For all δ_{n+1} either $\delta_{n+1} \in \Delta$ or $\sim\delta_{n+1} \in \Delta$.

(11) If $\exists x \delta_{n+1}(x) \in \Delta$ then for some u in L_{n+1} $\delta_{n+1}(u) \in \Delta$.

One can obtain such a theory Δ since L_{n+1} has λ more individual constants than L_n.

We now proceed to the actual construction of the model. First extend Θ_0 to a complete and saturated theory Θ_1 in the language M_1.

LEMMA 12. The following theory is consistent: $\Delta_0 \cup \{\psi_1 \mid \psi_1 \in L_1 \cap M_1$ and $\psi_1 \in \Theta_1\}$, i.e. we can add consistently to Δ_0 all the sentences ψ_1 of Θ_1 which are in the common language.

Proof. Assume otherwise, then we have

$$\vdash \delta_0 \rightarrow \sim\psi_1$$

for some $\delta_0 \in \Delta_0$ and $\psi_1 \in L_1 \cap M_1 \cap \Theta_1$. (We do not have to take conjunctions since both theories are complete!)

Now if $c_1 \ldots$ are the constants of C_0 occurring in ψ_1 we have:

$$\vdash \delta_0 \longrightarrow (\forall c_1 \ldots)\sim\psi_1(c_1, \ldots)$$

where we use now the constants as variables.

Now, since $\delta_0 \in \Delta_0$ we get that $(\forall c_1 \ldots)\sim\psi_1 \in \Delta_0$. But the latter formula is in the language $L_0 \cap M_0$ and so $(\forall c_1 \ldots)\sim\psi_1 \in \Theta_0$ which is a contradiction.

We now extend the theory of Lemma 12 to a complete and saturated theory Δ_2 in the language L_2.

LEMMA 13. The following theory is consistent:

$$\Theta_1 \cup \{\psi_2 \mid \psi_2 \in L_2 \cap M_2 \text{ and } \psi_2 \in \Delta_2\}.$$

Proof. Otherwise, for some $\epsilon_1 \in \Theta_1$ and some $\psi_2 \in L_2 \cap M_2 \cap \Delta_2$ we have:

$$\vdash \epsilon_1 \longrightarrow \sim\psi_2.$$

Now, if we again use the new constants of C_1 as variables, and quantify over them, we get

$$\vdash \epsilon_1 \longrightarrow (\forall c_1 \ldots)\sim\psi_2.$$

Now since $\epsilon_1 \in \Theta_1$ we get $(\forall c_1 \ldots)\sim\psi_2 \in \Theta_1$ since this sentence is in $L_1 \cap M_1$, but by our construction $(\forall c_1 \ldots)\sim\psi_2 \in \Delta_2$ which contradicts $\psi_2 \in L_2 \cap M_2 \cap \Delta_2$.

Now extend this theory to a complete and saturated theory Θ_3 in the language M_3.

We continue in this manner to get the sequences:

(14)
$$\Delta_0 \subseteq \Delta_2 \subseteq \Delta_4 \ldots$$
$$\Theta_0 \subseteq \Theta_1 \subseteq \Theta_3 \ldots$$

in which the following holds:

(15a) For $n \geqslant 1$, $\psi_{2n} \in \Delta_{2n} \Rightarrow \psi_{2n} \in \Theta_{2n+1}$.

(15b) $\psi_{2n+1} \in \Theta_{2n+1} \Rightarrow \psi_{2n+1} \in \Delta_{2n+2}$.

(16) Each one of the theories Δ_n, Θ_n with $n \geqslant 1$ is complete and saturated in its respective language.

Now let Δ_ω and Θ_ω be the respective unions of the sequences. Then

(17) Δ_ω, Θ_ω are complete and saturated.

(18) $\Delta_\omega \cap \Theta_\omega$ is a complete and saturated theory of the common language $L_\omega \cap M_\omega$. (Since $L_\omega \cap M_\omega \cap \Theta_1 \subseteq L_\omega \cap M_\omega \cap \Delta_2 \subseteq L_\omega \cap M_\omega \cap \Theta_3 \subseteq \ldots$)

(19) Since $\Delta_0 \subseteq \Delta_\omega$, $\Theta_0 \subseteq \Theta_\omega$ we may therefore assume that our initial theories themselves (namely Δ_0 and Θ_0) have already properties (17) and (18).

We now begin to construct two models $\underline{A}$ and $\underline{B}$ of Δ_0 and Θ_0 respectively which shall eventually be proved to be isomorphic.

We assume now that Δ_0 and Θ_0 are in the languages L_0 and M_0 and have properties (17) and (18).

LEMMA 20. Let $\sim\Box\phi \in \Theta_0$, then there exists a complete and saturated consistent theory $\Theta_1(\phi)$ in the language M_1 with the following properties:

(21) $\sim\phi \in \Theta_1(\phi)$.

(22) For all β, $\Box\beta \in \Theta_0 \Rightarrow \beta \in \Theta_1(\phi)$.

<u>Proof</u>. This is well known.

LEMMA 23. The following theory is consistent:

$$\{\gamma \mid \Box\gamma \in \Delta_0\} \cup \{\psi_1 \mid \psi_1 \in L_1 \cap M_1 \text{ and } \psi_1 \in \Theta_1(\phi)\}.$$

<u>Proof</u>. Otherwise $\vdash (\bigwedge_i \gamma_i \rightarrow \sim\bigwedge_j \psi_{1,j})$ for $\Box\gamma_i \in \Delta_0$ and $\psi_{1,j} \in L_1 \cap M_1 \cap \Theta_1(\phi)$, and since $\Box(\gamma \wedge \gamma') \leftrightarrow \Box\gamma \wedge \Box\gamma'$ and since Δ_0 and $\Theta_1(\phi)$ are complete and consistent we have for $\gamma = \bigwedge_i \gamma_i$ and $\psi_1 = \bigwedge_j \psi_{1,j}$, $\Box\gamma \in \Delta_0$, $\psi_1 \in L_1 \cap M_1 \cap \Theta_1(\phi)$ that

$$\vdash \gamma \rightarrow \sim\psi_1.$$

So if ψ_1 is $\psi_1(c_1,\ldots,c_n)$ where $c_1, \ldots, c_n$ are the constants of C_0 occurring in ψ_1 we have:

$$\vdash \gamma \rightarrow (\forall c_1 \ldots)\sim\psi_1.$$

Therefore $\vdash \Box\gamma \rightarrow \Box(\forall c_1 \ldots)\sim\psi_1$. Since $\Box\gamma \in \Delta_0$ we get $\Box(\forall c_1 \ldots)\sim\psi_1 \in \Delta_0$. But this formula is in $L_0 \cap M_0$ and so $\Box(\forall c_1 \ldots)\sim\psi_1 \in \Theta_0$ and so $(\forall c_1 \ldots)\sim\psi_1 \in \Theta_1(\phi)$ which contradicts $\psi_1(c_1,\ldots,c_n) \in \Theta_1(\phi)$.

Now extend the theory of Lemma 23 to a complete and saturated consistent theory Δ_2^ϕ in the language L_2.

LEMMA 24. The following theory is consistent:

$$\Theta_1(\phi) \cup \{\psi_2 \mid \psi_2 \in L_2 \cap M_2 \text{ and } \psi_2 \in \Delta_2^\phi\}.$$

Proof. Otherwise, by the completeness and consistency of $\Theta_1(\phi)$, Δ_2^ϕ, we have for some $\gamma \in \Theta_1(\phi)$ and some $\psi_2 \in \Delta_2^\phi$ where $c_1, \ldots, c_n$ are the new constants of L_2 which occur in ψ_2: $\vdash \gamma \rightarrow \sim\psi_2(c_1,\ldots,c_n)$ and hence $\vdash \gamma \rightarrow (\forall c_1 \ldots)\sim\psi_2$.

Now since $\gamma \in \Theta_1(\phi)$ we get that $(\forall c_1 \ldots)\sim\psi_2 \in \Theta_1(\phi)$. But the latter formula is in the language $L_1 \cap M_1$ and so by the construction of Δ_2^ϕ $(\forall c_1 \ldots)\sim\psi_2 \in \Delta_2^\phi$ which is a contradiction.

Now enlarge the theory of Lemma 24 to a complete consistent and saturated theory $\Theta_3(\phi)$ in the language M_3.

We now continue in this manner and get the following two sequences:

$$\Delta_2^{\phi} \subseteq \Delta_4^{\phi} \subseteq \ldots$$
$$\Theta_1(\phi) \subseteq \Theta_3(\phi) \subseteq \ldots .$$

Now let Δ^{ϕ} and $\Theta(\phi)$ be the unions of the respective sequences, then we have:

(25) For all y, $\Box y \in \Delta_0 \rightarrow y \in \Delta^{\phi}$.

(26) For all y, $\Box y \in \Theta_0 \rightarrow y \in \Theta(\phi)$.

(27) Δ^{ϕ} and $\Theta(\phi)$ are complete and saturated in the languages L_{ω} and M_{ω} respectively.

(28) $\Delta^{\phi} \cap \Theta(\phi)$ is a complete theory in the common language.

(29) Δ^{ϕ} and $\Theta(\phi)$ are in a language with λ more constants than the language of Δ_0 and Θ_0.

Now we notice two facts:

(30) Similar construction can be carried out for any other β such that $\sim\Box\beta \in \Theta_0$ and we get Δ^{β} and $\Theta(\beta)$ with properties (25), (26), (27), (28) and (29). Similarly for any β such that $\sim\Box\beta \in \Delta_0$ we can construct $\Delta(\beta)$ and Θ^{β} with similar properties as before.

All the theories which we construct in this process are saturated!!

We shall now define S_A and S_B, and an isomorphism between S_A and S_B simultaneously. The members of S_A and S_B will be of the form (Γ, x) where Γ is a saturated and complete theory and x is a finite sequence of formulas.

Our definition is by induction and its first step is as follows:

Let $(\Delta_0, \langle A\rangle) \in S_A$, $(\Theta_0, \langle B\rangle) \in S_B$ where Δ_0 and Θ_0 are the saturated and complete theories fulfilling the requirements of Theorem 9 and (18). Let

$$f((\Delta_0, \langle A\rangle)) = (\Theta_0, \langle B\rangle).$$

The n+1-st induction step is: We assume that at this stage f is defined on the part of S_A which has already been constructed. f is one-one and onto, and whenever $f((\Gamma,x)) = (E,y)$ then the following holds:

(a) Γ and E are in the languages L_ω, M_ω (for some ω) respectively and they fulfill (17) and (18).

(b) If we replace $\langle B\rangle$ which is the first letter in the sequence comprising y by $\langle A\rangle$ then we get x.

CONSTRUCTION 31. To continue the construction let (Γ,x) be constructed at the n-th (the preceding) step and let $f((\Gamma,x)) = (E,y)$. Let $\sim\Box\phi \in \Gamma$ construct the theories $\Gamma(\phi)$, E^ϕ. (This is possible to do since (17) and (18) hold.) Put $(\Gamma(\phi), x\hat{}\langle\phi\rangle) \in S_A$ and $(E^\phi, y\hat{}\langle\phi\rangle) \in S_B$ where ^ is concatenation of sequences, and write $f(\Gamma(\phi), x\hat{}\langle\phi\rangle) = (E^\phi, y\hat{}\langle\phi\rangle)$. Similarly let $\sim\Box\psi \in E$, construct the theories $E(\psi)$, Γ^ψ and put $(E(\psi), y\hat{}\langle\psi\rangle) \in S_B$ and $(\Gamma^\psi, x\hat{}\langle\psi\rangle) \in S_A$ and write $f(\Gamma^\psi, x\hat{}\langle\psi\rangle) = (E(\psi), y\hat{}\langle\psi\rangle)$. Clearly f and the new theories fulfill the assumptions of the n-th step and so we can proceed to the n+2-th step.

Now, having obtained, after ω steps, the sets S_A and S_B we define $(\Gamma,x)\ R_A\ (\Gamma',x')$ iff for some ϕ $x' = x\hat{}\langle\phi\rangle$. Similarly define $(E,y)\ R_B\ (E',x')$; we get that f is an isomorphism between (S_A,R_A) and (S_B,R_B). The following holds for S_A:

(32) $\Box\gamma \in \Gamma \Rightarrow \gamma \in \Gamma_1$ if there exists an x and x' such that $(\Gamma,x)\ R_A\ (\Gamma_1,x')$.

(33) If $\sim\Box\phi \in \Gamma$ and $(\Gamma,x) \in S_A$ then $\sim\phi \in \Gamma(\phi)$ and $(\Gamma(\phi),x^\frown\langle\phi\rangle) \in S_A$ and by definition $(\Gamma,x)\ R_A\ (\Gamma(\phi),x^\frown\langle\phi\rangle)$.

Let $(\Delta_0,\langle A\rangle)$ be our O.

Now for $(\Gamma,x) \in S_A$ define $\underline{A}_{(\Gamma,x)}$ as follows: the set of constants of $\underline{A}_\Gamma$ is the set of constants of the language of Γ. $[P(x_1,\ldots,x_n)] = T$ iff $P(x_1,\ldots,x_n) \in \Gamma$ for $x_1, \ldots, x_n$ in the language of Γ.

LEMMA 34. $[\phi]_{(\Gamma,x)} = T$ iff $\phi \in \Gamma$.

Proof. The proof is by induction: $\sim$, $\wedge$ present no difficulties. The case of $\exists$ goes through since our theories are all saturated and complete. Now the case of $\Box\phi$ also makes no trouble because of (32) and (33).

In a similar manner we construct the model $\underline{B}$ using all the Θ-theories.

Now we have that $\underline{A}$ is a model of Δ_0 and $\underline{B}$ is a model of Θ_0.

LEMMA 35. $\underline{A}$ and $\underline{B}$ are isomorphic in the common language $L_0 \cap M_0$.

Proof. We define an isomorphism g between S_A and S_B, using the function f. Let $A_{(\Gamma,x)}$ be given then $B_{f(\Gamma,x)}$ has the same domain as $A_{(\Gamma,x)}$ by properties (a) and (b) mentioned in step n+1 and so we take the isomorphism to be identity. This is indeed an isomorphism since the intersection of the left hand components of (Γ,x) and $f((\Gamma,x))$ is a complete and saturated theory of the common language.

Thus we conclude that $\underline{A}$ is a model of $\Delta_0 \cup \Theta_0$ and thus Theorem 9 is proved.

COROLLARY 36. To obtain the joint consistency theorem for T (or S4) just take R_A^* to be the reflexive (or the reflexive and transitive) closure of R_A.

Since $\vdash \Box\phi \longrightarrow \phi$ (or this and $\vdash \Box\phi \longrightarrow \Box^2\phi$) and since in the construction, all theories were taken to be saturated in the new system T (or S4), we get that:

$$(\Gamma,x)\ R_A^*\ (\Gamma',x') \wedge \Box\phi \in \Gamma \Rightarrow \phi \in \Gamma'.$$

Similarly for R_B^*.

COROLLARY 37. To obtain the theorem for $K + \Box\phi \longrightarrow \Box^{m+1}\phi$ define $(\Gamma,x)\ R_A^*\ (\Gamma',x')$ iff for some $k \geq 0$

$$(\Gamma,x)\ R_A^{km+1}\ (\Gamma',x').$$

Now since $\vdash \Box\phi \longrightarrow \Box^{m+1}\phi$ we get that if $\Box\phi \in \Gamma$ and $(\Gamma,x)\ R_A^*\ (\Gamma',x')$ then $\phi \in \Gamma'$. (S_A,R_A^*) is a model of $K + \Box\phi \longrightarrow \Box^{m+1}\phi$ since $(\Gamma,x)\ R_A^{*(m+1)}\ (\Gamma',x') \longrightarrow (\Gamma,x)\ R_A^*\ (\Gamma',x')$ can be proved to hold. Similarly for (S_B,R_B^*).

COROLLARY 38. C2 is like K except that the rule $\vdash \phi \Rightarrow \vdash \Box\phi$ is replaced by $\vdash \phi \longrightarrow \psi \Rightarrow \vdash \Box\phi \longrightarrow \Box\psi$. C2 is complete for semantics of the form (N,S,R,0) where $N \subseteq S$ is the set of normal worlds.

In the definition of satisfaction we change (7) to read:

$$[\Box\phi]_t = T \text{ iff } t \in N \text{ and } \forall s\ (tRs \longrightarrow [\phi]_s = T).$$

So if $t \notin N$ no formula of the form $\Box\phi$ can hold at t.

Let ψ be a sentence of the common language and let Δ_0 and Θ_0 be two theories fulfilling properties (17) and (18) we then have

$$\Box(\psi \longrightarrow \psi) \in \Delta_0 \quad \text{iff} \quad \Box(\psi \longrightarrow \psi) \in \Theta_0.$$

In case that $\Box(\psi \longrightarrow \psi) \in \Theta_0$ we continue the construction of $\Theta_1(\phi)$ etc.

as in (30). In case $\Box(\psi \dashrightarrow \psi) \not\in \Theta_0$, this means that in the model formed from Δ_0 (see lemma 34) Δ_0 is not normal and so we declare Δ_0 and Θ_0 to be non-normal worlds and stop dealing with them.

We now define S_A and S_B as in the case of K except that (see 31) certain worlds may be declared non-normal. Let N_A and N_B be the sets of normal worlds. Clearly $N_B = f``N_A$.

Thus the joint consistency theorem is proved for C2 as well.

COROLLARY 39. To obtain the theorem for D2, E2, $S2^m$, S3, note that appropriate semantics has to do either with normality conditions or reflexivity or transitivity.

COROLLARY 40. Let us consider the system (d) of section 0. It is complete for the semantics with the condition

$$uR^m x \wedge uR^{m+1} y \longrightarrow xRy.$$

Note that in the propositional case $\Diamond^m \Box\phi \dashrightarrow \Box^{m+1}\phi$ is sufficient to axiomatize the semantics. The predicate case needs the entire conjunction. The semantics fulfills (proof by induction on k)

$$uR^{m+k} x \wedge uR^{m+k+1} y \longrightarrow xRy.$$

To obtain the joint consistency theorem we construct (S_A, R_A) and (S_B, R_B) up to stage m. We can also assume that in S_A and S_B whenever (Γ, x) and (Γ', x') are not $\bar{R}_A$ comparable ($\bar{R}_A$ is the transitive closure of R_A) then if (Γ_0, x_0) is the highest point below both,

$$(41) \qquad (L_\Gamma - L_{\Gamma_0}) \cap (L_{\Gamma'} - L_{\Gamma_0}) = 0$$

where L_Γ is the language of the theory Γ.

LEMMA 42. Let $(\Delta_0,\langle A\rangle)\ R^m_A\ (\Delta,x)$ and let $\sim\Box\phi \in \Delta$ then

$$\{\sim\phi\} \cup \{\alpha \mid \Box\alpha \in \Delta' \text{ and } (\Delta_0,\langle A\rangle)\ R^m_A\ (\Delta',x')\}$$

is consistent.

Proof. Otherwise for some $\alpha_1, \ldots, \alpha_n$ we have $\vdash \alpha_1\wedge\ldots\wedge\alpha_n \longrightarrow \phi$, $\alpha_i \in \Delta_i$. Let $x_1 \ldots$ be the constants in Δ_1 and not in the rest, $x_2 \ldots$ be the constants in Δ_2 and not in $\Delta_3, \ldots, \Delta_n, \Delta$ etc., then:

$$\vdash (\exists x_1\ldots)\alpha_1 \longrightarrow ((\exists x_2\ldots)\alpha_2 \longrightarrow \ldots((\exists x_n\ldots)\alpha_n \longrightarrow \forall y\phi)\ldots).$$

Let us take the typical case of $n = 2$.

$$\vdash \Box(\exists x_1\ldots)\alpha_1 \longrightarrow \Box((\exists x_2\ldots)\alpha_2 \longrightarrow \forall y\phi).$$

Since $\Box\alpha_1 \in \Delta_1$ we get that

$$\Box((\exists x_2\ldots)\alpha_2 \longrightarrow \forall y\phi) \in \Delta_1.$$

Let $(\Delta_{1,2}, x_{1,2})$ be that highest point in S_A that is below both (Δ_1,x_1) and (Δ_2,x_2). By (41) $\Box(\exists x_2\alpha_2 \longrightarrow \forall y\phi)$ is in the language of $\Delta_{1,2}$ and so if $(\Delta_{1,2}, x_{1,2})$ has height k we get that $\Diamond^{m-k}\Box(\exists x_2\alpha_2 \longrightarrow \forall y\phi) \in \Delta_{1,2}$ and so by the axiom $\Box^{m+1-k}(\exists x_2\alpha_2 \longrightarrow \forall y\phi)$ is in $\Delta_{1,2}$ and so $\Box(\exists x_2\alpha_2 \longrightarrow \forall y\phi)$ is in Δ_2 and so $\Box\forall y\phi \in \Delta_2$.

Let $(\Delta_{2,*}, x_{2,*})$ be the highest element below both (Δ_2,x_2) and (Δ,x). Clearly $\forall y\phi$ is in the language of $\Delta_{2,*}$. Repeating the process we get that $\Box\forall y\phi \in \Delta$, a contradiction.

Extend the theory of (42) to a saturated theory $\Delta^{\phi,1}$ in a language $L_\Delta + C_\phi$ (where C_ϕ is a set of additional constants).

LEMMA 43. $\{\psi \mid \psi \in \Delta^{\phi,1} \cap (L_{f(\Delta)} + C_{\phi})\} \cup \{\beta \mid \Box\beta \in E$ for (E,y) such that $(\Theta_0,\langle B\rangle)\ R_B^m\ (E,y)\}$ is consistent.

<u>Proof</u>. Otherwise for some $\beta_i \in E_i$ we have

$$\vdash \bigwedge_i \beta_i \longrightarrow \sim\psi.$$

From this point we repeat the proof of (42) and obtain that $\Box\forall y\sim\psi \in E_n$. Let $(E_n,y_n) = f(\Delta_n,x_n)$ then since ψ is in the common language $\Box\forall y\sim\psi \in \Delta_n$ and so $\Box\forall y\sim\psi \in \Delta^{\phi,1}$ by (42) which is a contradiction.

The two lemmas (42) and (43) allow us, as in (13) - (17), to construct two theories $\Delta(\phi)$ and E^{ϕ} such that (17) and (18) hold and whenever $(\Delta_0,\langle A\rangle)\ R_A^m\ (\Delta',x')$ and $\Box\alpha \in \Delta'$ then $\alpha \in \Delta(\phi)$ and similarly for E^{ϕ}. Let $(\Delta(\phi), x\hat{}\langle\phi\rangle) \in S_A$ and $(E^{\phi}, f(x)\hat{}\langle\phi\rangle) \in S_B$ and define R_A and R_B to hold between every pair (Γ,x) of height m and our new pair. We do this for every $\sim\Box\phi \in \Delta$ for (Δ,x) of height m. We treat (S_B,R_B) similarly. Thus stage m + 1 is completed. Stage m + k + 1 is like stage m + 1, we treat points constructed at stage m + k.

Thus the joint consistency is proved for (d).

COROLLARY 44. To obtain the joint consistency theorem for S4.1 notice that S4.1 is complete for the following condition (besides those of S4):

$$(45) \qquad \forall x\ \exists y\ (xRy \wedge \forall z\ (yRz \longrightarrow y = z)).$$

To construct a (S_A,R_A) that fulfills this condition we modify (31) as follows: from each pair (Δ,x) and (E,y) we also construct a new pair $(\Delta^{\infty}, x\hat{}\langle\infty\rangle)$, $(E^{\infty}, y\hat{}\langle\infty\rangle)$, where Δ^{∞} is a 'classical' extension of Δ (i.e. $\Box\phi \longleftrightarrow \phi$ holds for all ϕ) and similarly E^{∞}. This can be done using the following lemma:

LEMMA 46. $\{\alpha \mid \Box\alpha \in \Delta\} \cup \{\forall x(\alpha \longrightarrow \Box\alpha)\}$ is consistent.

This holds since Δ is consistent and so has a model fulfilling (45).

This theory can be extended to a saturated theory Δ_1^∞ such that $\alpha \longleftrightarrow \Box\alpha \in \Delta_1^\infty$ for all α. A corresponding E_2^∞ can be found. Now we can continue as in (23) - (29), and get a Δ^∞ and E^∞. We do not construct any theories of the form $\Delta^\infty(\phi)$. Nevertheless we still have that (34) holds, because $\sim\Box\phi \longleftrightarrow \sim\phi \in \Delta^\infty$. Thus an (S_A, R_A) can be found.

S4.2 is complete for the condition (besides those of S4):

$$\forall y\ \forall x\ \exists z\ (xRz \wedge yRz).$$

To obtain the joint consistency theorem for S4.2 we modify the construction of (S_A, R_A) (see (31)) as follows: At stage n of the form $n = 2^m$ we construct, for each two pairs (Δ_i, x_i) and $f((\Delta_i, x_i)) = (E_i, y_i)$, $i = 1, 2$, previously defined another pair $(\Delta_{1,2}, x_{1,2})$ and $(E_{1,2}, y_{1,2})$ such that $(\Delta_i, x_i)\ R_A\ (\Delta_{1,2}, x_{1,2})$ and $(E_i, y_i)\ R_B\ (E_{1,2}, y_{1,2})$ using the lemmas below. We assume (41) to hold for the part of (S_A, R_A^*) constructed up to this stage. R_A^* is a partial order (not a tree) such that for every two points there exists a unique maximal point below both (see (36) for R_A^*).

LEMMA 47. Let (Δ_i, x_i) be given then

$$\{\alpha \mid \Box\alpha \in \Delta_1\} \cup \{\beta \mid \Box\beta \in \Delta_2\}$$

is consistent.

Proof. Otherwise for some α, β

$$\vdash \alpha \longrightarrow \sim\beta.$$

Let (Δ_0, x_0) be the maximal point below both. Let $x_1 \ldots, y_1 \ldots$ be the new variables in α and β respectively then

$$\vdash (\exists x_1 \ldots)\alpha \longrightarrow (\forall y_1 \ldots)\sim\beta,$$

$$\vdash \Box(\exists x_1 \ldots)\alpha \longrightarrow \Box(\forall y_1 \ldots)\sim\beta.$$

$\Box(\forall y_1 \ldots)\sim\beta \in \Delta_1$ and is in the language of Δ_0. So since $\Box\Diamond\alpha \vee \Box\Diamond\sim\alpha \in \Delta_0$ for all α, we get $\Box\Diamond(\forall y_1 \ldots)\sim\beta \in \Delta_0$ so $\Diamond(\forall y_1 \ldots)\sim\beta \in \Delta_1$, a contradiction.

Extend this theory to a saturated theory Δ_{*1}.

LEMMA 48. The following theory is consistent:

$$\{\alpha \mid \Box\alpha \in E_1\} \cup \{\beta \mid \Box\beta \in E_2\} \cup$$
$$\{\psi \mid \psi \in \Delta_{*1} \text{ and } \psi \text{ in the common language}\}$$

<u>Proof</u>. Simple and similar to (43).

These lemmas allow us to construct a $\Delta_{1,2}$ and $E_{1,2}$ as needed. We can give indexing in such a manner that $x_{1,2}$ may be defined e.g. $x_{1,2} = \{x_1, x_2\}$ and modify R_A such that $(\Delta_1, x_1)\ R_A^*\ (\Delta_{1,2}, x_{1,2})$. At stage $2^m + 1$ we continue as in the case of stage 3 (see (31)) treating both theories constructed at stage $2^m - 1$ and 2^m. At stage $2^m + 2$ we treat theories constructed at stage $2^m + 1$, and so on.

The system (e) of section 0 is complete for the condition

$$uR^m v \wedge uR^m v' \longrightarrow v = v'.$$

The proof of the joint consistency proceeds like in (40). We assume (41) and prove

LEMMA 49. The following is consistent:

$$\{\alpha \mid \Box\alpha \in \Delta \text{ and } (\Delta_o, \langle A\rangle)\ R_A^m\ (\Delta, x) \text{ for some } \Delta\}.$$

The proof goes like (42) since the axioms of (40) are available. Now to complete the proof notice that if $\sim\Box\phi \in \Delta$ for such Δ then from the axioms $\Box\sim\phi \in \Delta$, and so level $m + 1$ can be taken to consist of one theory alone. We leave the details to the reader.

The method of (36), (37) can be used to give the joint consistency theorem for $K + \Diamond^m\Box\phi \longrightarrow \Box^n\phi$ (the propositional part). It is complete for $uR^m x \wedge uR^n y \longrightarrow xRy$.

A NOTE ON MODELS AND SUBMODELS OF ARITHMETIC[1]

Haim Gaifman

The Hebrew University of Jerusalem

Introduction

The aim of this note is to point out certain applications of the equality of recursively enumerable and diophantine relations for the theory of models of arithmetic. The equality itself is easily seen to be equivalent to a model theoretic statement (Theorem 1). Roughly speaking, it amounts to the absoluteness of recursive relations in models of true arithmetic. Observing also the effective way in which it has been proved, one can derive the absoluteness in all models of Peano's arithmetic of formulas which are provably recursive (Theorem 2). The main result of this note is Theorem 4 which asserts the following:

---

[1]The results of this paper have been obtained in the summer of 1970. They do not constitute the subject matter of the invited talk given in the London Conference which was a survey-type lecture. They are offered here as a substitute. [Haim Gaifman's invited talk was on 'Probabilities for logical calculi', and surveyed the area described in: H. Gaifman, Concerning measures in first order calculi, Israel Journal of Math. Vol. 2 (1964) 1-18; D. Scott and P. Krauss, Assigning probabilities to logical formulas, Aspects of inductive logic (Hintikka, Suppes eds.), North-Holland (1966) 219-264; together with some recent unpublished work of J. Stavi. (Editor)]

After submitting the paper, in February 1971, the author's attention has been called to a paper by A. Adler, in Zeitschrift für Math. Logik und Grundlagen der Math., 1969 pp. 289-290, in which a result which is essentially Theorem 4 for the case of models of true arithmetic is stated. That paper had been unknown to the author before that. At a crucial point there is, however, a mistake in the proof which is given by Adler, and it seems to the author that any way of filling in the gap would amount in essence to a proof of the implication (3) $\Rightarrow$ (1) of Theorem 3, as given in the present paper. For more details see the postscript.

If M_1 is a submodel of M_2 and both satisfy the Peano axioms, then the initial segment, $\bar{M}_1$, of M_2, which is determined by M_1 (i.e. $x \in \bar{M}_1$ iff for some $y \in M_1$, $x \leq y$) is also a model of Peano's arithmetic and, in fact, M_1 is an elementary submodel of $\bar{M}_1$. Equivalently stated, it asserts that every extension of a model of Peano's arithmetic to another model of this theory can be obtained in two steps: first take an elementary cofinal extension, then take an end-extension.

In §2 this is applied to an old result of Rabin which asserts that every non-standard model of true arithmetic has an extension, which is also a model of true arithmetic, in which a diophantine equation, which is unsolvable in the original model, is solvable. One can get now the second model as an end-extension of a cofinal elementary extension of the first. Whether one can always make the second model an end-extension of the first is still open. The result is further generalized to all complete extensions of Peano's arithmetic and to Σ_n-formulas, by applying the theorem of Matijasevič and its effective proof to other old results of Rabin. The application is straightforward and is stated here mostly for the sake of the record and because of the great simplification which it brings. Finally, it is shown that one can have a model of true arithmetic with two submodels of true arithmetic, of which one is an elementary submodel and the other is a n-elementary submodel (i.e. $\Sigma_n \cup \Pi_n$-formulas are invariant) whose intersection violates a Peano axiom. The proof follows the same method of the proof of a weaker result in this direction of Rabin. From Theorem 4 it follows, however, that whenever two submodels determine the same initial segment their intersection is an elementary submodel of both.

Notations and terminology

We use '0', '1', '+', '.' and '=' to denote both the symbols of the formal first-order language of arithmetic, and their interpretations in the particular model under discussion. The same applies to '$\leq$', whether it is taken as an additional symbol or as a short-hand for the formula which defines this relation in terms of +.

By '<u>the language of arithmetic</u>' we mean the first-order language with 0, 1, +, · as non-logical symbols.

<u>Peano's arithmetic</u> is the theory axiomatized by the usual first-order formulation of Peano's axioms. It is denoted by '<u>P</u>'.

<u>True arithmetic</u> is the complete theory of the standard model whose domain is $\{0,1,2,\ldots,n,\ldots\}$.

If M is a model then '$x \in M$' means that x belongs to the universe of M.

'$M_1 \subset M_2$' means that M_1 is a submodel of M_2.

'$M_1 \prec M_2$' means that M_1 is an elementary submodel of M_2.

We shall use freely the members of a model as names of themselves and add them to the language, whenever we wish to speak about satisfaction of certain formulas for certain evaluations of their free variables. Thus, '$M \models \phi(a)$', where $a \in M$ and $\phi = \phi(v_0)$, means what is usually denoted by '$M \models \phi[a]$'.

Strings of variables are denoted by '$\vec{u}$', '$\vec{v}$', ...

Strings of members of a model are denoted by '$\vec{a}$', '$\vec{b}$', ...

M_2 is an <u>end-extension</u> of M_1 if $x \leq y$ for every $x \in M_1$, $y \in M_2 - M_1$.

M_2 is a <u>cofinal extension</u> of M_1 if for every $y \in M_2$ there exists $x \in M_1$ such that $y \leq x$.

§1.

A formula $\phi(v_0,\ldots,v_{n-1})$ is said to be <u>persistent</u> in a class of models K if, for every $M_1, M_2 \in K$ such that $M_1 \subset M_2$ and every $\vec{a} \in M_1{}^n$, $M_1 \models \phi(\vec{a}) \Rightarrow M_2 \models \phi(\vec{a})$.

A formula is said to be <u>absolute</u> in K if an analogous condition holds, with "$\Rightarrow$" replaced by "$\Leftrightarrow$".

Obviously, ϕ is absolute iff ϕ and $\neg\phi$ are persistent.

THEOREM 1. The assertion that every recursively enumerable relation is diophantine is equivalent to each of the following:

(1) Every formula (of arithmetic) which defines (in the standard model) a recursively enumerable relation is persistent in models of true arithmetic.

(2) Every formula which defines a recursive relation is absolute in models of true arithmetic.

<u>Proof</u>. If $\phi(\vec{v})$ defines a recursively enumerable relation, then, assuming its equality with a diophantine relation, we have a diophantine equation $\delta(\vec{u},\vec{v})$ such that the following is a theorem of true arithmetic:

$$\phi(\vec{v}) \leftrightarrow \exists\vec{u}\ \delta(\vec{u},\vec{v}).$$

Since the formula on the right side is existential it is persistent.

(1) $\Rightarrow$ (2), because ϕ defines a recursive relation iff ϕ and $\neg\phi$ define recursively enumerable relations.

(2) $\Rightarrow$ (1) because every ϕ which defines a recursively enumerable relation is equivalent, in true arithmetic, to a formula of the form $\exists\vec{v}\ \psi$, where ψ defines a recursive relation. If ψ is absolute $\exists\vec{v}\ \psi$ is persistent.

Finally, assume (1) and consider any recursively enumerable relation, X. There is a formula of arithmetic ϕ which defines X. This formula is persistent, and, by a characterization of Tarski of persistent formulas, [2], ϕ is equivalent in true arithmetic to $\exists\vec{v}\ \psi$ where ψ is quantifier-free. By well-known devices every quantifier-free formula of arithmetic can be easily put in form $\exists\vec{u}\ \delta$ where δ is a diophantine equation. Hence X is diophantine.

q.e.d.

Consider now a standard way of presenting recursively enumerable relations. For example, let them be presented through machines of the Turing type which enumerate the relation on an infinite tape. The description of such a machine can be formalized in a standard way in the language of arithmetic, leading to a formula $\phi(\vec{x})$ which asserts that $\vec{x}$ belongs to the particular recursively enumerable relation. Such a formula will be referred to as a <u>standard description</u> of the recursively enumerable relation.

DEFINITION. $\phi(\vec{v})$ is <u>provably recursively enumerable</u> if there is a standard description, $\psi(\vec{v})$, of a recursively enumerable relation, such that

$$\underline{P} \vdash \forall\vec{v}\ (\phi(\vec{v}) \longleftrightarrow \psi(\vec{v})).$$

$\phi(\vec{v})$ is <u>provably recursive</u> if both $\phi(\vec{v})$ and $\neg\phi(\vec{v})$ are provably recursively enumerable.

THEOREM 2. Every formula which is provably recursively enumerable is persistent and every formula which is provably recursive is absolute in the models of $\underline{P}$.

<u>Proof</u>. A survey of the proof of Matijasevič, that every recursively enumerable relation is diophantine, [8], as well as the proofs

of the results of J. Robinson, [1], and Davis-Putnam-J. Robinson, [3], on which it relies, shows that, for any given recursively enumerable relation, they can be carried out using means which are provided by the Peano axioms.[2] If one formalizes them in Peano's arithmetic one gets, for every standard description, $\phi(\vec{u})$, of a recursively enumerable relation, a diophantine equation $\psi(\vec{u},\vec{v})$, such that $\underline{P} \vdash \phi(\vec{u}) \leftrightarrow \exists\vec{v}\ \psi(\vec{u},\vec{v})$. Since $\exists\vec{v}\ \psi(\vec{u},\vec{v})$ is persistent in models of $\underline{P}$, so is $\phi(\vec{u})$, and so is every formula which is provably equivalent to $\phi(\vec{u})$. The absoluteness of provably recursive formulas follows now immediately.

q.e.d.

Note: The occurrence of the prefix "provable" in the formulation of the theorem is unavoidable, because there are two formulas defining in the standard model the same recursive set, whose equivalence is not provable in $\underline{P}$.

A bounded quantifier is a quantifier of the form $\forall u \leq v$ or of the form $\exists u \leq v$. Its meaning is given by the rules:

$$\forall u{\leq}v\ \phi \text{ is to be rewritten as } \forall u\ [u \leq v \rightarrow \phi]$$
$$\exists u{\leq}v\ \phi \text{ is to be rewritten as } \exists u\ [u \leq v \wedge \phi].$$

Here, $u \leq v$ is defined in the way which is usual in Peano's arithmetic, namely, by $\exists z\ [u + z = v]$. A more convenient way is to add $\leq$ as a binary predicate to the language of arithmetic and to include among Peano's axioms the biconditional $u \leq v \leftrightarrow \exists z\ [u + z = v]$.

The set of all bounded formulas is the smallest set which includes the formulas $t_1 = t_2$, $t_1 \leq t_2$, where t_1 and t_2 are terms, and is closed

---

[2] The author wishes to thank Mr. A. Pridor for carrying out the detailed checking which is necessary to verify this claim. Any mistake should, however, be attributed to the author.

under combinations by means of sentential connectives and under bounded quantification.

COROLLARY to Theorem 2. Every bounded formula is absolute in models of $\underline{P}$.

<u>Proof</u>. By Theorem 2 and the well-known observation that every bounded formula is provably recursive.

In the formulation of the next theorem it is assumed that the language of arithmetic contains also the binary predicate $\leqslant$. $\underline{P}$ will contain the axiom $u \leqslant v \leftrightarrow \exists z\,(u + z = v)$, so that $\leqslant$ will be interpreted in a model of $\underline{P}$ as the usual ordering.

THEOREM 3. Let M_1 be a model of $\underline{P}$. Let M_2 be a model of the language of arithmetic such that $M_1 \subset M_2$, $\leqslant$ is in M_2 a total ordering, and M_1 is cofinal in M_2 (i.e. for each $x \in M_2$ there exists $y \in M_1$ such that $x \leqslant y$). The following are then equivalent:

(1) $M_1 \prec M_2$

(2) M_2 is a model of $\underline{P}$

(3) If $\psi(\vec{v})$ is of the form $\forall_{\leqslant} \exists_{\leqslant} \phi$, where '$\forall_{\leqslant}$' denotes a block of bounded universal quantifiers, '$\exists_{\leqslant}$' - a block of bounded existential quantifiers, and ϕ is quantifier free, then

$$M_1 \models \psi(\vec{c}) \Rightarrow M_2 \models \psi(\vec{c}), \quad \text{for all } \vec{c} \in M_1^{\,n}.$$

(4) The same as (3) except that now ψ is of the form $\forall\exists\phi$, where '$\forall$' and '$\exists$' denote blocks of unbounded universal and existential quantifiers, respectively.

<u>Note</u>: The implication (2) $\Rightarrow$ (1), with '$\underline{P}$' replaced everywhere by 'true arithmetic', appears essentially in [6]. At that time Matijasevič's result was unknown, hence Rabin formulated it with respect to

any family of predicates from which all recursively enumerable relations could be existentially defined. The main point, however, of Theorem 3 is the implication (3) $\Rightarrow$ (1).

Proof. Obviously (1) $\Rightarrow$ (2). The implication (2) $\Rightarrow$ (3) follows from the corollary to Theorem 2. (It is true without assuming the cofinality of M_1 in M_2.) It remains to prove that (3) $\Rightarrow$ (4) and (4) $\Rightarrow$ (1).

(3) $\Rightarrow$ (4): Assume that $M_1 \models \forall\vec{u}\ \exists\vec{v}\ \phi(\vec{u},\vec{v},\vec{c})$, with $\vec{c} \in M_1{}^n$ and ϕ quantifier-free. For the sake of simplicity consider the case where $\vec{u} = u$, $\vec{v} = v$, and $\vec{c} = c$. To show $M_2 \models \forall u\ \exists v\ \phi(u,v,c)$, let $a \in M_2$ and show the existence of b such that $M_2 \models \phi(a,b,c)$. Let $a' \in M_1$ be such that $a \leqslant a'$. We have $M_1 \models \forall u{\leqslant}a'\ \exists v\ \phi(u,v,c)$. Now, in Peano's arithmetic, the following is provable for every formula ψ:

$$\forall u{\leqslant}u'\ \exists v\ \psi \longrightarrow \exists v'\ (\forall u{\leqslant}u'\ \exists v{\leqslant}v'\ \psi).$$

Hence, there exists $b' \in M_1$ such that $\forall u{\leqslant}a'\ \exists v{\leqslant}b'\ \phi(u,v,c)$ holds in M_1. Hence, assuming (3), it holds in M_2. Since $a \leqslant a'$, there is $b \in M_2$ such that $b \leqslant b'$ and $M_2 \models \phi(a,b,c)$. The proof for general $\vec{u}$, $\vec{v}$, $\vec{c}$ is completely analogous.

(4) $\Rightarrow$ (1): Encode all the finite sequences of natural numbers by natural numbers, in one of the standard ways, and let $(y)_x$ be the xth member of the sequence encoded by y. The relation $(y)_x = z$, as a relation in x, y, z, is recursive. If one formalizes the encoding in the language of arithmetic, in a standard way, one gets a formula which asserts that the xth member of the sequence encoded by y is z. This formula which we also denote by "$(y)_x = z$" is provably recursive. Arguing as in the proof of Theorem 2 we get two diophantine equations, σ_1 and σ_2, for which the following is provable in P.

$$(y)_x = z \leftrightarrow \exists\vec{u}\ \sigma_1(x,y,z,\vec{u})$$

$$(y)_x \neq z \leftrightarrow \exists\vec{v}\ \sigma_2(x,y,z,\vec{v}).$$

Consequently the following sentences hold in M_1:

(i) $$\forall x,y,z\ [\exists\vec{u}\ \sigma_1(x,y,z,\vec{u}) \leftrightarrow \neg\exists\vec{v}\ \sigma_2(x,y,z,\vec{v})]$$

(ii) $$\forall x,y\ \exists z\ \exists\vec{u}\ \sigma_1(x,y,z,\vec{u})$$

(iii) $$\forall x,y,z,z_1\ (\exists\vec{u}\ \sigma_1(x,y,z,\vec{u}) \wedge \exists\vec{u}\ \sigma_1(x,y,z_1,\vec{u}) \rightarrow z = z_1)$$

Each one of these can be put in the form $\forall\exists$. Hence, assuming (4), it follows that they hold in M_2. The satisfaction of (ii) and (iii) means that $\exists\vec{u}\ \sigma_1(x,y,z,\vec{u})$ defines a function, so that z is the function's value for the arguments x, y. The satisfaction of (i) means that the same function is defined also by $\neg\exists\vec{v}\ \sigma_2(x,y,z,\vec{v})$. This implies the absoluteness of the function for the submodel M_1, that is, the function defined by $\exists\vec{u}\ \sigma_1(x,y,z,\vec{u})$ in M_1 is the restriction of the function defined in M_2.

For the sake of technical convenience, consider an extension of the language of arithmetic by adding to it a function sign of two arguments ( ), with the understanding that $(x)_y$ is to define in any model satisfying (ii) and (iii) the function defined by $\exists\vec{u}\ \sigma_1(x,y,z,\vec{u})$. M_1 is a submodel of M_2, where both are considered as models of the extended language, and hence $M_1 \models \phi(\vec{a})$ implies $M_2 \models \phi(\vec{a})$ whenever ϕ is quantifier-free. We will show, by induction on k, that the same is true for every ϕ in the extended language whose prenex normal form has k quantifiers.

If the assertion is true for $\phi(v_0,\ldots,v_{n-1})$, then, obviously, it is true for $\exists v_0\ \phi(v_0,\ldots,v_{n-1})$. The only point to be established is the passage from k to k + 1 quantifiers, where the left-most added quantifier is $\forall$.

Using the equivalence $(x)_y = z \leftrightarrow \exists\vec{u}\ \sigma_1(x,y,z,\vec{u})$, one can translate quantifier-free formulas of the extended language into existential formulas which do not involve terms of the form $(x)_y$. For instance, $((x)_y)_z \neq (x')_{y'} + z'$, can be translated as $\exists x_1, x_2, x_3\ \exists \vec{u}_1, \vec{u}_2, \vec{u}_3$ $(\sigma_1(x,y,x_1,\vec{u}_1) \wedge \sigma_1(x_1,z,x_2,\vec{u}_2) \wedge \sigma_1(x',y',x_3,\vec{u}_3) \wedge x_2 \neq x_3 + z')$. In this manner the new function symbol can be eliminated and one gets an existential formula in the original language. From (4) it follows that $M_1 \models \forall\vec{u}\ \exists\vec{v}\ \psi(\vec{a},\vec{u},\vec{v})$ implies $M_2 \models \forall\vec{u}\ \exists\vec{v}\ \psi(\vec{a},\vec{u},\vec{v})$, whenever ψ is a quantifier free formula of the extended language. Since M_1 is a model for $\underline{P}$, we have:

$$M_1 \models \forall u_0,\dots,u_{k-1}\ \exists v\ [(v)_0 = u_0 \wedge (v)_1 = u_1 \wedge \dots \wedge (v)_k = u_k].$$

The same sentence holds in M_2. This enables one to contract quantifiers in the same manner both in M_1 and in M_2, that is, one can replace '$Qv_0,v_1\ \phi(v_0,v_1)$' by '$Qv\ \phi((v)_0,(v)_1)$' where Q is either $\exists$ or $\forall$. It remains therefore to show that the satisfaction in M_1 of a formula with $k + 1$ alternating quantifiers starting with $\forall$, where $k + 1 > 2$, implies its satisfaction in M_2.

Assume $M_1 \models \forall u_0\ \exists u_1\ \phi(\vec{a},u_0,u_1)$ where ϕ is with $k - 1$ quantifiers.

Let $b \in M_2$ and let $b' \in M_1$ be such that $b \leq b'$. Obviously, $M_1 \models \forall u_0{\leq}b'\ \exists u_1\ \phi(\vec{a},u_0,u_1)$. Now, the following is provable in $\underline{P}$:

$$\forall u_0{\leq}v_0\ \exists u_1\ \phi(u_0,u_1) \rightarrow \exists v\ \forall u{\leq}v_0\ \phi(u_0,(v)_u).$$

Hence: $M_1 \models \exists v\ \forall u{\leq}b'\ \phi(\vec{a},u,(v)_u)$. Let $d \in M_1$ be such that $M_1 \models \forall u{\leq}b'\ \phi(\vec{a},u,(d)_u)$. By the induction hypothesis $M_2 \models \forall u{\leq}b'\ \phi(\vec{a},u,(d)_u)$. Hence $M_2 \models \phi(\vec{a},b,(d)_b)$.

q.e.d.

THEOREM 4. Let M_1 and M_2 be models of $\underline{P}$ such that $M_1 \subset M_2$. Let $\bar{M}_1$ be the submodel of M_2 consisting of all x such that $x \leq y$ for some $y \in M_1$. Then $M_1 \prec \bar{M}_1$.

Proof. By the corollary to Theorem 2, we have, for every bounded formula $\phi(\vec{v})$ and every $\vec{a} \in {M_1}^n$,

$$M_1 \models \phi(\vec{a}) \Leftrightarrow M_2 \models \phi(\vec{a}).$$

Since M_2 is an end-extension of $\bar{M}_1$ we also have:

$$\bar{M}_1 \models \phi(\vec{a}) \Leftrightarrow M_2 \models \phi(\vec{a}).$$

Hence

$$M_1 \models \phi(\vec{a}) \Leftrightarrow \bar{M}_1 \models \phi(\vec{a}).$$

Now apply the implication (3) $\Rightarrow$ (1) of Theorem 3.

q.e.d.

COROLLARY 1. If M, M_1, M_2 are models of $\underline{P}$, M_1 and M_2 are submodels of M and they are cofinal in each other (i.e., for each $x_1 \in M_1$ there exists $x_2 \in M_2$ such that $x_1 \leq x_2$, and vice versa), then M_1 and M_2 have a common elementary extension which is a submodel of M. In particular, they are elementary equivalent and their intersection is an elementary submodel of each.

Proof. $M' = \{x \in M : x \leq x_1 \text{ for some } x_1 \in M_1\}$ is, by the last theorem, a common elementary extension of M_1 and M_2. Since all Skolem functions are definable in Peano's arithmetic, it follows that an intersection of two elementary submodels of a given model is also an elementary submodel of this model, which concludes the proof.

Theorem 4 can be also stated as follows:

Every extension of a model, M, of $\underline{P}$ to a model, M^*, of $\underline{P}$ can be decomposed into two successive extensions: an elementary cofinal extension, $\bar{M}$, of M, and an end-extension M^*, of $\bar{M}$.

Consequently, if Q is any property of a model, M, of $\underline{P}$, such that Q does not hold in an extension of M, but Q is preserved under elementary extensions, then there is a cofinal elementary extension of M having the property Q, for which a certain end-extension does not have the property Q.

§2.

Rabin has shown in [5] that there exists a diophantine equation $\delta(u,\vec{v})$, such that in every non-standard model of true arithmetic, M, there is an element c for which $M \models \neg\exists\vec{v}\ \delta(c,\vec{v})$, but M has an extension M^* which is a model of true arithmetic such that $M^* \models \exists\vec{v}\ \delta(c,\vec{v})$.

Since $\neg\exists\vec{v}\ \delta(c,\vec{v})$ will hold in every elementary extension of M, we get using the last observation of §1:

COROLLARY 2. There is a diophantine equation $\delta(u,\vec{v})$ such that every non-standard model of true arithmetic, M, has an element c and a cofinal elementary extension $\bar{M}$, such that $\delta(c,\vec{v})$ is unsolvable in $\bar{M}$ but is solvable in an end-extension of $\bar{M}$, which is a model of true arithmetic.

The question remains whether for the given $\delta(c,\vec{v})$, or for another diophantine equation which is unsolvable in M, there is always a model, which is an end-extension of M, in which a solution exists. This amounts to asking whether one can always choose $\bar{M} = M$.

The result which is weaker than Corollary 2 and which asserts the existence of models of true arithmetic, M_1, M_2, with M_2 an end-extension of M_1, and a diophantine equation unsolvable in M_1 and solvable in M_2, can be derived, in a straightforward way, from Matijasevič's result and the characterization given by Feferman in [7] for formulas preserved under end-extensions. This has been pointed out to the author by Macintyre.

Using a second paper of Rabin, [6], one can, with the help of Matijasevič's result and the observation used in the proof of Theorem 2, generalize Corollary 2 to any complete theory which extends <u>P</u>, as well as in other directions, as follows.

In [6], using the methods of [5], Rabin gets a similar result for a system of arithmetic which includes additional predicates to denote certain recursive relations. For this system the result generalizes to any complete theory which includes Peano's arithmetic. It can be also generalized to Σ_n-formulas, where a Σ_n-<u>formula</u> is one which in prenex form has n alternating blocks of quantifiers starting with an existential one. The additional predicate symbols are needed in order to make possible an enumeration of all k-ary Σ_n-relations by a universal (k+1)-ary Σ_n-relation.

An analysis of the proof shows that, arguing as in the proof of Theorem 2, one can dispense with the additional symbols. One is able to get the following result in a straightforward way from Rabin's:

Say that M_1 is a <u>n-elementary submodel</u> of M_2 if $M_1 \subset M_2$ and, for every $\phi(\vec{v})$, which in prenex normal form has at most n alternating blocks of quantifiers, and every $\vec{a} \in M_1^k$, one has: $M_1 \models \phi(\vec{a})$ iff $M_2 \models \phi(\vec{a})$.

THEOREM 5. For every $n \geq 0$, there exists a Σ_{n+1}-formula, $\phi(v)$, such that every non-standard model, M, of <u>P</u> has a member c, such that $M \models \neg\phi(c)$, but there exists a n-elementary extension, M^*, of M, which is elementary equivalent to M, satisfying $M^* \models \phi(c)$. Again, M^* is an end-extension of a cofinal elementary extension of M.

Rabin used his result and an additional device to construct an ascending sequence of models of true arithmetic $M_0 \subset M_1 \subset \ldots \subset M_i \subset \subset M_{i+1} \subset \ldots$ such that each M_i is a n-elementary submodel of M_{i+1} but $\cup M_i$ is not a model of $\underline{P}$. The same works for any complete extension of $\underline{P}$. As before, one is now able to get the same result without having to increase the language of arithmetic.

In [3 p. 80] Robinson shows how a union of an ascending sequence, $M_0 \subset M_1 \subset \ldots \subset M_i \subset M_{i+1} \subset \ldots$ of models of a given theory can be represented as an intersection of two models of that theory, M', M'' which are submodels of a common extension M^* which is a model of the theory. Using his construction with a slight modification one can show that in addition one can get:

(i) One of the models, M' say, is such that $M' \prec M^*$.

(ii) If, for each i, M_i is a n-elementary submodel of M_{i+1}, then M'' and $M' \cap M''$ are n-elementary submodels of M^*.

Putting all these together one gets, for every fixed n:

THEOREM 6. Every complete extension, T, of $\underline{P}$ has a model M which has an elementary submodel M_1 and a n-elementary submodel M_2, which is a model of T, such that $M_1 \cap M_2$, which is a n-elementary submodel of M, violates one of Peano's axioms.

This strengthens and generalizes Theorem 8 in [5].

We conclude by posing a problem. As we have seen, if $M_1 \subset M_2$ and both satisfy Peano's axioms then the recursive functions in M_1 are the restrictions of those in M_2. In particular, M_1 is closed under the recursive functions of M_2. The converse is not true. One can have a model, M_2, of true arithmetic, or any other complete extension of $\underline{P}$, and a submodel M_1 which is closed under the recursive functions of M_2 and is not a model of $\underline{P}$. The same is true if "recursive functions" is replaced by "Σ_n-definable functions", where n is any fixed natural number. This is implied by Theorem 6, and can be also easily derived from the known result that Peano's arithmetic is not implied by any consistent set of Σ_k-sentences, where k is fixed. The question, however, remains open if M_1 is assumed to be an initial segment of M_2. The problem therefore is:

Is every initial segment of a model, M, of $\underline{P}$ (or true arithmetic) which is closed under the recursive functions of M (or the Σ_n-definable functions) also a model of $\underline{P}$?

REFERENCES

[1] J. Robinson, Existential definability in arithmetic, Trans. Am. Math. Soc. 72 (1952).

[2] Tarski, Contributions to the theory of models, Indg. Math. 16 (1954).

[3] A. Robinson, Introduction to model theory and to the metamathematics of algebra, North-Holland, Amsterdam (1963).

[4] M. Davis, H. Putnam and J. Robinson, The decision problem for exponential diophantine equations, Ann. of Math. (2) 74 (1961).

[5] M. O. Rabin, Diophantine equations and non-standard models of arithmetic, Proceedings of the 1960 International Congress in Logic, Methodology and Philosophy of Science, Stanford University Press.

[6] M. O. Rabin, Non standard models and the independence of the induction axiom, Essays on the foundations of mathematics dedicated to A. A. Fraenkel, The Magnes Press, Jerusalem (1961).

[7] Solomon Feferman, Persistent and invariant formulas for outer extensions, Compositio Mathematica 20 (1968) p. 29.

[8] Ju. V. Matijasevič, Enumerable sets are diophantine, Soviet mathematics, vol. 11, number 2 (1970). (Translated from Russian)

POSTSCRIPT

In Lemma 2 of his paper Adler produces an argument which, if true, would imply that if M_1 is a cofinal submodel of M_2 and M_1 is a model of true arithmetic and if bounded formulas persist when passing from M_1 to M_2 then $M_1 \prec M_2$. The following is a quotation:

"Lemma 2. Let α_i, $\beta_i \in M$. If for all α_{k+1} there exists β_{k+1} such that

$$\forall x_1 \leq \alpha_1 \ \exists y_1 \leq \beta_1 \ \ldots \ \forall x_{k+1} \leq \alpha_{k+1} \ \exists y_{k+1} \leq \beta_{k+1} \ P(\vec{x}, \vec{y})$$

is true in $M * N$, then

$$\forall x_1 \leq \alpha_1 \ \exists y_1 \leq \beta_1 \ \ldots \ \forall x_{k+1} \ \exists y_{k+1} \ P(\vec{x}, \vec{y})$$

is true in $M * N$.

Proof. Immediate for to every $x_{k+1} \in M * N$, there exists $\alpha_{k+1} \in M$ such that $x_{k+1} \leq \alpha_{k+1}$."

Here M is a model of true arithmetic, N is a model of true arithmetic which extends M and $M * N = \{y : \text{for some } x \in M,\ y \leq x\}$. $P(\vec{x}, \vec{y})$ is any formula. The α_i and β_i range over M.

The argument is wrong. Consider for instance the simple case where, for every $\alpha_2 \in M$ the formula:

$$\forall x_1 \leq \alpha_1 \ \exists y_1 \leq \beta_1 \ \forall x_2 \leq \alpha_2 \ P(x_1, x_2, y_1)$$

holds in $M * N$.

One <u>cannot</u> directly deduce that $\forall x_1 \leq \alpha_1 \ \exists y_1 \leq \beta_1 \ \forall x_2 \ P(x_1, x_2, y_1)$ holds in $M * N$, for, it could happen that for some $x_1 \in M * N$, such that $x_1 \leq \alpha_1$, the y_1 such that $M * N \models \forall x_2 \leq \alpha_2 \ P(x_1, x_2, y_1)$ depends on

α_2, so that as α_2 ranges over M, y_1 keeps changing. One cannot deduce that there is a single y_1 such that $M * N \models P(x_1, x_2, y_1)$ for all x_2. If it is known that M * N is a model of arithmetic then one can argue inside M * N that since the y_1's are bounded by β_1 there is at least one y_1 which is good for an infinite set of α_2's. But this can be done only if M * N is assumed to be a model of arithmetic, which is to be proved! Actually, the result which assumes in addition that the cofinal extension of M is a model of true arithmetic, was proved by Rabin in [5].

It seems to the author that any way of passing from bounded to unbounded quantifiers would involve a device which is similar to the one employed in the proof of Theorem 3 of the present paper, in which $\forall\exists$ is changed into $\exists\forall$ using a coding of a partial Skolem function.

[Added in proof: A. Wilkie has provided partial answers to two of the problems raised above. His results are:

(i) Every non-standard countable model of P has an end-extension in which a diophantine equation, not solvable in the original model, has a solution. We can, in fact, take the new model to be isomorphic to the original one. (The proof uses a device of Harvey Friedman.)

(ii) Every model of P contains a proper initial segment, which is closed under all provably-recursive functions, but is not a model of P.]

AN APPLICATION OF ULTRA-PRODUCTS TO PRIME RINGS WITH POLYNOMIAL IDENTITIES

Arnulf Hirschelmann

Bonn, West Germany

A ring R is said to be prime if $aRb = (0)$ $(a, b \in R)$ implies $a = 0$ or $b = 0$. The centroid of R is the set of elements in the endomorphism ring of the additive group of R which commute elementwise with the multiplication ring of R which is generated (in the endomorphism ring) by linear transformations caused by multiplication. A ring R is said to satisfy a polynomial identity over its centroid if there is an $f \neq 0$ in the free algebra over the centroid in the noncommutative variables $x_1, \ldots, x_n$ for some n, such that $f(a_1,\ldots,a_n) = 0$ for all $a_1, \ldots, a_n$ in R. An obvious example is the following: a commutative ring satisfies $f(x_1,x_2) = x_1x_2 - x_2x_1$. So the just defined rings generalize somewhat the notion of commutativity. In the sequel a special but extremely significant polynomial identity (because it is multilinear) is needed, the so-called standard identity of degree n:

$$* \qquad [x_1,\ldots,x_n] = \Sigma_{\sigma\in\mathcal{S}_n} (-1)^{\sigma} x_{\sigma(1)}\cdots x_{\sigma(n)}.$$

R is said to be a (left-)order in S if S is a ring of (left) quotients for R (cf. [3]).

In his paper "Problems in Ring Theory" Kaplansky asked whether any ring with a polynomial identity can be embedded in a matrix ring. The answer is in general negative. But on the affirmative side there is Posner's theorem:

Let R be a prime ring satisfying a polynomial identity over its centroid. Then R can be embedded as an order in D_n where D is a division algebra finite-dimensional over its center.

Another version of Amitsur's proof [1] will be given indicating the main steps. Under the assumptions made above R turns out to be a Goldie ring, which means that R satisfies the ascending chain condition on left annihilators and R contains no infinite direct sums of left ideals. By Goldie's theory it follows that R has a simple Artinian quotient ring Q(R) which by Wedderburn's theorem has the form D_n for some division ring D. Just the polynomial identity forces D to be finite-dimensional as will now be seen.

Primeness of R implies that R has no nilpotent ideals. It can be shown [3, p. 184] that in fact R has no nil ideals. Therefore a theorem of Amitsur [3, p. 150] can be applied to prove that R[t] is semisimple. Furthermore it can be proved that R satisfies a standard identity, because R has no nil ideals. Hence R[t] satisfies the same identity as R; we need only to consider the case of R semisimple. Now every semisimple ring is a subdirect sum of primitive (having a faithful irreducible module) rings,

$$R = \Sigma_{\alpha \in A} \oplus_s R_\alpha$$

for some index set A. This set A will be important later in constructing an ultra-product.

Each R_α is a homomorphic image of R and therefore satisfies *. Were R primitive the proof would be complete. For a primitive ring is prime and a theorem of Kaplansky [3, p. 157] in connection with the density theorem says that a primitive ring satisfying a polynomial identity is a simple ring, finite-dimensional over its center. In particular R has the form D_n for some division ring D and hence is its own quotient ring.

If R is not primitive (Goldie [2] then avoids completely the construction of an ultra-product but proceeds analogously) let

$$T_a = \{\alpha \in A \mid a(\alpha) \neq 0\} \quad \text{for } 0 \neq a \in R.$$

Since R is prime, given $a \neq 0$, $b \neq 0$ in R there exists $x \in R$ with $axb \neq 0$. Hence

$$\emptyset \neq T_{axb} \subset T_a \cap T_b.$$

From this it is clear that the set $\{T_a \mid 0 \neq a \in R\}$ can be enlarged to a filter and moreover to an ultra-filter $\mathcal{F}$ on A. Now $r \neq 0$ implies $r \not\equiv 0 \pmod{\mathcal{F}}$ which shows that R is <u>isomorphically embedded</u> in

$$S = \Pi R_\alpha/\mathcal{F}.$$

S has a ring of quotients which by generalizing an idea of Santosuosso [4] (who treated the commutative case) is canonically isomorphic to the ultra-product of quotient rings of the R_α:

$$Q(S) \xrightarrow{\sim} \Pi Q(R_\alpha)/\mathcal{F}.$$

But in this case one has $S = Q(S)$. Hence also $Q(R)$ is isomorphically embedded in S.

In a letter to the author from G. Sabbagh (Paris) a general argument which explains why the class of primitive rings is closed under ultra-products is given (without settling the question about the elementary character of the class of primitive rings). Namely let $\mathfrak{M}$ be the class of all couples (R,ρ) where R is a primitive ring and ρ is a maximal regular right ideal in R such that $(\rho:R) = (0)$ (cf. Herstein [3 p. 40]). $\mathcal{M}$ is then elementary (in the wider sense). Hence the class of primitive rings is a PC_Δ-class (e.g. Tarski) and it is well-known that any PC_Δ-class is closed under ultra-products.

So S is primitive and hence after Kaplansky again it is finite-dimensional and satisfies *. Thus Q(R) does.

REFERENCES

[1] S. A. Amitsur, Prime rings having polynomial identities with arbitrary coefficients, Proc. London Math. Soc. III 17 (1967) 470-486.

[2] A. W. Goldie, A note on prime rings with polynomial identities, Journal London Math. Soc. II 1 (1969) 606-608.

[3] I. N. Herstein, Noncommutative rings, Carus Mathematical Monographs, John Wiley (1968).

[4] G. Santosuosso, Sul trasporto ad un ultraprodotto di anelli di proprietà dei suoi fattori, Rend. Math. (6) vol. 1 (1968) 82-99.

EMBEDDING NONDISTRIBUTIVE LATTICES IN THE RECURSIVELY ENUMERABLE DEGREES

A. H. Lachlan

Simon Fraser University

This paper is directed at the problem of deciding which countable lattices may be embedded in the recursively enumerable (r.e.) degrees. In [3] Thomason has shown that any finite distributive lattice may be embedded in the r.e. degrees.[1] The method that he used is a generalisation of the "minimal pair" construction of Yates [4]. We first observe that by a very simple change in Thomason's construction one can see that in fact any countable distributive lattice can be embedded in the r.e. degrees. All one has to do is to let α and β in the quadruples $\langle e,\alpha,f,\beta\rangle$ of [3] range through an r.e. class $\mathcal{R}$ of recursive sets, where $\mathcal{R}$ is uniformly recursive, has $\emptyset$ as its only finite member and forms an atomless Boolean algebra when the operations of union, intersection, and complementation are defined in the usual way. Thomason's construction then yields an embedding of $\mathcal{R}$ considered as a lattice in the r.e. degrees. Since any countable distributive lattice can be embedded in $\mathcal{R}$ we have the result.

Below we shall show that the two five element nondistributive lattices M_5, the modular one, and N_5, the nonmodular one, are both embeddable in the r.e. degrees. In order to avoid recalling a lot of

---

[1] Lerman has given an independent proof of the same result.

notations and terminology we shall assume that the reader is familiar with the first eight pages of [1].

When embedding a distributive lattice one can choose very simple functionals to ensure that the joins in the embedding all have the correct values. Indeed the Thomason embedding is simultaneously an embedding in the m-degrees. Some of the functionals that must be employed for M_5 and N_5 cannot even yield truth-table reducibility. To be specific let A, A_0, B be r.e. sets with r.e. degrees $\mathbf{a}$, $\mathbf{a}_0$, $\mathbf{b}$ respectively such that $\{\mathbf{0}, \mathbf{a}_0, \mathbf{a}, \mathbf{b}, \mathbf{a} \cup \mathbf{b}\}$ constitutes an embedding of N_5 in which $\mathbf{a}_0 < \mathbf{a}$. Let $A_0 \oplus B$ denote $\{2x \mid x \in A_0\} \cup \{2x+1 \mid x \in B\}$; then since $\mathbf{a}_0 \cup \mathbf{b} = \mathbf{a} \cup \mathbf{b}$ we must have A recursive in $A_0 \oplus B$. The point we wish to make is that A cannot be truth-table reducible to $A_0 \oplus B$ because of:

LEMMA. If X, Y_0, Y_1 are r.e. sets such that X is truth-table reducible to $Y_0 \oplus Y_1$ then there are disjoint r.e. sets X_0, X_1 such that $X_0 \cup X_1 = X$, X_0 is recursive in Y_0, and X_1 is recursive in Y_1.

Since the lemma is straightforward we shall not prove it. Applied to the situation above it would yield r.e. degrees $\mathbf{a}_0^* \leq \mathbf{a}_0$ and $\mathbf{b}^* \leq \mathbf{b}$ such that $\mathbf{a} = \mathbf{a}_0^* \cup \mathbf{b}^*$. Then $\mathbf{b}^* \neq \mathbf{0}$ contradicts $\mathbf{a} \cap \mathbf{b} = \mathbf{0}$, and $\mathbf{b}^* = \mathbf{0}$ contradicts $\mathbf{a}_0 < \mathbf{a}$. Similarly, if A_0, A_1, and A_2 are nonrecursive r.e. sets with degrees $\mathbf{a}_0$, $\mathbf{a}_1$, and $\mathbf{a}_2$ respectively such that $\{\mathbf{0}, \mathbf{a}_0, \mathbf{a}_1, \mathbf{a}_2, \mathbf{a}_0 \cup \mathbf{a}_1 \cup \mathbf{a}_2\}$ constitutes an embedding of M_5 then A_0 cannot be truth-table reducible to $A_1 \oplus A_2$, and likewise with the sets permuted. This makes it at least plausible that some of the complications in the constructions below are necessary.

That N_5 can be embedded is an immediate consequence of the following:

THEOREM 1. There exist nonrecursive r.e. degrees $\mathbf{a}_0$, $\mathbf{a}$, $\mathbf{b}$ such that $\mathbf{a}_0 < \mathbf{a}$, $\mathbf{a} \cap \mathbf{b} = \mathbf{0}$, and $\mathbf{a} \leq \mathbf{a}_0 \cup \mathbf{b}$.

Proof. Rather than write out the proof in full we shall show how to modify the proof of Theorem 1 of [1]. We shall assume that the reader is already familiar with that proof. We give $\Phi_i(A) \neq B$ order $3i$ rather than $2i$, and $\Phi_i(B) \neq A$ order $3i + 1$ rather than $2i + 1$. Let $A_0 = \{x \mid 2x \in A\}$ and $A_{0,s} = \{x \mid 2x \in A_s\}$. We give order $3i + 2$ to the condition $\Phi_i(A_0 \oplus \emptyset) \neq A$ where we use $A_0 \oplus \emptyset$ instead of A_0 for technical convenience. Followers will be appointed for the condition $\Phi_i(A_0 \oplus \emptyset) \neq A$ just as for the other conditions but they must be odd numbers. In step $s + 1$ for each odd number x we define an even number $f(x,s) \notin A_s \cup B_s$ such that $f(x,s)$ is increasing in s and strictly increasing in x. Within the constraints and any further specific one made in step s, $f(x,s)$ is to be given the least possible value. It will be a feature of the construction that in any step s where an odd number x is enumerated in A, $f(x,s)$ will also be enumerated in A. Further, $f(x,s+1) \neq f(x,s)$ only if some even number $\leq f(x,s)$ is enumerated in $A \cup B$ at step $s + 1$. By insisting that when a follower e of order i is appointed at step $s + 1$ then $e > f(x,s)$ for every odd $x < i$, we ensure that $\lim_s f(x,s)$ exists for every odd x. This yields $\underset{\sim}{a} \leq \underset{\sim}{a}_0 \cup \underset{\sim}{b}$. To discover whether an odd number x is in A, knowing the membership of A_0 and B, we effectively compute s_0 such that no even number enters $A \cup B$ at a step $\geq s_0$. Then $f(x,s_0) = \lim_s f(x,s)$ and so $x \in A$ if and only if $x \in A_s$. Of course, to discover whether an even number is in A, knowing A_0, is trivial. To ensure that $\underset{\sim}{a}_0 > \underset{\sim}{0}$ we restrict the followers of $\Phi_i(B) \neq A$ to be even for each i. For $\Phi_i(A) \neq B$ and $\Phi_i(B) \neq A$ the notions of "satisfaction" and "requiring attention" are to be exactly the same as in [1]. Similarly $\Phi_i(A_0 \oplus \emptyset) \neq A$ is said to be <u>satisfied at step</u> s if at the end of step s it has a realized follower e such that $\Phi_{i,s}(A_{0,s} \oplus \emptyset;\ e)$ is defined and different from $A_s(e)$. If $\Phi_i(A_0 \oplus \emptyset) \neq A$ is unsatisfied at step s, suppose that at the end of step s its followers in order of appointment are $e_0, e_1, \ldots, e_p$. Then at the end of step s there will be associated with each realized

e_y some potential requirement $m \leq L(n,x)$ with $n < i$.

The condition $\Phi_i(A_0 \oplus \emptyset) \neq A$ is said to <u>require attention at step s + 1</u> if it is unsatisfied at step s and if one of the following three possibilities obtains just after step s:

1. The potential requirement of one of the followers of $\Phi_i(A_0 \oplus \emptyset) \neq A$ is satisfied by s.

2. $\Phi_i(A_0 \oplus \emptyset) \neq A$ has no unrealized follower.

3. $\Phi_{i,s}(A_s;e)$ is defined where e is one of the unrealized followers of $\Phi_i(A_0 \oplus \emptyset) \neq A$.

As in the proof of Theorem 1 of [1] in step 0 we do nothing. In step s + 1 we look for the condition of least order which requires attention. If that condition is $\Phi_i(A) \neq B$ or $\Phi_i(B) \neq A$ then step s + 1 is to be exactly the same as before except for the changes indicated above such as replacing "2i" by "3i" and insisting that followers of $\Phi_i(B) \neq A$ be even. Now suppose that $\Phi_i(A_0 \oplus \emptyset) \neq A$ is the condition of least order which requires attention. Then we cancel all followers and requirements of order $> 3i + 2$. There are now three cases corresponding to the three possibilities just mentioned. We adopt the case that corresponds to the first of the three possibilities which holds.

<u>Case 1</u>. Let e_q be the first follower of $\Phi_i(A_0 \oplus \emptyset) \neq A$ in order of appointment whose potential requirement, $m \leq L(n,x)$ say, is satisfied by s. Then we appoint requirements of order $3i + 2$ which preserve all the values $\Psi_{n,s}(A_s;y)$, $\Theta_{n,s}(B_s;y)$ for $y < m$. These values will all be defined because $m \leq L(n,s)$. Further, suppose that e_q was appointed at step $s_q + 1$; then we seek the greatest number z, $z < n$, such that $M(z,s_q) < M(z,s)$. Suppose no such z exists. If enumerating $f(e_q,s)$ in A does not injure the value $\Phi_{i,s}(A_{0,s} \oplus \emptyset;\ e_q)$ then enumerate both e_q and $f(e_q,s)$ in A, and if enumerating $f(e_q,s)$ in A would injure the value $\Phi_{i,s}(A_{0,s} \oplus \emptyset;\ e_q)$ then enumerate $f(e_q,s)$ in B and nothing in A.

In the latter case the follower e_q is to become unrealised (so that it has no associated potential requirement), and $f(e_q, s+1)$ is to be defined large enough so that enumerating it in A would not injure the value $\Phi_{i,s}(A_{0,s} \oplus \emptyset; e_q)$. Finally, suppose that z does exist, then associate with e_q the potential requirement $M(z,s) \leq L(z,x)$ in place of $m \leq L(n,x)$.

<u>Case 2</u>. Appoint requirements of order $3i + 2$ to preserve all the values of the finite functions $\Psi_{x,s}(A_s)$ and $\Theta_{x,s}(B_s)$ for each $x < i$. We choose the least odd number e which has not yet been appointed as a follower, which is not in any existing requirements and which satisfies $e > f(x,s)$ for every odd $x < i$. We appoint e to follow $\Phi_i(A_0 \oplus \emptyset) \neq A$.

<u>Case 3</u>. Let e_q be the first unrealized follower of $\Phi_i(A_0 \oplus \emptyset) \neq A$ in order of appointment such that $\Phi_{i,s}(A_{0,s} \oplus \emptyset; e_q)$ is defined; e_q now becomes realized and a requirement of order $3i + 2$ is appointed to preserve the value $\Phi_{i,s}(A_{0,s} \oplus \emptyset; e_q)$. If $\Phi_{i,s}(A_{0,s} \oplus \emptyset; e_q) \neq 1$ do nothing more. If $\Phi_{i,s}(A_{0,s} \oplus \emptyset; e_q) = 1$ let e_q have been appointed at step $s_q + 1$. If $M(x,s_q) = M(x,s)$ for all $x < i$ then proceed as in Case 1 when z in Case 1 does not exist. Otherwise associate with e_q the potential requirement $M(z,s) \leq L(z,x)$ where z is the greatest number $< i$ such that $M(z,s_q) < M(z,s)$.

This completes the description of the construction. The proof that the construction works is much the same as in [1]. In addition to the lemmas of [1] we need ones showing that the condition $\Phi_i(A_0 \oplus \emptyset) \neq A$ requires attention at only a finite number of steps and that $\Phi_i(A_0 \oplus \emptyset) \neq A$ is satisfied. The proofs of these lemmas are very like those of Lemmas 1 and 2 of [1], so we leave them as an exercise for the reader. We have designed the construction so as to obtain $\underset{\sim}{b} \not\leq \underset{\sim}{a}$, $\underset{\sim}{a}_0 \not\leq \underset{\sim}{b}$, $\underset{\sim}{a}_0 \leq \underset{\sim}{a}$, $\underset{\sim}{a} \leq \underset{\sim}{a}_0 \cup \underset{\sim}{b}$, and $\underset{\sim}{a} \cap \underset{\sim}{b} = \underset{\sim}{0}$. This is clearly sufficient for the conclusion of the theorem.

It turns out that embedding M_5 is a much more difficult problem. Our terminology and notation will undergo a few slight changes. What were "conditions" in Theorem 1 will now be referred to as "requirements". Also, instead of writing functionals with suffixes such as $\Psi_{i,j}$, Φ_i we shall write $\Psi(i,j)$, $\Phi(i)$. The class of all r.e. sets will be denoted by $\mathcal{W}$. The set $\{0,1,2\}$ will be denoted by T. If σ, τ are sequences we write $\sigma \subseteq \tau$ to mean that τ extends σ, and we use $\sigma * \tau$ to denote the sequence obtained by juxtaposing τ to the right of σ. The empty sequence like the empty set will be denoted by $\emptyset$.

THEOREM 2. There exist nonrecursive r.e. degrees $\underset{\sim}{a}_0$, $\underset{\sim}{a}_1$, and $\underset{\sim}{a}_2$ such that $\underset{\sim}{a}_0 \cup \underset{\sim}{a}_1 = \underset{\sim}{a}_1 \cup \underset{\sim}{a}_2 = \underset{\sim}{a}_2 \cup \underset{\sim}{a}_0$ and $\underset{\sim}{a}_0 \cap \underset{\sim}{a}_1 = \underset{\sim}{a}_1 \cap \underset{\sim}{a}_2 = \underset{\sim}{a}_2 \cap \underset{\sim}{a}_0 = \underset{\sim}{0}$.

Proof. Effectively enumerate all quadruples of the form (i,j,Θ,Φ) where i, j are distinct members of T and Θ, Φ are p.r. functionals. Let the enumeration be $\langle(i(n),j(n),\Theta(n),\Phi(n)) \mid n = 0,1,\ldots\rangle$. We suppose that there is given an s.r.e. double sequence $\langle\Theta(n,s) \mid n = 0,1,\ldots;\ s = 0,1,\ldots\rangle$ of finite functionals increasing in s such that for each n, $\Theta(n) = \lim_s \Theta(n,s)$, and similarly for Φ in place of Θ. With each natural number n associate in an effective manner either (i) a positive integer $\pi(n)$, or (ii) a pair $(\kappa(n),W(n))$ where $\kappa(n) \in T$ and $W(n)$ is an r.e. set, or (iii) a pair of natural numbers $(\eta(n),\xi(n))$ with $\eta(n) < n$ and $\xi(n) > 0$. We shall say that n is of type 0, 1, or 2 according as n falls under case (i), (ii), or (iii) respectively. The association is to be exhaustive in the sense that

$$\text{rng } \pi = N - \{0\} \ \& \ \{(\kappa(n),W(n)) \mid 0 \leq n\} = T \times \mathcal{W} \ \& \ \{(\eta(n),\xi(n)) \mid 0 \leq n\} = N \times (N - \{0\}).$$

Below we shall describe a construction consisting of stages 0, 1, ... in which we shall effectively enumerate sets A(0), A(1), and

A(2) having the desired degrees $\underset{\sim}{a}_0$, $\underset{\sim}{a}_1$, and $\underset{\sim}{a}_2$ respectively. Simultaneously we shall be effectively enumerating W(0), W(1), ... and also defining certain auxiliary functions. Let Σ be the set of all finite sequences each member of which is either ω or a natural number. In stage s for each σ and $\tau \in \Sigma$ such that $\tau \subseteq \sigma$, and each $i \in T$ and $x \in N$ we shall define numbers $f(i,x,s)$, $c(\sigma,s)$, $\ell(\sigma,s)$, $r(\sigma,s)$, $r(\sigma,\tau,i,s)$ and also a finite sequence $\alpha(\sigma,\tau,s)$ of members of $T \times N$. For $i \in T$ we shall denote by $A(i,s)$ the finite set of numbers which have been enumerated in A(i) by the end of stage s.

Before stating the construction we shall attempt to describe the intuitive ideas underlying it. The function f is directly concerned with the need to ensure that $\underset{\sim}{a}_0 \cup \underset{\sim}{a}_1 = \underset{\sim}{a}_1 \cup \underset{\sim}{a}_2 = \underset{\sim}{a}_2 \cup \underset{\sim}{a}_0$. The construction will be such that $f(i,s,x)$ is $> x$, increasing in s, and strictly increasing in x, that is,

(1) $i \in T \rightarrow f(i,x,s) \leq f(i,x,s+1)$

and

(2) $i \in T \rightarrow x < f(i,x,s) < f(i,x+1,s)$.

Further, we shall have

(3) $i,\ j \in T \ \&\ i \neq j \rightarrow f(i,x,s) \notin A(j,s)$,

(4) $\forall i\ [i \in T. \rightarrow. A(i,s) = A(i,s+1)] \vee \exists i \exists j \exists k \exists x\ [\{i,j,k\} = T \ \&$
$A(i,s) = A(i,s+1) \ \&\ A(j,s+1) - A(j,s) = \{x\} \ \&\ A(k,s+1) - A(k,s)$
$= \{f(j,x,s)\}]$,

and

(5) $i \in T \ \&\ f(i,x,s) \neq f(i,x,s+1). \rightarrow. \exists j \exists y\ [j \in T \ \&\ j \neq i \ \&$
$y \leq f(i,x,s) \ \&\ A(j,s+1) - A(j,s) = \{y\}]$.

In the construction we shall ensure that $\lim_s f(i,x,s)$ exists for each

(i,x) in $T \times N$. Denote the value of $\lim_s f(i,x,s)$ by $f(i,x)$. Suppose that $\{i,j,k\} = T$ and that assuming knowledge of the membership of $A(j)$ and $A(k)$ we wish to compute whether or not $x \in A(i)$. We can effectively find a step s such that neither $A(j) - A(j,s)$ nor $A(k) - A(k,s)$ has a member $\leq f(i,x,s)$. From (5) $f(i,x) = f(i,x,s)$. We claim that $x \in A(i)$ if and only if $x \in A(i,s)$. If not, then $x \in A(i,t+1) - A(i,t)$ for some $t \geq s$. From (4) either $f(i,x,t) = f(i,x) \in A(j,t+1) \cup A(k,t+1)$ which contradicts (3), or there exists y such that either $A(j,t+1) - A(j,t) = \{y\}$ and $f(j,y,t) = x$, or $A(k,t+1) - A(k,t) = \{y\}$ and $f(k,y,t) = x$. In either of the latter cases we have $y \leq x \leq f(i,x)$ by (2) which contradicts the choice of s since either $y \in A(j) - A(j,s)$ or $y \in A(k) - A(k,s)$. This establishes the claim which in turn demonstrates that $A(i)$ is recursive in $A(j)$ and $A(k)$ together.

Having reduced the problem of ensuring that $\underset{\sim}{a}_0 \cup \underset{\sim}{a}_1 = \underset{\sim}{a}_1 \cup \underset{\sim}{a}_2 = \underset{\sim}{a}_2 \cup \underset{\sim}{a}_0$ to that of ensuring that $\lim_s f(i,x,s)$ exists for every (i,x) in $T \times N$ we shall now attempt to describe the whole scheme of priorities that will be employed. Until further notice we shall assume that in $c(\sigma,x)$, $\ell(\sigma,x)$, etc. the argument σ has been replaced by its length n. This will enable us to give a first approximation to the picture of the construction that we wish to present. It is clear that for the construction to be successful there must be for each n an algorithm A_n which computes the common value of $\Theta(n,A(i(n)))$ and $\Phi(n,B(j(n)))$ if these turn out to be the same total function a_n. The algorithm A_n which is only implicit in the construction will consist of an enumeration of the graph of a_n in order of magnitude of the argument. If $\ell(n,s) = m$ this means that at the end of stage s the enumeration A_n is complete just for arguments $< m$. The number $r(n,i(n),s)$ will be defined so that, if $\Theta(n,s,A(i(n),s))$ agrees with A_n at the end of stage s and $A(i(n),s+1) - A(i(n),s)$ has no member $< r(n,i(n),s)$ then $\Theta(n,s,A(i(n),s+1))$ will agree with A_n at the end of stage $s + 1$. Similarly for $j(n)$ and Φ in place of $i(n)$ and $\circledast$ respectively. We give priority $1/2n+1$ to our desire that no errors be introduced in A_n. If n is of type 0 we give priority $1/2n+2$ to our desire that $\lim_s f(i,x,s)$ exist whenever $x < \pi(n)$. If n is of type 1 we give priority $1/2n+2$ to our desire that $N - A(\kappa(n)) \not\subseteq W(n)$. If n is of type 2 we give priority $1/2n+2$ to our desire that the enumeration $A_{\eta(n)}$ be completed for arguments $< \xi(n)$. In stage s the number $r(n,s)$ is chosen so that if for each $i \in T$, $A(i,s+1) - A(i,s)$ contains no member $< r(n,s)$ then no requirement with priority $1/2m+2$ where $m \leq n$ will be injured in stage $s + 1$.

As the word "priority" suggests the principal feature of our construction is that requirements of greater priority take precedence over those of lower priority. The requirements regarding f which correspond to numbers of type 0 play an almost trivial role in the construction since the satisfaction of any particular one of them only necessitates restraining a finite set of numbers from each $A(i)$. The requirements corresponding to numbers of type 2, which necessitate extending the algorithms A_n to further arguments play a somewhat more important but still minor role. The main conflict in the construction is between our desire not to introduce errors in the algorithms A_n and the need to make $A(i)$ nonrecursive for each $i \in T$. (The latter need has been decomposed into the requirements corresponding to numbers of type 1.) We shall now explore the means by which this serious conflict is overcome. Consider a fixed number n of type 1. The associated requirement is $N - A(\kappa(n)) \not\subseteq W(n)$ which must be satisfied without injuring any of the requirements given higher priority. Let $W(n,s)$ denote the finite set of numbers that have been enumerated in $W(n)$ by the end of stage s. We shall ensure that $r(n,s) \notin A(\kappa(n),s)$ if $A(\kappa(n),s) \cap W(n,s) = \emptyset$. Further $r(n,s)$ will be increasing in n and s and such that $\lim_s r(n,s)$ exists for each n, the limiting value being denoted by $r(n)$. The requirement that no errors be introduced in the enumeration A_m will be denoted R_m. R_m is said to be <u>persistent at stage</u> $s + 1$ if at the end of stage s the current approximations to $\Theta(A(i(m)))$ and $\Phi(A(j(m)))$ are defined and equal on an initial segment of length at least $\ell(m,s)$. In line with the replacement of σ by its length made above, we shall assume that $\mathcal{R}_m$ is persistent for each $m \leq n$, where $\mathcal{R}_m$ is said to be <u>persistent</u> if it is persistent at infinitely many stages s + 1.

Suppose now that the construction has reached the point at which $r(n,s)$ attains its limiting value $r(n)$. If $r(n) \notin W(n)$ then certainly $N - W(n) \not\subseteq A(\kappa(n))$ because as stated above we shall ensure that

$$A(\kappa(n),s) \cap W(n,s) = \emptyset \rightarrow r(n,s) \notin A(\kappa(n),s).$$

Thus we may assume that $r(n)$ is eventually enumerated in $W(n)$. It is now sufficient to show that $r(n)$ can be enumerated in $A(\kappa(n))$ without any requirement of priority $> 1/2n+2$ being injured. By restraining numbers $< r(n)$ from being enumerated in $A(0) \cup A(1) \cup A(2)$ we ensure that no requirement with priority $1/2m+2$, where $m < n$, is injured. That leaves only the problem of not injuring $\mathcal{R}_0, \mathcal{R}_1, \ldots, \mathcal{R}_n$ which is solved by:

INFORMAL PROPOSITION. Let $m \leq n$ and suppose that the following is true of m: if $\{i,j,k\} = T$ and r_i, r_j, r_k, and x are numbers such that $f(i,x,s) \geq r_j$ and $f(j,f(i,x,s),s) \geq r_i$, then without injuring any $\mathcal{R}_p$, $p < m$, it is possible to ensure that for some $t > s$, $x \in A(i,t)$ and

$$y \in T \ \& \ z < r_y \ \& \ z \in A(y,t) - A(y,s) . \rightarrow . y = i \ \& \ z = x.$$

Then the same is true of m + 1.

We first recall that (1) - (5) have to be satisfied throughout the construction. For m = 0 the statement following the colon is clearly true. Indeed we may choose t to be s + 1 in this case by enumerating x in A(i) and f(i,x,s) in A(j) at stage s + 1. Thus given the truth of the proposition we see by induction that the statement following the colon is true for m = n. Taking $r_i = r_j = r_k = x = r(n)$ and $i = \kappa(n)$ it is clear that r(n) can eventually be enumerated in $A(\kappa(n))$ without any of $\mathcal{R}_0, \mathcal{R}_1, \ldots, \mathcal{R}_n$ being injured. We now turn to the proof of the proposition. Since we are supposing that r(n,s) has achieved its limiting value r(n) we may suppose that $\ell(m,s)$ remains fixed while we are attempting to enumerate r(n) in $A(\kappa(n))$ and thus fixed within the context of the proposition. This means that the enumeration A_m will contain exactly the same $\ell(m,s)$ pairs at all stages we shall be considering. We satisfy $\mathcal{R}_m$ by ensuring that at each stage u

(6) either $\Theta(m,u,A(i(m),u))$ or $\Phi(m,u,A(j(m),u))$ agrees with A_m

where a partial function ϕ agrees with A_m provided that all the pairs in the enumeration A_m belong to the graph of ϕ. Assume the hypothesis of the proposition, i.e. that m has the property cited. Let r_i, r_j, r_k, and x be given satisfying the stated conditions at the end of stage s. We suppose that (6) is satisfied at stage s. We have to show that it is possible to ensure that for some t > s, $x \in A(i,t)$, and

(7) $y \in T \;\&\; z < r_y \;\&\; z \in A(y,t) - A(y,s) . \rightarrow . y = i \;\&\; z = x.$

At the same time no $\mathcal{R}_p$ with $p \leq m$ may be injured at a stage > s and $\leq t$. Let k be the unique member of $T - \{i,j\}$. Let $u_0 + 1$ be the least stage > s at which $\mathcal{R}_m$ is persistent, then by restraining appropriate finite sets of numbers from A(i), A(j), and A(k) at stages > s and $\leq u_0$ we may assume that neither (7) nor any $\mathcal{R}_p$ with $p \leq m$ has been injured by the end of stage u_0. Let α' be the sequence $\langle (i_0,x_0), \ldots ,(i_p,x_p)\rangle$ in $N \times T$ and i_{p+1} be the number such that $(i_0,x_0) = (i,x)$ and

$$(8) \quad z \leq p \rightarrow \{i_{z+1}\} = \{i,j\} - \{i_z\}. \& .z < p \rightarrow x_{z+1} = f(i_z,x_z,u_o)$$

where p is the least number such that $f(i_p,x_p,u_0) \geq r(m,i_{p+1},u_0)$ and $f(i_{p+1},f(i_p,x_p,u_0),u_0) \geq r(m,i_p,u_0)$. For $i \in T$ let $r_i' = \max\{r_i, r(m,i,u_0), x_p+1\}$ then it should be clear from (1) that $x_0 < x_1 < \ldots < x_p$, $f(i_p,x_p,u_0) \geq r'_{i_{p+1}}$, and $f(i_{p+1},f(i_p,x_p,u_0),u_0) \geq r'_{i_p}$. Now apply the hypothesis of the proposition to r'_{i_p}, $r'_{i_{p+1}}$, r_k' and x_p with $s = u_0$. We conclude that we can ensure there exists $u_1 > u_0$ such that $x_p \in A(i_p,u_1)$,

$$(9) \quad y \in T \ \& \ z < r_y' \ \& \ z \in A(y,u_1) - A(y,u_0). \rightarrow .y = i_p \ \& \ z = x_p$$

and no $\mathcal{R}_p$ with $p < m$ is injured. From (9) and the definition of r_i' it follows that (6) is satisfied at every stage u, $u_0 \leq u \leq u_1$, since m is persistent at stage $u_0 + 1$. If $p = 0$ there is no more to prove, so assume $p = q + 1$. From (5), (9), and the fact that r_i' was chosen $\geq x_p + 1$, it is clear that for $z < q$, $x_{z+1} = f(i_z,x_z,u_1)$. Let α'' be the sequence obtained from α' by deleting its last member. From (9) we may suppose that $r(m,i,u_1) = r(m,i,u_0)$ for $i = i_q$ and $i = k$. Further, since at some stage $> u_0$ and $\leq u_1$, x_p was enumerated in $A(i_p)$ we may suppose that $f(i_q,x_q,u_1) \geq r_k'$ and that $f(k,f(i_q,x_q,u_1),u_1) \geq r'_{i_q}$. Finally, we may suppose by simply waiting long enough that $\mathcal{R}_m$ is persistent at stage $u_1 + 1$. Defining r_i'' at stage u_1 as r_i' was defined at stage u_0 we see that $r''_{i_q} \leq r'_{i_q}$ and $r''_{i_k} \leq r'_{i_k}$. Thus we are in exactly the same situation at stage u_1 with respect to α'' and the numbers r_i'' as we were in at stage u_0 with respect to α' and the numbers r_i'. Thus we can apply the hypothesis of the proposition repeatedly and eventually reach a stage $t = u_{p+1}$ such that $x = x_0 \in A(i,t)$, (7) holds, and such that no $\mathcal{R}_p$, $p \leq m$, is injured at a stage $> s$ and $\leq t$. This completes the proof of the informal proposition.

In the actual construction the sequences $\alpha(\sigma,\tau,s)$ play a role similar to that of α' and α'', and the numbers $r(\sigma,\tau,i,s)$ play a role similar to that of r_i' and r_i''. The additional argument τ, whose only significant values are those $\subseteq \sigma$, is required because of the inductive nature of our reasoning.

The motivation that we have given so far would be perfectly satisfactory were it not for the fact that some of R_0, R_1, ... will turn out to be not persistent. We overcome this final difficulty as follows. For simplicity we shall first discuss the problem as though R_0 was the only member of the sequence that might not be persistent. In this case at stages where R_0 is persistent we carry out the construction in the manner outlined above. Between stages at which R_0 is persistent we protect, through finite restraints, whatever has been done in the earlier stages at which R_0 was persistent, and at the same time we carry out the same construction but ignoring R_0 completely except for the finite restraints just mentioned. It should be clear that this modification of the original idea will suffice if R_0 is the only member of the sequence which might not be persistent. In the general case this modification must be nested within itself repeatedly, the depth of nesting being n when all of R_0, R_1, ... , R_n are being considered. We associate with R_0 the ordinal ω if R_0 is persistent and the number n otherwise, where n is the number of stages at which R_0 is persistent. For each m, within each possibility regarding the persistence of the requirements R_0, ... , R_m there is one possibility regarding the persistence of R_{m+1} corresponding to each member of $N \cup \{\omega\}$. It is for this reason that the set Σ is introduced. This concludes our attempt to draw an intuitive picture of the construction. It only remains to specify the construction and justify it.

Stage 0. Let $c(\sigma,0) = \ell(\sigma,0) = r(\sigma,0) = r(\sigma,i,0) = r(\sigma,\tau,i,0) = 0$ and $f(i,x,0) = x + 1$ for each σ, $\tau \in \Sigma$, $i \in T$, and $x \in N$. For each σ and τ in Σ let $\alpha(\sigma,\tau,0)$ be the empty sequence. (Note that as indicated above the values of $r(\sigma,\tau,i,s)$ and $\alpha(\sigma,\tau,s)$ are only significant in case $\tau \subseteq \sigma$.)

Before stage $s + 1$ can be described certain preliminary definitions are necessary. Σ is linearly ordered by: $\sigma \lessdot \tau$ if either (i) there exist $\alpha, \beta \in N \cup \{\omega\}$ and a common initial segment ν of σ and τ such that $\nu * \langle\alpha\rangle \subseteq \sigma$, $\nu * \langle\beta\rangle \subseteq \tau$, and $\alpha < \beta$, or (ii) $\tau \subseteq \sigma$. For any p.r. functional Θ and set A we say that r <u>preserves</u> $\Theta(A)$ <u>up to</u> ℓ if for any set B such that

$$y \in (A - B) \cup (B - A) \rightarrow r \leqslant y$$

we have

$$x < \ell \ \&\ \Theta(A;x) \text{ defined } .\rightarrow. \ \Theta(B;x) \text{ defined } \&\ \Theta(B;x) = \Theta(A;x).$$

Suppose that r does preserve $\Theta(A)$ up to ℓ and that $\Theta(A;x)$ is defined for each $x < \ell$. Below we shall need to use the fact that if $\Theta \subseteq \Theta'$, $A \subseteq A'$, and $A' - A$ has no member $< r$, then $\Theta(A)$ and $\Theta'(A')$ agree on arguments $< \ell$ and moreover r preserves $\Theta'(A')$ up to ℓ. Notice also that if a finite functional Θ and a finite set A are given then we can effectively determine the least number r which preserves $\Theta(A)$ up to ℓ.

Suppose now that stage s has been completed and that $\sigma \in \Sigma$ has length n, σ is said to be <u>persistent at stage</u> s + 1 if for each $x < \ell(\sigma,s)$ we have $\Theta(n,s,A(i(n),s);x)$ and $\Phi(n,s,B(j(n),s);x)$ defined and equal. Also, σ is said to <u>require attention at stage</u> s + 1 if one of the following four cases holds:

<u>Case 1</u>. $r(\sigma,s) = 0$.

<u>Case 2</u>. n is of type 0 and there exist $i \in T$ and $x \in N$ such that

$f(i,x,s) \geq r(\sigma,s)$ and $x < \pi(n)$.

Case 3. n is of type 1 and $r(\sigma,s) \in W(n,s) - A(\kappa(n),s)$.

Case 4. n is of type 2, $\ell(\tau,s) < \xi(n)$, $\tau * \langle\omega\rangle \subseteq \sigma$, and $\Theta(\eta(n),s,A(i(\eta(n)),s))$ and $\Phi(\eta(n),s,A(j(\eta(n)),s))$ are both defined and equal for each argument $< \xi(n)$, where τ is the initial segment of σ of length $\eta(n)$.

Stage s + 1. Define a number n and a sequence $\sigma_0, \sigma_1, \ldots, \sigma_n$ in Σ as follows. Let $\sigma_0 = \emptyset$, and suppose that $\sigma_0, \sigma_1, \ldots, \sigma_m$ have already been defined. If σ_m requires attention at stage $s + 1$ let $n = m$. Otherwise, if σ_m is persistent at stage $s + 1$ define σ_{m+1} to be $\sigma_m * \langle\omega\rangle$, and if σ_m is not persistent at stage $s + 1$ define σ_{m+1} to be $\sigma_m * \langle c(\sigma_m,s)\rangle$. Once the construction has been stated completely it will be easy to see that this inductive definition terminates with the definition of n in a finite number of steps because for only finitely many σ is $r(\sigma,s) \neq 0$. Denote σ_n by σ below. Call σ the index of stage $s + 1$ and n the length of stage $s + 1$.

Let $c(\sigma',s+1)$, $\sigma' \in \Sigma$, be defined as follows. If $\sigma' = \sigma_m$ for some $m < n$ then let $c(\sigma',s+1)$ be $c(\sigma',s) + 1$ if $\sigma_{m+1} = \sigma_m * \langle\omega\rangle$ and be $c(\sigma',s)$ otherwise. If $\sigma' \leq \sigma$ let $c(\sigma',s+1) = 0$, and if $\sigma' > \sigma$ but not $\sigma' \subseteq \sigma$ let $c(\sigma',s+1) = c(\sigma',s)$.

If $\sigma' < \sigma$ let $\ell(\sigma',s+1) = r(\sigma',s+1) = r(\sigma',\tau',i,s+1) = 0$ and $a(\sigma',\tau',s+1)$ be empty. If $\sigma' > \sigma$ let $r(\sigma',\tau',i,s+1) = r(\sigma',\tau',i,s)$ and $a(\sigma',\tau',s+1) = a(\sigma',\tau',s)$. If $\sigma' > \sigma$ and $\sigma' \not\subseteq \sigma$ let $\ell(\sigma',s+1) = \ell(\sigma',s)$. The remaining definitions that have to be made in stage $s + 1$ depend on the case under which σ requires attention at stage $s + 1$. For all m let $k(m)$ be such that $T = \{i(m),j(m),k(m)\}$.

Case 1. $r(\sigma,s) = 0$. Let $f(i,x,s+1) = f(i,x,s)$ for all $i \in T$ and $x \in N$. Let $\ell(\sigma_m,s+1) = \ell(\sigma_m,s)$ for all $m \leq n$. Let $r(\sigma,\tau,i,s+1) = 0$

and $\alpha(\sigma,\tau,s+1)$ be empty for all $i \in T$ and all τ. For $\sigma' \in \Sigma$ of length m let $r(\sigma',i(m),s+1)$ be the least number preserving $\Theta(m,s+1,A(i(m),s+1))$ up to $\ell(\sigma',s+1)$. Similarly for $j(m)$ and Φ in place of $i(m)$ and Θ respectively. Let $r(\sigma',k(m),s+1) = 0$. Finally, let $r(\sigma,s+1)$ be the least number > 0 satisfying

$$(10)\ r(\sigma,s) \leq r(\sigma,s+1).\&.r(\sigma,s+1) \notin A(0,s+1) \cup A(1,s+1) \cup A(2,s+1),$$

$$(11)\ \sigma' > \sigma.\rightarrow.r(\sigma',s+1) < r(\sigma,s+1) \ \&\ r(\sigma',\tau',i,s+1) < r(\sigma,s+1),$$

$$(12)\ \sigma' > \sigma \text{ and } (i,x) \text{ is a member of } \alpha(\sigma',\tau',s+1).\rightarrow.x < r(\sigma,s),$$

$$(13)\ \sigma' * \langle\omega\rangle > \sigma \ \&\ \sigma' * \langle\omega\rangle \nsubseteq \sigma.\rightarrow.r(\sigma',i,s+1) \leq r(\sigma,s+1).$$

<u>Case 2</u>. n is of type 0. Proceed as in Case 1 but with the additional condition

$$(14)\ x < \pi(n) \longrightarrow f(i,x,s+1) < r(\sigma,s+1).$$

<u>Case 3</u>. n is of type 1. Then $r(\sigma,s) \in W(n,s) - A(\kappa(n),s)$ and our aim is to hasten the advent of a stage $t > s$ at which $r(\sigma,s)$ can be enumerated in $A(\kappa(n))$. Let $\ell(\sigma_m,s+1) = \ell(\sigma_m,s)$ for all $m < n$ and let $\ell(\sigma,s+1) = 0$ and $r(\sigma,s+1) = r(\sigma,s)$. For the rest we consider four subcases:

<u>Case 3.1</u>. $\alpha(\sigma,\sigma,s)$ is empty. Let $f(i,x,s+1) = f(i,x,s)$ for all $i \in T$ and $x \in N$. If $\tau \neq \sigma$ let $r(\sigma,\tau,i,s+1) = 0$ and $\alpha(\sigma,\tau,s+1)$ be empty. Let $\alpha(\sigma,\sigma,s+1)$ be $\langle(\kappa(n),r(\sigma,s))\rangle$, and $r(\sigma,\sigma,i,s+1) = r(\sigma,s)$ for each $i \in T$.

<u>Case 3.2</u>. There exists $m < n$ such that $\alpha(\sigma,\sigma_{m+1},s)$ is nonempty, $r(\sigma,\sigma_{m+1},i,s) > 0$ for each $i \in T$ and $\alpha(\sigma,\sigma_m,s)$ is empty. Let $f(i,x,s+1) = f(i,x,s)$ for all $i \in T$ and $x \in N$. If $\tau \nsubseteq \sigma$ or $\sigma_m \nsubseteq \tau$ let $r(\sigma,\tau,i,s+1) = 0$ and $\alpha(\sigma,\tau,s+1)$ be empty. If $\sigma_{m+1} \subseteq \tau \subseteq \sigma$ let

$r(\sigma,\tau,i,s+1) = r(\sigma,\tau,i,s)$ and $\alpha(\sigma,\tau,s+1) = \alpha(\sigma,\tau,s)$. Let (i^*,x^*) be the last member of $\alpha(\sigma,\sigma_{m+1},s+1)$. We shall verify later that there exists $j^* \in T$, $j^* \neq i^*$, such that

$$f(i^*,x^*,s) \geq r(\sigma,\sigma_{m+1},j^*,s) \;\&\; f(j^*,f(i^*,x^*,s),s) \geq r(\sigma,\sigma_{m+1},i,s).$$

There are now two subsubcases:

Case 3.2.1. $\sigma_{m+1} \neq \sigma_m * \langle\omega\rangle$. Let $r(\sigma,\sigma_m,i,s+1) = r(\sigma,\sigma_{m+1},i,s+1)$ for each $i \in T$. Let $\alpha(\sigma,\sigma_{m+1},s+1)$ be $\langle(i^*,x^*)\rangle$.

Case 3.2.2. $\sigma_{m+1} = \sigma_m * \langle\omega\rangle$. Let $\alpha(\sigma,\sigma_m,s+1)$ be the sequence $\langle(i_0,x_0),\ (i_1,x_1),\ \ldots\ ,\ (i_p,x_p)\rangle$ where $(i_0,x_0) = (i^*,x^*)$, where for $q > 0$ $\{i_q\} = \{i^*,j^*\} - \{i_{q-1}\}$ and $x_q = f(i_{q-1},x_{q-1},s)$ and where p is the least number such that $f(i_p,x_p,s) \geq r(\sigma_m,i_{p+1},s)$ and $f(i_{p+1},f(i_p,x_p,s),s) \geq r(\sigma_m,i_p,s)$. The reader should note that if $0 < q \leq p$ then

$$x_q = f(i_{q-1},x_{q-1},s) \geq r(\sigma,\sigma_{m+1},i_q,s).$$

Let $r(\sigma,\sigma_m,i,s+1) = \max\{r(\sigma,\sigma_{m+1},i,s),\ r(\sigma_m,i,s)\}$. Notice that

$$f(i_p,x_p,s+1) \geq r(\sigma,\sigma_m,i_{p+1},s+1) \;\&$$
$$f(i_{p+1},f(i_p,x_p,s+1),s+1) \geq r(\sigma,\sigma_m,i_p,s+1).$$

Case 3.3. $\alpha(\sigma,\sigma_0,s)$ is nonempty and $r(\sigma,\sigma_0,i,s) > 0$ for each $i \in T$. Let (i^*,x^*) be the last member of $\alpha(\sigma,\sigma_0,s)$. As we shall verify later there exists $j^* \in T$, $j^* \neq i^*$, such that $f(i^*,x^*,s) \geq r(\sigma,\tau,j^*,s)$ for all τ. Let m be the greatest number $\leq n$ such that (i^*,x^*) is the last member of $\alpha(\sigma,\sigma_m,s)$. Enumerate x^* in $A(i^*)$ and $f(i^*,x^*,s)$ in $A(j^*)$. If $m < p \leq n$ let $\alpha(\sigma,\sigma_p,s+1) = \alpha(\sigma,\sigma_p,s)$ and $r(\sigma,\sigma_p,i,s+1) = r(\sigma,\sigma_p,i,s)$ for each $i \in T$. Let $\alpha(\sigma,\sigma_m,s+1)$ be the sequence obtained from $a(\sigma,\sigma_m,s)$ by deleting the last member (i^*,x^*). Let $r(\sigma,\sigma_m,i^*,s+1) = 0$ and $r(\sigma,\sigma_m,i,s+1) = r(\sigma,\sigma_m,i,s)$ for $i \in T - \{i^*\}$. For values of τ not yet considered let $a(\sigma,\tau,s+1)$ be empty and $r(\sigma,\tau,i,s+1) = 0$. There are two subsubcases:

Case 3.3.1. $m < n$. As we shall verify later $a(\sigma,\sigma_m,s+1)$ is nonempty and has last member (i_q,x_q) with $f(i_q,x_q,s) = x^*$ and $i_q \neq i^*$. Let j_q be the unique member of $T - \{i^*,i_q\}$. Let $f(i,x,s+1) = f(i,x,s)$ if $f(i,x,s) < x^*$, and otherwise define $f(i,x,s+1)$ in a manner consistent with (1) to (5) and such that

$$f(i_q,x_q,s+1) \geqslant r(\sigma,\sigma_m,j_q,s+1) \ \& \\ f(j_q,f(i_q,x_q,s+1),s+1) \geqslant r(\sigma,\sigma_m,i_q,s+1).$$

Case 3.3.2. $m = n$. Then as will be seen later (i^*,x^*) is $(\kappa(n),r(\sigma,s))$. Define $f(i,x,s+1)$ in any manner consistent with (1) to (5).

Case 3.4. There exist $m < n$ and a unique $i^* \in T$ such that $a(\sigma,\sigma_m,s)$ is nonempty and $r(\sigma,\sigma_m,i^*,s) = 0$. Let $f(i,x,s+1) = f(i,x,s)$ for all $i \in T$ and $x \in N$. Let $a(\sigma,\tau,s+1) = a(\sigma,\tau,s)$ for all τ. Let $r(\sigma,\tau,i,s+1) = r(\sigma,\tau,i,s)$ unless $\tau = \sigma_m$ and $i = i^*$. Let $r(\sigma,\sigma_m,i^*,s+1) = \max\{r(\sigma,\sigma_{m+1},i^*,s), r(\sigma_m,i^*,s)\}$.

Finally to complete Case 3 define $r(\sigma',i,s+1)$ as in Case 1.

Case 4. n is of type 2. Then $\ell(\tau,s) < \xi(n)$, $\tau * \langle\omega\rangle \subseteq \sigma$, and $\Theta(\eta(n),s,A(i(\eta(n)),s))$ and $\Phi(\eta(n),s,A(j(\eta(n)),s))$ are both defined and equal for each argument $< \xi(n)$ where τ is the initial segment of σ of length $\eta(n)$. Let $f(i,x,s+1) = f(i,x,s)$ for all $i \in T$ and $x \in N$. Let $\ell(\sigma_m,s+1) = \ell(\sigma_m,s)$ if $m \leqslant n$ and $\sigma_m \neq \tau$. Let $\ell(\tau,s+1) = \xi(n)$, and $r(\sigma,s+1) = r(\sigma,s)$. For the rest proceed as in Case 1.

This completes the construction. In stage $s + 1$ we adopt the first of Cases 1 to 4 which holds. Our immediate aim is to show that the construction is self consistent.

PROPOSITION 1. Let $\sigma_0, \sigma_1, \ldots, \sigma_n = \sigma$ be a strictly increasing sequence in Σ with $\sigma_0 = \emptyset$ and σ_n of length n, then

(i) for every t and $p < n$

$$r(\sigma,\sigma_p,i,t) \geq r(\sigma,\sigma_{p+1},i,t) \vee r(\sigma,\sigma_p,i,t) = 0$$

and if $\tau \not\subseteq \sigma$ then $r(\sigma,\tau,i,t) = 0$;

(ii) for every s there exists an integer m, $-1 \leq m \leq n$, such that

(15) $m < p \leq n \rightarrow a(\sigma,\sigma_p,t)$ is nonempty. $\&.0 \leq p \leq m \rightarrow$ $a(\sigma,\sigma_p,t)$ is empty

(16) $m < p < n$ & (i,x) is a member of $a(\sigma,\sigma_p,s)$ but not the first member. $\rightarrow .x \geq r(\sigma,\sigma_{p+1},i,s)$

(17) $m < p < n \rightarrow$ the last member of $a(\sigma,\sigma_{p+1},t)$ is the first member of $a(\sigma,\sigma_p,t)$

(18) $m + 1 < p \leq n \rightarrow r(\sigma,\sigma_p,i,t) > 0. \&. r(\sigma,\sigma_p,i,t) > 0 \rightarrow$ $m + 1 \leq p \leq n$

(19) $m + 1 \leq p \leq n$ & (i^*,x^*) is the last member of $a(\sigma,\sigma_p,t)$. $\rightarrow .\exists j^* \, [j^* \in T \, \& \, j^* \neq i^* \, \& \, f(i^*,x^*,t) \geq r(\sigma,\sigma_p,j^*,t) > 0$ $\& \, f(j^*,f(i^*,x^*,t),t) \geq r(\sigma,\sigma_p,i^*,t) > 0]$

and

(20) $m < n. \rightarrow .n$ is of type 1 & $a(\sigma,\sigma,t) = \langle(\kappa(n),r(\sigma,t))\rangle$ & $r(\sigma,t) > 0$;

(iii) for every t and τ if (i,x) and (i',x') are consecutive members of $a(\sigma,\tau,s)$ then $i' \neq i$ and $x' = f(i,x,s)$.

<u>Proof</u>. For proof by contradiction let $t = s + 1$ be the least value of t which witnesses the falsity of the proposition and let σ_0, $\sigma_1, \ldots, \sigma_n = \sigma$ be the corresponding sequence in Σ. Consider first of

all the case in which the index of stage $s + 1$ is $\sigma' \neq \sigma$. Then $\sigma' < \sigma$, otherwise we should have $\alpha(\sigma,\tau,s+1)$ empty and $r(\sigma,\tau,i,s+1) = 0$ for every $\tau \in \Sigma$ and $i \in T$. Since $\sigma' < \sigma$ we have $r(\sigma,s+1) = r(\sigma,s)$, $\alpha(\sigma,\tau,s+1) = \alpha(\sigma,\tau,s)$ and $r(\sigma,\tau,i,s+1) = r(\sigma,\tau,i,s)$ for every $\tau \in \Sigma$ and $i \in T$. Thus since the proposition fails in respect of σ for $t = s + 1$ it must be the case that there are consecutive members (i,x), (i',x') of some $\alpha(\sigma,\tau,s)$ such that $x' = f(i,x,s) \neq f(i,x,s+1)$. From (5) it follows that some number $\leq x'$ is enumerated in $A(0) \cup A(1) \cup A(2)$ at stage $s + 1$ whence Case 3.3 holds at that stage. Since the proposition is true for $t = s$ we have from (2), (17), (20) and (iii) that $x' \geq r(\sigma',s) > 0$ because any number x^* enumerated in $A(0) \cup A(1) \cup A(2)$ at stage $s + 1$ must be such that for some $i^* \in T$, (i^*,x^*) is a member of $\alpha(\sigma',\emptyset,s)$. Let s' be the greatest stage $\leq s$ which has index σ then either $\alpha(\sigma,\tau,u) = \alpha(\sigma,\tau,s')$ for every $\tau \in \Sigma$ and u in $s' \leq u \leq s + 1$ or there is some stage u, $s' < u \leq s$, of index $> \sigma$. The latter is impossible because it would mean that $\alpha(\sigma,\tau,s+1)$ was empty and that $r(\sigma,\tau,i,s+1) = 0$ for every $\tau \in \Sigma$ and $i \in T$. By the same token s' must exist. Since $r(\sigma',s') = 0$ it is clear that there is a least stage s'', $s' < s'' < s$ such that $r(\sigma',u) = r(\sigma',s)$ for all u in $s'' \leq u \leq s$. By (12) at stage s'' we have $x' < r(\sigma',s'') = r(\sigma',s)$ which contradicts the inequality $x' \geq r(\sigma',s)$ found above.

It only remains to consider the case in which stage $s + 1$ has index σ. If $r(\sigma,s) = 0$ or the type of n is not 1 it is almost immediate that (i), (ii), and (iii) hold for $t = s + 1$. Otherwise Case 3 obtains at stage $s + 1$ and it is a routine matter to check through the subcases and subsubcases in order to verify that (i), (ii), and (iii) hold for $t = s + 1$ since they hold for $t = s$.

If we observe that in the above argument (11) yields $r(\sigma,s'') < r(\sigma',s'')$ we can draw the following conclusion. Let stage s' have index σ and all the stages $> s'$ and $\leq s + 1$ have index $< \sigma$ then any number enumerated in $A(0) \cup A(1) \cup A(2)$ at stage $s + 1$ is $> r(\sigma,s') = r(\sigma,s)$ and also $> x'$ if (i',x') is any member of one of the sequences $\alpha(\sigma,\tau,s') = \alpha(\sigma,\tau,s)$. We shall make frequent use of this fact below. We now prove a series of propositions from which the success of the construction may be deduced.

PROPOSITION 2. Let $\sigma_0, \sigma_1, \ldots, \sigma_n = \sigma$ be a strictly increasing sequence in Σ and σ have length n where n has type 0. Suppose there are infinitely many stages with index $\supseteq \sigma$ but only a finite number with index $> \sigma$. Then there are only a finite number of stages with index σ, and $\lim_s f(i,x,s)$ exists for all $i \in T$ and $x < \pi(n)$.

<u>Proof</u>. Let s_0 be the first stage with index $\supseteq \sigma$ such that no stage $> s_0$ has index $> \sigma$ then $r(\sigma,s_0-1) = 0$, since $r(\sigma,t) = 0$ when $t = 0$ or stage t has index $< \sigma$, and since $r(\sigma,t) = r(\sigma,t-1)$ if stage t has index $< \sigma$. Thus stage s_0 has index σ through Case 1 and $r(\sigma,s_0) > 0$. Let stage $s + 1$ be the first after stage s_0 which has index $\supseteq \sigma$, then $r(\sigma,s) = r(\sigma,s_0) > 0$ and from Case 2 we see that either $f(i,x,s) < r(\sigma,s)$ for all $x < \pi(n)$, or $f(i,x,s+1) < r(\sigma,s+1)$ for all $x < \pi(n)$. Without loss suppose $f(i,x,s) < r(\sigma,s)$ for all $x < \pi(n)$. Then it is easy to show by induction on t that for all $t > s$: $r(\sigma,t) = r(\sigma,s)$, any number enumerated in $A(0) \cup A(1) \cup A(2)$ at stage t is $> r(\sigma,s)$, stage t has index $< \sigma$, and $f(i,x,t) = f(i,x,s)$ for all $i \in T$ and $x < \pi(n)$. The conclusion follows.

PROPOSITION 3. Under the hypothesis of Proposition 2 but with n now of type 1, there are only a finite number of stages with index σ and $N - A(\kappa(n)) \not\subseteq W(n)$.

<u>Proof</u>. Let s_0 be the first stage with index $\supseteq \sigma$ such that no stage $> s_0$ has index $> \sigma$ then as above $r(\sigma,s_0) > 0$ and $r(\sigma,s_0) \not\in A(i,s_0)$ for all $i \in T$ through Case 1. Suppose that $r(\sigma,s_0) \not\in W(n)$ then it is easy to show by induction on t that for all $t > s_0$: $r(\sigma,t) =$

$r(\sigma,s_0) \not\in A(i,t)$ for all $i \in T$, and stage t has index $< \sigma$. The conclusion follows immediately in this case. Thus we may suppose that $r(\sigma,s_0) \in W(n)$. Since every stage $> s_0$ has index $\leq \sigma$ and n has type 1 we have $r(\sigma,t) = r(\sigma,s_0) > 0$ for all $t > s_0$. Let s_1 be the least stage $> s_0$ with index $\supseteq \sigma$ and such that $r(\sigma,s_0) \in W(n,s_1-1)$. Then $r(\sigma,s_0)$ $= r(\sigma,s_1-1) \not\in A(\kappa(n),s_1-1)$ since every stage $> s_0$ and $< s_1$ must have index $< \sigma$, and $a(\sigma,\tau,s_1-1)$ is empty for all τ since Case 1 holds at stage s_0. It follows that stage s_1 has index σ and that $a(\sigma,\sigma,s_1) =$ $<(\kappa(n),r(\sigma,s_0))>$ by Case 3 at that stage. Suppose for contradiction that $r(\sigma,s_0) \not\in A(\kappa(n))$. It is clear that every stage $\geq s_1$ with index $\supseteq \sigma$ has index σ and at every such stage Case 3 obtains. By induction on t we see that $a(\sigma,\sigma,t) = <(\kappa(n),r(\sigma,s_0))>$ for all $t \geq s_1$, because this could only fail through a stage $t = s + 1 > s_1$, with index σ, at which Case 3.3 obtains with $(i^*,x^*) = <(\kappa(n),r(\sigma,s_0))>$. Suppose that for some m, $0 \leq m < n$, there exists $s_2 \geq s_0$ such that $a(\sigma,\sigma_{m+1},t)$ is a fixed nonempty sequence for all $t \geq s_2$. We claim that for some $s_3 \geq$ s_2, $a(\sigma,\sigma_m,s_3)$ is nonempty. If not, let s_4 be the least stage $> s_2$ with index σ, then Case 3.4 holds at stage s_4 because Case 3.2 would make $a(\sigma,\sigma_m,s_4)$ nonempty. From Case 3.4 we see that $r(\sigma,\sigma_{m+1},i,s_4)$ > 0 for each $i \in T$. Let s_3 be the least stage $> s_4$ with index σ, then $r(\sigma,\sigma_{m+1},i,s_3-1) = r(\sigma,\sigma_{m+1},i,s_4) > 0$ for each $i \in T$ and so $a(\sigma,\sigma_m,s_3)$ is nonempty. This establishes the claim. Consider any stage $s + 1 >$ s_3 such that $a(\sigma,\sigma_m,s)$ is nonempty and $a(\sigma,\sigma_m,s+1) \neq a(\sigma,\sigma_m,s)$ then Case 3.3 must hold at stage $s + 1$. Thus $a(\sigma,\sigma_m,s+1)$ is obtained from $a(\sigma,\sigma_m,s)$ by deleting its last member which cannot be the same as its first member because $a(\sigma,\sigma_{m+1},s+1) = a(\sigma,\sigma_{m+1},s)$ since $s \geq s_2$. It follows that for all $t \geq s_3$ we have $a(\sigma,\sigma_m,t+1) \subseteq a(\sigma,\sigma_m,t)$ and $a(\sigma,\sigma_m,t+1)$ nonempty. Hence for all sufficiently large t, $a(\sigma,\sigma_m,t)$ is a fixed nonempty sequence. By descending induction on m it follows that for every $m \leq n$, $a(\sigma,\sigma_m,t)$ is a fixed nonempty sequence for all sufficiently large t. Let s_5 be a stage of index σ such that

$\alpha(\sigma,\sigma_0,t) = \alpha(\sigma,\sigma_0,s_5-1)$ for all $t \geq s_5$. At stage s_5 Case 3.4 must hold so that $r(\sigma,\sigma_0,i,s_5) > 0$ for each $i \in T$. At the first stage $> s_5$ with index σ, Case 3.3 will hold yielding a contradiction. It follows that $r(\sigma,s_0) \in A(\kappa(n))$ and the proposition is thereby proved.

PROPOSITION 4. Under the hypothesis of Proposition 3 but with n now of type 2, there are only a finite number of stages with index σ and if $\Theta(\eta(n),A(i(\eta(n))))$ and $\Phi(\eta(n),A(j(\eta(n))))$ are the same total function then $\ell(\sigma_{\eta(n)},t) \geq \xi(n)$ for all sufficiently large t.

<u>Proof</u>. Let s_0 be the first stage with index $\supseteq \sigma$ such that no stage $> s_0$ has index $> \sigma$ then as above $r(\sigma,s_0) > 0$. Let s_1 be the least number $\geq s_0$ if any such that $\ell(\sigma_{\eta(n)},s_1) \geq \xi(n)$ and $r(\sigma,s_1) > 0$. By induction on t for all $t > s_1$ stage t has index $< \sigma$, $\ell(\sigma_{\eta(n)},t) \geq \xi(n)$, and $r(\sigma,t) = r(\sigma,s_1)$. If there is a stage $s_2 > s_0$ with index σ then s_1 certainly exists by Case 4 since for the least such s_2, $r(\sigma,s_2-1) = r(\sigma,s_0) > 0$. Suppose for contradiction that s_1 does not exist and that $\Theta(\eta(n),A(i(\eta(n))))$ and $\Phi(\eta(n),A(j(\eta(n))))$ are the same total function then certainly there is a stage $s_3 > s_0$ such that $r(\sigma,s_3-1) = r(\sigma,s_0) > 0$, $\ell(\sigma_{\eta(n)},s_3-1) < \xi(n)$, and $\Theta(\eta(n),s_3-1,A(i(\eta(n)),s_3-1))$ and $\Phi(\eta(n),s_3-1,A(j(\eta(n)),s_3-1))$ are both defined and equal for all arguments $< \xi(n)$. Thus stage s_3 has index σ which is the desired contradiction. This completes the proof of the proposition.

PROPOSITION 5. Let σ, $\tau \in \Sigma$ have lengths n, m respectively and suppose $\tau * \langle\omega\rangle \subseteq \sigma$ and $r(\sigma,t-1) \in W(n,t-1) - A(\kappa(n),t-1)$; then

(i) if $r(\sigma,\tau,i(m),t) > 0$ then $r(\sigma,\tau,i(m),t)$ preserves $\Theta(m,t,A(i(m),t))$ up to $\ell(\tau,t)$; similarly for $j(m)$ and Φ in place of $i(m)$ and Θ respectively;

(ii) if $r(\sigma,\tau,i(m),t)$ and $r(\sigma,\tau,j(m),t)$ are both > 0 then $\Theta(m,t,A(i(m),t))$ and $\Phi(m,t,A(j(m),t))$ are both defined and equal for arguments $< \ell(\tau,t)$, i.e. τ is persistent at stage $t + 1$.

<u>Proof</u>. For proof by contradiction suppose the proposition is false. Consider a counterexample in which $t = s + 1$ is chosen as small as possible. Suppose firstly that the proposition fails through (i). Without loss suppose that $r(\sigma,\tau,i(m),s+1) > 0$ and that $r(\sigma,\tau,i(m),s+1)$ does not preserve $\Theta(m,s+1,A(i(m),s+1))$ up to $\ell(\tau,s+1)$. Let s_0 be the greatest stage $\leq s$ such that $r(\sigma,\tau,i(m),s\ ) = 0$, then stage $s_0 + 1$ has index σ. We claim that

$$\tau' \subseteq \sigma \ \& \ \tau' \neq \sigma \ \& \ \tau' * \langle\omega\rangle \not\subseteq \sigma \ \& \ s_0 \leq t \leq s. \rightarrow. \quad c(\tau',t) = c(\tau',s_0). \tag{21}$$

The only way (21) can fail is through a stage t, $s_0 < t \leq s$, which has index $\geq \tau' * \langle\omega\rangle$ whence $r(\sigma,\tau,i(m),t) = 0$. This would contradict the choice of s_0 and so (21) must be true. We now consider three cases:

<u>Case 1</u>. $s_0 = s$. Then stage $s + 1$ has index σ and since $0 = r(\sigma,\tau,i(m),s) < r(\sigma,\tau,i(m),s+1)$, in stage $s + 1$ $r(\sigma,\tau,i(m),s+1)$ is defined either through Case 3.2.2 or through Case 3.4 and hence is $\geq r(\tau,i(m),s)$. In either case no number is enumerated in $A(0) \cup A(1) \cup A(2)$ at stage $s + 1$ and $\ell(\tau,s+1) = \ell(\tau,s)$. Since $\tau * \langle\omega\rangle \subseteq \sigma$, $\Theta(m,s,A(i(m),s))$ and $\Phi(m,s,A(j(m),s))$ are defined and equal for arguments $< \ell(\tau,s)$. By definition $r(\tau,i(m),s)$ preserves $\Theta(m,s,A(i(m),s))$ up to $\ell(\tau,s)$. This is enough to show that this case cannot hold.

<u>Case 2</u>. $s_0 < s$ and $\ell(\tau,s) > \ell(\tau,s+1)$. Then $r(\sigma,\tau,i(m),t) = r(\sigma,\tau,i(m),s+1)$ for $s_0 < t \leq s$, because $r(\sigma,\tau,i(m),t) \neq r(\sigma,\tau,i(m),t+1)$ implies that one of $r(\sigma,\tau,i(m),t)$ and $r(\sigma,\tau,i(m),t+1)$ is zero. Since (i) holds for $t = s$ it must be the case that in stage $s + 1$ a number $x < r(\sigma,\tau,i(m),s)$ is enumerated in $A(i(m))$. We consider two subcases:

<u>Case 2.1</u>. Stage $s + 1$ has index $\sigma' \neq \sigma$. Then $\sigma' < \sigma$, for otherwise $r(\sigma,\tau,i(m),s+1) = 0$. Hence $r(\sigma',s_0+1) = 0$. Let s_1 be the greatest number $\leq s$ such that $r(\sigma',s\ ) = 0$. Since Case 1 occurs at

stage $s_1 + 1$ and Case 3.3 occurs at stage $s + 1$ we have $s_1 < s$. From stage $s_1 + 1$ we have $r(\sigma,\tau,i(m),s_1) \leq r(\sigma',s_1+1)$. By induction on t, $r(\sigma',t) \geq r(\sigma,\tau,i(m),s_1)$ for $s_1 < t \leq s$. From (1) and Proposition 1 it follows that $x \geq r(\sigma',s)$. Since $r(\sigma,\tau,i(m),s_1) = r(\sigma,\tau,i(m),s)$ from above, we can deduce that $x \geq r(\sigma,\tau,i(m),s)$ which contradicts another of our findings above. Hence this subcase cannot hold.

<u>Case 2.2</u>. Stage $s + 1$ has index σ. Then Case 3.3 holds at stage $s + 1$ and since $r(\sigma,\tau,i(m),s+1) \neq 0$ the 'm' of the statement of Case 3.3, denote it by m', satisfies $m' < m$. From Proposition 1 it follows that $x \geq r(\sigma,\tau,i(m),s)$. Hence this subcase cannot hold.

<u>Case 3</u>. $s_0 < s$ and $\ell(\tau,s) < \ell(\tau,s+1)$. Then Case 4 holds at stage $s + 1$, and stage $s + 1$ has index $\sigma' \supseteq \tau * \langle\omega\rangle$. Since $r(\sigma,\tau,i(m),s) \neq 0$ we have $\sigma' < \sigma$. We cannot have $\sigma' = \sigma$ because Case 4 holding at stage $s + 1$ means that the length of σ' has type 2 whereas from $r(\sigma,\tau,i(m),s) \neq 0$ we see by Proposition 1 that n has type 1. Let $\sigma'_0, \sigma'_1, \ldots, \sigma'_q = \sigma'$ be a strictly increasing sequence in Σ where q is the length of σ'. Since $\sigma' \supseteq \tau * \langle\omega\rangle$ we have $\sigma'_u = \sigma_u$ for $u \leq m + 1$. We shall prove that for $m + 1 < u \leq n$, $u \leq q$ and $\sigma'_u = \sigma_u$. For induction suppose that $m + 1 \leq u < n$, $u \leq q$, and $\sigma'_u = \sigma_u$, then certainly $u + 1 \leq q$ since $\sigma' < \sigma$. If $\sigma_u * \langle\omega\rangle \not\subseteq \sigma$ then $\sigma_u * \langle\omega\rangle \not\subseteq \sigma'$ again because $\sigma' < \sigma$. Hence $\sigma'_{u+1} = \sigma_u * \langle c(\sigma_u,s)\rangle$. But from (21), $c(\sigma_u,s) = c(\sigma_u,s_0)$ whence $\sigma'_{u+1} = \sigma_{u+1}$. If $\sigma_u * \langle\omega\rangle \subseteq \sigma$ then since $u \geq m + 1$ and $r(\sigma,\tau,i(m),s) \neq 0$ we have from (18) that $r(\sigma,\sigma_u,i,s) > 0$ for each $i \in T$. Since the proposition holds for $t = s$, σ_u is persistent at stage $s + 1$, whence $\sigma'_{u+1} = \sigma_u * \langle\omega\rangle = \sigma_{u+1}$. This completes the induction and shows that $\sigma \subseteq \sigma'$. Since $r(\sigma,s) \in W(n,s) - A(\kappa(n),s)$ by assumption, σ requires attention at stage $s + 1$ which contradicts our findings above that the index σ' of stage $s + 1$ satisfies $\sigma \subseteq \sigma'$ and $\sigma' \neq \sigma$. Thus this case cannot hold.

We have shown that $t = s + 1$ cannot be a counterexample to the proposition through the failure of (i). For (ii) we follow through the same case analysis and conclude that neither can (ii) fail for $t = s + 1$. This completes the proof of the proposition.

PROPOSITION 6. Suppose that every stage $\geq s_0$ has index $< \tau$ where τ has length m, that $\ell(\tau, s_0) > x$, that τ is persistent at stage s_0, and that

$$\Theta(m, s_0, A(i(m), s_0); x) = \Phi(m, s_0, A(j(m), s_0); x) = y.$$

Then for every $t \geq s_0$

(22) $$\Theta(m, s, A(i(m), t); x) = y \vee \Phi(m, s, A(j(m), t); x) = y.$$

Proof. Assume the hypothesis. By induction on t, $\ell(\tau, t) \geq \ell(\tau, s_0)$ for all $t \geq s_0$. For proof by contradiction let $t = s + 1 > s_0$ be the least value for which (22) fails. There are two cases to consider.

Case 1. Stage $s + 1$ has index σ, $\tau * \langle\omega\rangle \not\subseteq \sigma$. Let stage s_1 be the greatest stage $< s + 1$ with index $\supseteq \tau * \langle\omega\rangle$, then $s_0 \leq s_1 < s + 1$ and if the index of stage s_1 is σ', every stage t, $s_1 < t \leq s + 1$, has some index σ'' satisfying $\theta * \langle\omega\rangle > \sigma''$ and $\tau * \langle\omega\rangle \not\subseteq \sigma''$. Without loss suppose that $\Theta(i(m), s_1, A(i(m), s_1); x) = y$. We claim that

(23) $$s_1 \leq t \leq s + 1 \rightarrow \Theta(i(m), t, A(i(m), t); x) = y.$$

If not, then we might as well suppose that $t = s + 1$ is the least value for which (23) fails. Since $\sigma < \sigma'$, $r(\sigma, s_1) = 0$. Since Case 3.3 must occur at stage $s + 1$ we have $r(\sigma, s) > 0$. Let s_2 be the greatest number, $s_1 \leq s_2 < s$, such that $r(\sigma, s_2) = 0$, then stage $s_2 + 1$ has index σ and Case 1 occurs at that stage. From (13), $r(\tau,i(m),s_2+1) \leq r(\sigma,s_2)$. Also $r(\tau,i(m),s_2+1)$ preserves $\Theta(m,s_2+1, A(i(m),s_2+1))$ up to $\ell(\tau,s_2+1) > x$. By induction between $s_2 + 1$ and $s + 1$, and recalling the remark made after Proposition 1, any number enumerated in $A(i(m))$ at a stage $> s_2 + 1$ and $\leq s + 1$ is $\geq r(\sigma,s_2)$. Thus, since (23) holds for $t = s_2 + 1$ it must also hold for $t = s + 1$. This completes the first case.

Case 2. Stage $s + 1$ has index σ, $\tau * \langle\omega\rangle \subseteq \sigma$. Then Case 3.3 holds at stage $s + 1$ and $r(\sigma,\emptyset,i,s) > 0$ for each $i \in T$. Since $r(\sigma,\emptyset,i,s) > 0$ there must be a stage s_1 of index σ, where $s_1 < s$, such that every stage $> s_1$ and $\leq s + 1$ has index $\leq \sigma$. Clearly $r(\sigma,s_1) = r(\sigma,t)$ for $s_1 \leq t \leq s + 1$ and $r(\sigma,s_1) \in W(n,s_1)$. Also from stage $s + 1$, $r(\sigma,s) \notin A(\kappa(n),s)$. Thus $r(\sigma,s-1) \in W(n,s-1) - A(\kappa(n),s-1)$. From Proposition 1, $r(\sigma,\tau,i,s) > 0$ for each $i \in T$. Applying (i) of Proposition 5 we see that $r(\sigma,\tau,i(m),s)$ preserves $\Theta(m,s,A(i(m),s))$ up to $\ell(\tau,s)$ which is $> x$, and that $r(\sigma,\tau,j(m),s)$ preserves $\Phi(m,s,A(j(m),s))$ up to $\ell(\tau,s)$. Moreover

$$\Theta(m,s,A(i(m),s));x) = \Phi(m,s,A(j(m),s);x) = y$$

since τ is persistent at stage $s + 1$. From the statement of Case 3.3, at stage $s + 1$ either no number $< r(\sigma,\tau,i(m),s)$ is enumerated in $A(i(m))$ or no number $< r(\sigma,\tau,j(m),s)$ is enumerated in $A(j(m))$. In either case we get (22) for $t = s + 1$. This completes the proof of the proposition.

PROPOSITION 7. For each n there exists σ of length n such that there are infinitely many stages with index $\geq \sigma$ but only a finite number with index $> \sigma$.

Proof. Note that the result is trivially true for $n = 0$ since $\emptyset$ is the greatest member of Σ. For induction assume the result for n. According as n is of type 0, 1, or 2 we see by Proposition 2, 3, or 4 respectively that there are only a finite number of stages with index σ. Let s_0 be chosen such that no stage $\geq s_0$ has index $\geq \sigma$. If $t + 1 \geq s_0$ and stage $t + 1$ has index $\sigma' \not\geq \sigma$ then $c(\sigma,t+1) = c(\sigma,t)$ since $\sigma' < \sigma$. If $t + 1 \geq s_0$ and stage $t + 1$ has index $\sigma' \supseteq \sigma$ then either $c(\sigma,t+1) = c(\sigma,t) + 1$ or $c(\sigma,t+1) = c(\sigma,t)$ according as $\sigma' \supseteq \sigma * \langle\omega\rangle$ or $\sigma' \supseteq \sigma * \langle c(\sigma,t)\rangle$. If there are infinitely many stages with index $\supseteq \sigma * \langle\omega\rangle$ then it is clear that the proposition is true for $n + 1$ because amongst all the members of Σ which are $< \sigma$, $\sigma * \langle\omega\rangle$ is the greatest. Otherwise, $\lim_t c(\sigma,t)$ exists with value $c(\sigma)$ say and all but a finite number of stages with index $\supseteq \sigma$ have index $\supseteq \sigma * \langle c(\sigma)\rangle$. Thus again the proposition is true for $n + 1$. This completes the proof.

From Propositions 7 and 2, $\lim_s f(i,x,s)$ exists for every $i \in T$ and $x \in N$. Thus by the reasoning given at the beginning of the proof $\underset{\sim}{a}_0 \cup \underset{\sim}{a}_1 = \underset{\sim}{a}_1 \cup \underset{\sim}{a}_2 = \underset{\sim}{a}_2 \cup \underset{\sim}{a}_0$. From Propositions 7 and 3, for each $i \in T$ and $W \in \mathcal{W}$ we have $N - A(i) \not\leq W$. Hence $A(0)$, $A(1)$, and $A(2)$ are all nonrecursive. Let $\tau \in \Sigma$ have length m and suppose that infinitely many stages have index $\supseteq \tau$ but only a finite number have index $> \tau$. From Proposition 7 for each m such τ can be found. Suppose that $\Theta(m,A(i(m)))$ and $\Phi(m,A(j(m)))$ are the same total function. Let t_0 be such that no stage $\geq t_0$ has index $\geq \tau$. Let x be given and choose n such that $\eta(n) = m$ and $\xi(n) > \max\{x, \ell(\tau,t_0)\}$. By Proposition 7 there exists σ of length n such that there are infinitely many stages with index $\supseteq \sigma$ but only a finite number with index $> \sigma$. Then $\tau \subseteq \sigma$ from the way Σ is ordered. It follows by Proposition 4 that we can find $s_0 > t_0$ such that $\ell(\tau,s_0) \geq \xi(n)$ and $\ell(\tau,s_0-1) < \xi(n)$. Since $\ell(\tau,s_0) > \ell(\tau,s_0-1)$ τ is persistent at stage s_0 and $\Theta(m,s_0,A(i(m),s_0))$ and $\Phi(m,s_0,A(j(m),s_0))$ agree for arguments $< \ell(\tau,s_0)$. Let their common value at x be y. Then Proposition 6 shows that

$$\Theta(m,A(i(m));x) = \Phi(m,A(j(m));x) = y.$$

Since t_0 can be fixed and since n and s_0 can be found effectively from x we have shown that if $\Theta(m,A(i(m)))$ and $\Phi(m,A(j(m)))$ are the same total function then their common value is recursive. This completes the proof of Theorem 2.

REFERENCES

[1] A. H. Lachlan, Lower bounds for pairs of recursively enumerable degrees, Proc. London Math. Soc. 16 (1966) 537-569.

[2] S. K. Thomason, Sublattices of the r.e. degrees, Amer. Math. Soc. Notices 16 (1969) 423.

[3] S. K. Thomason, Sublattices of the recursively enumerable degrees, Z. Math. Logik Grundlagen Math., to appear.

[4] C. E. M. Yates, A minimal pair of recursively enumerable degrees, J. Symbolic Logic 31 (1966) 159-168.

DIRECT POWERS WITH DISTINGUISHED DIAGONAL

Angus Macintyre
King's College, University of Aberdeen

1. Introduction

An important group of results in model theory relates the first-order properties of product structures to the first-order properties of the factor structures. The classical sources are [12] and [3], and significant refinements and generalizations may be found in [1,5,6,17, 18].

In the case of direct powers $\mathfrak{M}^I$, the classical investigations ignore an important feature. There is a canonical embedding $\Delta: \mathfrak{M} \longrightarrow \mathfrak{M}^I$, where for $x \in \mathfrak{M}$, $\Delta(x)$ is the function on I with constant value x. Thus $\mathfrak{M}^I$ contains a canonical copy $\Delta(\mathfrak{M})$ of $\mathfrak{M}$. When $\mathfrak{M}$ is a ring, $\mathfrak{M}^I$ is not just a ring, but has a natural structure of algebra over $\mathfrak{M}$.

Let $\mathcal{L}$ be a first-order logic, $\mathfrak{M}$ an $\mathcal{L}$-structure, and I an index set. The image $\Delta(\mathfrak{M})$ is called the diagonal in $\mathfrak{M}^I$. We adjoin to $\mathcal{L}$ a new 1-ary predicate symbol P to correspond to the distinguished subset $\Delta(\mathfrak{M})$. Let $\mathcal{L}_\Delta$ be the resulting logic. Let $\mathfrak{M}^I_\Delta$ be the $\mathcal{L}_\Delta$-structure $(\mathfrak{M}^I, \Delta(\mathfrak{M}))$. We study the $\mathcal{L}_\Delta$-theory of $\mathfrak{M}^I_\Delta$, in search of analogues of known results about $\mathfrak{M}^I$.

It turns out that if I is finite we can prove analogues of the known results, and if I is infinite there are counterexamples to the analogues of the known results. When $\mathfrak{M}$ is finite, and I is infinite, we prove weakened versions of known results.

2. Preliminaries

2.1. We work throughout with first-order logics $\mathcal{L}$ with equality. $\mathcal{L}$ may have relation-symbols, operation-symbols and individual constants. As connectives, $\mathcal{L}$ has $\neg$, $\wedge$, and $\vee$. $\mathcal{L}$ has the usual quantifiers $\exists$ and $\forall$. For convenience we require that the variables of $\mathcal{L}$ are x_μ $(\mu < \lambda)$ for some infinite cardinal λ.

2.2. $\mathcal{L}_\Delta$ is got from $\mathcal{L}$ simply by adjoining a new 1-ary relation-symbol P.

2.3. If $\mathcal{M}$ is an $\mathcal{L}$-structure, Th($\mathcal{M}$), the theory of $\mathcal{M}$, is the set of all $\mathcal{L}$-sentences Φ such that $\mathcal{M} \models \Phi$.

2.4. Let Form($\mathcal{L}$) be the set of formulas of $\mathcal{L}$. For $n < \omega$, let $\mathrm{Form}_n(\mathcal{L})$ be the set of formulas of $\mathcal{L}$ with fewer than n free variables. Thus $\mathrm{Form}_0(\mathcal{L})$ is the set of sentences of $\mathcal{L}$.

2.5. Suppose T is an $\mathcal{L}$-theory, and $n < \omega$. The Ryll-Nardzewski algebra $F_n(T)$ (see [14]) is the Boolean Algebra of $\mathcal{L}$-formulas, with free variables among $x_0, \ldots, x_{n-1}$, which are inequivalent over T. $S_n(T)$ is the Stone space of $F_n(T)$. By Stone's Theorem, $S_n(T)$ is finite if and only if $F_n(T)$ is finite.

2.6. Suppose $\mathcal{M}$ is an $\mathcal{L}$-structure, and $a_0, \ldots, a_k \in \mathcal{M}$. Let $\Phi(x_{i_0}, \ldots, x_{i_k})$ be an $\mathcal{L}$-formula with $x_{i_0}, \ldots, x_{i_k}$ as its only free variables. We write $(\mathcal{M}, a_0, \ldots, a_k) \models \Phi(a_0, \ldots, a_k)$ to mean that the (k+1)-tuple $\langle a_0, \ldots, a_k \rangle$ satisfies Φ in $\mathcal{M}$.

2.7. If $\mathcal{M}$ is an $\mathcal{L}$-structure, and A is a subset of $\mathcal{M}$, we form a new structure $(\mathcal{M}, a)_{a \in A}$ by distinguishing the members of A. $(\mathcal{M}, a)_{a \in A}$ is an $\mathcal{L}(A)$-structure, where $\mathcal{L}(A)$ is got from $\mathcal{L}$ by adjoining constants $\bar{a}$ for $a \in A$.

2.8. If $\mathcal{M}$ is an $\mathcal{L}$-structure, and I an index set, we define projections $\pi_i^{\mathcal{M},I}: \mathcal{M}^I \longrightarrow \mathcal{M}$, for $i \in I$, by $\pi_i^{\mathcal{M},I}(f) = f(i)$. If the sense is clear we drop the superscripts $\mathcal{M}$, I.

2.9. If $\mathcal{M}$ is a structure, $\overline{\overline{\mathcal{M}}}$ is the cardinality of $\mathcal{M}$.

2.10. We drop brackets in repeated conjunctions and disjunctions. Thus we write $\Phi_1 \wedge \Phi_2 \wedge \Phi_3$ instead of $(\Phi_1 \wedge \Phi_2) \wedge \Phi_3$, and so on.

3. Positive Results

Most of our positive results are consequences of Lemma 1 below.

Suppose $\mathcal{M}$ is an $\mathcal{L}$-structure. The lemma says, roughly speaking, that, for finite n, any $\mathcal{L}_\Delta$-statement about elements of $\mathcal{M}_\Delta^n$ is effectively equivalent to an $\mathcal{L}$-statement about n-tuples of elements of $\mathcal{M}$.

3.1. LEMMA 1. Suppose $n < \omega$. Then there exists a recursive map τ_n: Form($\mathcal{L}_\Delta$) $\longrightarrow$ Form($\mathcal{L}$) such that

a) If Φ is a sentence, $\tau_n(\Phi)$ is a sentence;

b) If the free variables of Φ are exactly $x_{i_0}, \ldots, x_{i_k}$ then the free variables of $\tau_n(\Phi)$ are exactly $x_{n.i_j+\ell}$ for $0 \leqslant j \leqslant k$, and $0 \leqslant \ell < n$;

c) $\tau_n(\neg\Phi) = \neg\tau_n(\Phi)$;

d) $\tau_n(\Phi_1 \wedge \Phi_2) = \tau_n(\Phi_1) \wedge \tau_n(\Phi_2)$;
$\tau_n(\Phi_1 \vee \Phi_2) = \tau_n(\Phi_1) \vee \tau_n(\Phi_2)$;

e) If $\tau_n(\Phi(x_{i_0}, \ldots, x_{i_k})) = \Psi(x_{n.i_0}, \ldots, x_{n.i_0+(n-1)}, x_{n.i_1} \ldots$
$\ldots x_{n.i_k+(n-1)})$ and $\mathcal{M}$ is an $\mathcal{L}$-structure, and $f_0, \ldots, f_k$
$\in \mathcal{M}_\Delta^n$, then
$(\mathcal{M}_\Delta^n, f_0, \ldots, f_k) \models \Phi(f_0, \ldots, f_k)$ if and only if

$$(\mathcal{M}, \pi_0^{\mathcal{M},n}(f_0), \pi_1^{\mathcal{M},n}(f_0), \ldots, \pi_{n-1}^{\mathcal{M},n}(f_0), \pi_0^{\mathcal{M},n}(f_1), \ldots \pi_{n-1}^{\mathcal{M},n}(f_k))$$
$$\models \Psi(\pi_0^{\mathcal{M},n}(f_0), \pi_1^{\mathcal{M},n}(f_0), \ldots, \pi_{n-1}^{\mathcal{M},n}(f_0), \pi_0^{\mathcal{M},n}(f_1), \ldots \pi_{n-1}^{\mathcal{M},n}(f_k)).$$

__Proof__. For simplicity of notation we confine ourselves to the case where $\mathcal{L}$ has, in addition to =, just one relation-symbol R, and R is binary. We assume $\mathcal{L}$ has no operation-symbols and no individual constants. We indicate at the end of the proof how to handle operation-symbols and individual constants. No new ideas are required for the general case.

Suppose we have defined τ_n for atomic formulas of $\mathcal{L}_\Delta$. Then we define τ_n inductively on Form($\mathcal{L}_\Delta$) by:

$$\tau_n(\neg\Phi) = \neg\tau_n(\Phi);$$
$$\tau_n(\Phi_1 \wedge \Phi_2) = \tau_n(\Phi_1) \wedge \tau_n(\Phi_2);$$
$$\tau_n(\Phi_1 \vee \Phi_2) = \tau_n(\Phi_1) \vee \tau_n(\Phi_2);$$
$$\tau_n((\exists x_i)\Phi) = (\exists x_{n.i})(\exists x_{n.i+1})\ldots(\exists x_{n.i+(n-1)})\tau_n(\Phi);$$
$$\tau_n((\forall x_i)\Phi) = (\forall x_{n.i})(\forall x_{n.i+1})\ldots(\forall x_{n.i+(n-1)})\tau_n(\Phi).$$

Suppose τ_n maps atomic formulas of $\mathcal{L}_\Delta$ to formulas of $\mathcal{L}$. Then τ_n maps Form($\mathcal{L}_\Delta$) into Form($\mathcal{L}$). Recall that the variables of $\mathcal{L}$ and $\mathcal{L}_\Delta$ are x_μ ($\mu < \lambda$), where λ is an infinite cardinal. If $j < n$, and $\mu < \lambda$ then $n.\mu + j < \lambda$. Thus our inductive definition of τ_n gives a map from Form($\mathcal{L}_\Delta$) into Form($\mathcal{L}$).

It is trivial that properties (c) and (d) of the lemma hold.

Suppose property (b) holds for atomic Φ. We prove by induction that the property holds for all Φ. From the inductive definition of τ_n it suffices to prove that if the property holds for Φ then it holds for $(\exists x_i)\Phi$ and $(\forall x_i)\Phi$. The two cases are alike, so we just consider the former. So, suppose the free variables of $(\exists x_i)\Phi$ are exactly

$x_{i_0}, \ldots, x_{i_k}$. We distinguish two cases. In the first case, x_i is not free in Φ. Then by inductive assumption, the free variables of $\tau_n(\Phi)$ are exactly $x_{n.i_j+\ell}$ for $0 \leq j \leq k$ and $0 \leq \ell < n$. Now, $\tau_n((\exists x_i)\Phi) = (\exists x_{n.i})(\exists x_{n.i+1})\ldots(\exists x_{n.i+(n-1)})\tau_n(\Phi)$. We claim that none of the variables $x_{n.i+r}$ $(0 \leq r < n)$ is free in $\tau_n(\Phi)$. For if $x_{n.i+r}$ $(0 \leq r < n)$ is free in $\tau_n(\Phi)$ then $x_{n.i+r} = x_{n.i_j+\ell}$ for some j, ℓ with $0 \leq j \leq k$ and $0 \leq \ell < n$. But then $n.i + r = n.i_j + \ell$, whence $i = i_j$ and $r = \ell$. But then $x_i = x_{i_j}$, so x_i is free in Φ, contrary to assumption. Thus none of the variables $x_{n.i+r}$ $(0 \leq r < n)$ is free in $\tau_n(\Phi)$, so the free variables of $\tau_n((\exists x_i)\Phi)$ are exactly $x_{n.i_j+\ell}$ for $0 \leq j \leq k$ and $0 \leq \ell < n$. This proves the result in the first case. In the second case, x_i is free in Φ. We leave this to the reader, since the argument is similar to the above. Thus if (b) holds for atomic Φ, then (b) holds for all Φ. A similar argument, or a special case of the above, proves that if (a) and (b) hold for atomic Φ then (a) holds for all Φ.

Now, suppose property (e) holds for atomic Φ. We prove by induction that property (e) holds for all Φ. As with property (b), it suffices to show that if the property (e) holds for Φ then it holds also for $(\exists x_i)\Phi$. Again there are two cases. This time we look just at the case where x_i is free in Φ. So, suppose without loss of generality that the free variables of Φ are $x_{i_0}, \ldots, x_{i_k}$, and $i = i_0$. Then the free variables of $(\exists x_i)\Phi$ are $x_{i_1}, \ldots, x_{i_k}$. Let $\mathcal{M}$ be an $\mathcal{L}$-structure, and $f_1, \ldots, f_k \in \mathcal{M}^n_\Delta$. Then

$$(\mathcal{M}^n_\Delta, f_1, \ldots, f_k) \models (\exists x_i)\Phi(x_i, f_1, \ldots, f_k)$$

$\iff$ there exists $f_0 \in \mathcal{M}^n_\Delta$ such that $(\mathcal{M}^n_\Delta, f_0, \ldots, f_k) \models \Phi(f_0, \ldots, f_k)$

$\iff$ (by induction) there exists $f_0 \in \mathcal{M}^n_\Delta$ such that

$$(\mathcal{M},\pi_0^{\mathcal{M},n}(f_0),\pi_1^{\mathcal{M},n}(f_0),\ldots,\pi_{n-1}^{\mathcal{M},n}(f_0),\pi_0^{\mathcal{M},n}(f_1),\ldots\pi_{n-1}^{\mathcal{M},n}(f_k))$$
$$\models \tau_n(\Phi)(\pi_0^{\mathcal{M},n}(f_0),\pi_1^{\mathcal{M},n}(f_0),\ldots,\pi_{n-1}^{\mathcal{M},n}(f_0),\pi_0^{\mathcal{M},n}(f_1),\ldots\pi_{n-1}^{\mathcal{M},n}(f_k))$$

$$\Leftrightarrow\quad (\mathcal{M},\pi_0^{\mathcal{M},n}(f_1),\ldots,\pi_{n-1}^{\mathcal{M},n}(f_k)) \models$$
$$(\exists x_{n.i})(\exists x_{n.i+1})\ldots(\exists x_{n.i+(n-1)})\tau_n(\Phi)(\pi_0^{\mathcal{M},n}(f_1),\ldots,\pi_{n-1}^{\mathcal{M},n}(f_k))$$

$$\Leftrightarrow\quad (\mathcal{M},\pi_0^{\mathcal{M},n}(f_1),\ldots,\pi_{n-1}^{\mathcal{M},n}(f_k)) \models \tau_n(\exists x_i\Phi)(\pi_0^{\mathcal{M},n}(f_1),\ldots,\pi_{n-1}^{\mathcal{M},n}(f_k))$$

as required.

It remains to define τ_n for atomic formulas, and verify that (a) - (e) hold for atomic formulas. In the case of a logic with operation-symbols and individual constants there are very many distinct types of atomic formulas, and it would be painful to list them all. Hence our restriction on $\mathcal{L}$.

The atomic formulas are of three types.

<u>Type 1</u>. $x_i = x_j$

<u>Type 2</u>. $R(x_i,x_j)$

<u>Type 3</u>. $P(x_i)$

We define

1) $$\tau_n(x_i = x_j) = (x_{n.i} = x_{n.j}) \wedge (x_{n.i+1} = x_{n.j+1}) \wedge \ldots \wedge (x_{n.i+(n-1)} = x_{n.j+(n-1)}).$$

2) $$\tau_n(R(x_i,x_j)) = R(x_{n.i},x_{n.j}) \wedge R(x_{n.i+1},x_{n.j+1}) \wedge \ldots \wedge R(x_{n.i+(n-1)},x_{n.j+(n-1)}).$$

3) $$\tau_n(P(x_i)) = (x_{n.i} = x_{n.i+1}) \wedge (x_{n.i+1} = x_{n.i+2}) \wedge \ldots \wedge (x_{n.i+(n-2)} = x_{n.i+(n-1)}).$$

It is trivial that (a), (b), (c), (d) hold for atomic formulae.

We now check that (e) holds for each of the three types. Let $\mathcal{M}$ be an $\mathcal{L}$-structure, and f_i, $f_j \in \mathcal{M}^n_\Delta$.

Type 1.

$$(\mathcal{M}^n_\Delta, f_i, f_j) \models f_i = f_j$$

$$\Longleftrightarrow \pi_k^{\mathcal{M},n}(f_i) = \pi_k^{\mathcal{M},n}(f_j) \text{ for } 0 \leq k < n$$

$$\Longleftrightarrow (\mathcal{M}, \pi_0^{\mathcal{M},n}(f_i), \pi_1^{\mathcal{M},n}(f_i), \ldots, \pi_{n-1}^{\mathcal{M},n}(f_i), \pi_0^{\mathcal{M},n}(f_j), \ldots \pi_{n-1}^{\mathcal{M},n}(f_j))$$
$$\models \Phi(\pi_0^{\mathcal{M},n}(f_i), \pi_1^{\mathcal{M},n}(f_i), \ldots, \pi_{n-1}^{\mathcal{M},n}(f_i), \pi_0^{\mathcal{M},n}(f_j), \ldots \pi_{n-1}^{\mathcal{M},n}(f_j))$$

where $\Phi(x_{n.i}, x_{n.i+1}, \ldots, x_{n.i+(n-1)}, x_{n.j}, \ldots, x_{n.j+(n-1)})$ is

$$(x_{n.i} = x_{n.j}) \wedge (x_{n.i+1} = x_{n.j+1}) \wedge \ldots \wedge (x_{n.i+(n-1)} = x_{n.j+(n-1)})$$

$$\Longleftrightarrow (\mathcal{M}, \pi_0^{\mathcal{M},n}(f_i), \pi_1^{\mathcal{M},n}(f_i), \ldots, \pi_{n-1}^{\mathcal{M},n}(f_i), \pi_0^{\mathcal{M},n}(f_j), \ldots, \pi_{n-1}^{\mathcal{M},n}(f_j))$$
$$\models \tau_n(x_i = x_j)(\pi_0^{\mathcal{M},n}(f_i), \pi_1^{\mathcal{M},n}(f_i), \ldots, \pi_{n-1}^{\mathcal{M},n}(f_i), \pi_0^{\mathcal{M},n}(f_j), \ldots$$
$$\ldots, \pi_{n-1}^{\mathcal{M},n}(f_j))$$

as required.

Type 2. The argument is similar to that for type 1, and is left to the reader.

Type 3.

$$(\mathcal{M}^n_\Delta, f_i) \models P(f_i)$$

$$\Longleftrightarrow f_i(k) = f_i(\ell) \text{ for all } k, \ell < n$$

$$\Longleftrightarrow \pi_0^{\mathcal{M},n}(f_i) = \pi_1^{\mathcal{M},n}(f_i) = \ldots = \pi_{n-1}^{\mathcal{M},n}(f_i)$$

$$\Longleftrightarrow (\mathcal{M}, \pi_0^{\mathcal{M},n}(f_i), \ldots, \pi_{n-1}^{\mathcal{M},n}(f_i)) \models (\pi_0^{\mathcal{M},n}(f_i) = \pi_1^{\mathcal{M},n}(f_i)) \wedge$$
$$(\pi_1^{\mathcal{M},n}(f_i) = \pi_2^{\mathcal{M},n}(f_i)) \wedge \ldots \wedge (\pi_{n-2}^{\mathcal{M},n}(f_i) = \pi_{n-1}^{\mathcal{M},n}(f_i))$$

$$\Longleftrightarrow (\mathcal{M}, \pi_0^{\mathcal{M},n}(f_i), \ldots, \pi_{n-1}^{\mathcal{M},n}(f_i)) \models$$
$$\tau_n(P(x_i))(\pi_0^{\mathcal{M},n}(f_i), \ldots, \pi_{n-1}^{\mathcal{M},n}(f_i))$$

as required.

[We now indicate how to handle operation-symbols and individual constants. Suppose $\mathcal{L}$ has, in addition to R, a binary operation symbol $\otimes$, and individual constants a and b. In the power $\mathfrak{M}^n_\Delta$, $\otimes$ is defined coordinatewise, and a, b correspond to constant functions. It ought to be obvious how to define τ_n. We give just one example. Let Φ be $R(x_i \otimes a, b \otimes x_j)$. Then $\tau_n(\Phi)$ is

$$R(x_{n.i} \otimes a, b \otimes x_{n.j}) \wedge R(x_{n.i+1} \otimes a, b \otimes x_{n.j+1})$$
$$\wedge \ldots \wedge R(x_{n.i+(n-1)} \otimes a, b \otimes x_{n.j+(n-1)}).$$

One may easily verify that (a) - (e) hold.]

Finally, it is clear that τ_n is recursive. This concludes the proof.

3.2. From the lemma one easily deduces analogues of two classical results. The classical results are:

a) If $\mathfrak{M} \equiv \mathfrak{N}$ then $\mathfrak{M}^I \equiv \mathfrak{N}^I$;

b) If $\mathfrak{M}$ is decidable, then $\mathfrak{M}^I$ is decidable.

We will see later that the analogues of these results for $\mathfrak{M}^I_\Delta$ and $\mathfrak{N}^I_\Delta$ fail, if I is infinite. However, we now prove exact analogues when I is finite.

THEOREM 1. Let $n < \omega$. If $\mathfrak{M} \equiv \mathfrak{N}$ then $\mathfrak{M}^n_\Delta \equiv \mathfrak{N}^n_\Delta$.

<u>Proof</u>. Suppose $\mathfrak{M}^n_\Delta \not\equiv \mathfrak{N}^n_\Delta$. Select a sentence Φ such that $\mathfrak{M}^n_\Delta \models \Phi$, and $\mathfrak{N}^n_\Delta \models \neg\Phi$. By Lemma 1, part (e), $\mathfrak{M} \models \tau_n(\Phi)$ and $\mathfrak{N} \models \tau_n(\neg\Phi)$. But $\tau_n(\neg\Phi) = \neg\tau_n(\Phi)$ by part (c).

$\therefore\ \mathfrak{M} \models \tau_n(\Phi)$ and $\mathfrak{N} \models \neg\tau_n(\Phi)$.

$\therefore\ \mathfrak{M} \not\equiv \mathfrak{N}$. This proves the theorem.

Remark. The proof shows that τ_n induces a continuous map $\hat{\tau}_n$ from the space of complete $\mathcal{L}$-theories to the space of complete $\mathcal{L}_\Delta$-theories. In fact $\hat{\tau}_n(T) = \{\Phi : \tau_n(\Phi) \in T\}$.

THEOREM 2. Let $n < \omega$. If $\mathfrak{M}$ is decidable then $\mathfrak{M}^n_\Delta$ is decidable.

Proof. $\mathfrak{M}^n_\Delta \models \Phi \iff \mathfrak{M} \models \tau_n(\Phi)$, and τ_n is recursive.

3.3. There are two other known results which belong with the classical results, but were discovered somewhat later. These are:

c) If Th($\mathfrak{M}$) and Th($\mathfrak{N}$) are ω_0-categorical, then Th($\mathfrak{M} \times \mathfrak{N}$) is ω_0-categorical;

d) If Th($\mathfrak{M}$) and Th($\mathfrak{N}$) are totally transcendental [11], then Th($\mathfrak{M} \times \mathfrak{N}$) is totally transcendental.

(c) is due to Grzegorczyk [6]. Another proof is in [17]. (d) was stated without proof by the present author in [7]. A proof may be given along the lines of [17], with a use of Frayne's Theorem as in Theorem 4 below.

We now prove analogues of (c) and (d). There is a common idea behind both proofs, and we now present this.

Let $n < \omega$, and let $\mathfrak{M}$ be an $\mathcal{L}$-structure. By part (b) of Lemma 1, τ_n maps formulas with free variables in the list $x_0, \ldots, x_k$ into formulas with free variables in the list $x_0, \ldots, x_{n.k+(n-1)}$. By parts (c) and (d) of Lemma 1, τ_n commutes with $\neg$, $\wedge$ and $\vee$. By part (e), Φ is satisfiable in $\mathfrak{M}^n_\Delta$ if and only if $\tau_n(\Phi)$ is satisfiable in $\mathfrak{M}$. Identifying $S_{k+1}(T)$, for a theory T, with the space of (k+1)-types over T, and using the Compactness Theorem, it follows directly from the preceding remarks that τ_n induces a continuous map

$$\hat{\tau}_n^{(k+1)} : S_{(k+1)n}(\mathrm{Th}(\mathfrak{M})) \rightarrow S_{k+1}(\mathrm{Th}(\mathfrak{M}^n_\Delta))$$

given by

$$\hat{\tau}_n^{(k+1)}(P) = \{\Phi : \tau_n(\Phi) \in P\}.$$

We claim $\hat{\tau}_n^{(k+1)}$ is surjective. Suppose $q \in S_{k+1}(\mathrm{Th}(\mathcal{M}_\Delta^n))$. Let $q^* = \{\tau_n(\Phi) : \Phi \in q\}$. We claim there exists $p \in S_{(k+1)n}(\mathrm{Th}(\mathcal{M}))$ such that $q^* \subseteq P$. It suffices to prove that any finite conjunction of members of q^* is satisfiable in $\mathcal{M}$, and this follows from the above-mentioned properties of τ_n. Then $\hat{\tau}_n^{(k+1)}(p) = q$. Thus $\hat{\tau}_n^{(k+1)}$ is surjective.

For the next two theorems, assume $\mathcal{L}$ countable, and $\mathcal{M}$ an $\mathcal{L}$-structure.

THEOREM 3. Let $n < \omega$. Suppose $\mathrm{Th}(\mathcal{M})$ is ω_0-categorical. Then $\mathrm{Th}(\mathcal{M}_\Delta^n)$ is ω_0-categorical.

Proof. By Ryll-Nardzewski's theorem [14], each $S_\ell(\mathrm{Th}(\mathcal{M}))$ is finite for $\ell < \omega$.

Since $\hat{\tau}_n^{(k+1)} : S_{(k+1)n}(\mathrm{Th}(\mathcal{M})) \to S_{k+1}(\mathrm{Th}(\mathcal{M}_\Delta^n))$ is surjective, $S_{k+1}(\mathrm{Th}(\mathcal{M}_\Delta^n))$ is finite. Again by Ryll-Nardzewski's theorem, $\mathrm{Th}(\mathcal{M}_\Delta^n)$ is ω_0-categorical. This completes the proof.

THEOREM 4. Let $n < \omega$. Suppose $\mathrm{Th}(\mathcal{M})$ is totally transcendental. Then $\mathrm{Th}(\mathcal{M}_\Delta^n)$ is totally transcendental.

Proof. Suppose $\mathrm{Th}(\mathcal{M}_\Delta^n)$ is not totally transcendental. Then there is a countable $\mathcal{L}_\Delta$-structure $\mathcal{N}$ such that $\mathcal{N} \equiv \mathcal{M}_\Delta^n$ and $S_1(\mathrm{Th}(\mathcal{N},a)_{a\in\mathcal{N}})$ is uncountable. Select $\mathcal{N}_1$ such that $\mathcal{N} \prec \mathcal{N}_1$, and every point of $S_1(\mathrm{Th}(\mathcal{N},a)_{a\in\mathcal{N}})$ is realized in $\mathcal{N}_1$.

We claim that there exists an $\mathcal{L}$-structure $\mathcal{M}_1$ such that $\mathcal{M}_1 \equiv \mathcal{M}$ and $\mathcal{N}_1 \prec (\mathcal{M}_1)_\Delta^n$. By Frayne's Theorem [4, Theorem 2.12], since $\mathcal{N}_1 \equiv \mathcal{N} \equiv \mathcal{M}_\Delta^n$, $\mathcal{N}_1$ is an elementary subsystem of an ultrapower $(\mathcal{M}_\Delta^n)^I/\mathcal{D}$. We claim that $(\mathcal{M}_\Delta^n)^I/\mathcal{D} \cong (\mathcal{M}^I/\mathcal{D})_\Delta^n$. Define a map

$\gamma : (\mathcal{M}^n_\Delta)^I/\mathcal{D} \longrightarrow (\mathcal{M}^I/\mathcal{D})^n_\Delta$ by

$$f/\mathcal{D} \rightsquigarrow (f/\mathcal{D})_*,$$

where, for $m < n$,

$$(f/\mathcal{D})_*(m) = f_*(m)/\mathcal{D},$$

where $f_*(m)(i) = f(i)(m)$, for $i \in I$. To prove that this is a genuine definition, we have to show that if $f, g \in (\mathcal{M}^n_\Delta)^I$ and $f \equiv g \bmod \mathcal{D}$, then $f_*(m) \equiv g_*(m) \bmod \mathcal{D}$, for each $m < n$. This is trivial, and doesn't need $\mathcal{D}$ to be an ultrafilter. It is also trivial to prove that γ is onto. γ is 1-1 because $\mathcal{D}$ is an ultrafilter. It is trivial to prove that γ is compatible with the primitive notions of $\mathcal{L}$, i.e. γ is an $\mathcal{L}$-homomorphism. Finally, we want to show that γ is an $\mathcal{L}_\Delta$-homomorphism, so it remains to show that if $x \in (\mathcal{M}^n_\Delta)^I/\mathcal{D}$ and $(\mathcal{M}^n_\Delta)^I/\mathcal{D} \models P(x)$ then $(\mathcal{M}^I/\mathcal{D})^n_\Delta \models P(\gamma(x))$.

Let $x = f/\mathcal{D}$. Suppose $(\mathcal{M}^n_\Delta)^I/\mathcal{D} \models P(x)$. Then $\{i \in I : \mathcal{M}^n_\Delta \models P(f(i))\} \in \mathcal{D}$, so

$$\{i \in I : \mathcal{M} \models (f(i)(0) = f(i)(1)) \wedge (f(i)(1) = f(i)(2)) \wedge \dots \wedge (f(i)(n-2) = f(i)(n-1))\} \in \mathcal{D},$$

so

$$\{i \in I : \mathcal{M} \models (f_*(0)(i) = f_*(1)(i)) \wedge (f_*(1)(i) = f_*(2)(i)) \wedge \dots \wedge (f_*(n-2)(i) = f_*(n-1)(i))\} \in \mathcal{D},$$

so

$$\mathcal{M}^I/\mathcal{D} \models (f_*(0)/\mathcal{D} = f_*(1)/\mathcal{D}) \wedge (f_*(1)/\mathcal{D} = f_*(2)/\mathcal{D}) \wedge \dots \wedge (f_*(n-2)/\mathcal{D} = f_*(n-1)/\mathcal{D}),$$

so $(\mathcal{M}^I/\mathcal{D})^n_\Delta \models P(f/\mathcal{D})$, i.e. $(\mathcal{M}^I/\mathcal{D})^n_\Delta \models P(\gamma(x))$, as required. Thus $(\mathcal{M}^n_\Delta)^I/\mathcal{D} \cong (\mathcal{M}^I/\mathcal{D})^n_\Delta$. Now take $\mathcal{M}_1 = \mathcal{M}^I/\mathcal{D}$, so $\mathcal{M}_1 \equiv \mathcal{M}$ and $\mathcal{N}_1 \prec (\mathcal{M}_1)^n_\Delta$.

Recall that

$$S_1(\mathrm{Th}(\mathcal{N},a)_{a\in\mathcal{N}}) \text{ is uncountable,}$$

and every point is realized in $\mathcal{N}_1$, and $\mathcal{N} \prec \mathcal{N}_1$. By the preceding paragraph, every point is realized in $(\mathcal{M}_1)^n_\Delta$, where $\mathcal{M}_1 \equiv \mathcal{M}$. Since $\mathcal{N} \subseteq (\mathcal{M}_1)^n_\Delta$, the elements of $\mathcal{N}$ are n-tuples of elements of $\mathcal{M}_1$. Let $A = \{\pi_m^{\mathcal{M}_1,n}(f) : f \in \mathcal{N},\ m < n\}$. Then A is countable, since $\mathcal{N}$ is countable. Now we work in the logic $\mathcal{L}(A)$. Consider the structure $\mathcal{M}_2 = (\mathcal{M}_1,a)_{a\in A}$. Since

$$S_1(\mathrm{Th}(((\mathcal{M}_1)^n_\Delta,a)_{a\in\mathcal{N}})) \text{ is uncountable,}$$

it is obvious that

$$S_1(\mathrm{Th}(((\mathcal{M}_1)^n_\Delta,a)_{a\in A^n})) \text{ is uncountable.}$$

But $((\mathcal{M}_1)^n_\Delta,a)_{a\in A^n} = ((\mathcal{M}_1,a)_{a\in A})^n_\Delta$, so

$$S_1(\mathrm{Th}((\mathcal{M}_2)^n_\Delta)) \text{ is uncountable.}$$

But we have a surjection $\hat{\tau}_n^{(1)}$ from $S_n(\mathrm{Th}(\mathcal{M}_2))$ to $S_1(\mathrm{Th}((\mathcal{M}_2)^n_\Delta))$, so $S_n(\mathrm{Th}(\mathcal{M}_2))$ is uncountable. Thus $S_n(\mathrm{Th}((\mathcal{M}_1,a)_{a\in A}))$ is uncountable, where $\mathcal{M}_1 \equiv \mathcal{M}$, and A is countable. Then, by [7a, Lemma 5], $\mathrm{Th}(\mathcal{M})$ is not totally transcendental. This completes the proof.

3.4. It is well-known that there exist $\mathcal{M}$ such that Th($\mathcal{M}$) is ω_1-categorical but Th($\mathcal{M}^2$) is not ω_1-categorical. An example is given in [17]. It does not follow directly from this that there exist $\mathcal{M}$ such that Th($\mathcal{M}$) is ω_1-categorical, but Th($\mathcal{M}^2_\Delta$) is not ω_1-categorical. However, examples do exist. Let $\mathcal{L}$ be the pure logic of identity theory. Let $\mathcal{M}$ be an infinite $\mathcal{L}$-structure. Then Th($\mathcal{M}$) and Th($\mathcal{M}^2$) are ω_1-categorical, but Th($\mathcal{M}^2_\Delta$) is not ω_1-categorical. In fact, Th($\mathcal{M}^2_\Delta$) has two non-isomorphic models in power ω_1.

4. Results for finite $\mathcal{M}$

The following is a special case of an important theorem of Feferman and Vaught [3, Theorem 6.6].

THEOREM. There is a recursive map f from $\mathrm{Form}_0(\mathcal{L})$ to ω such that for all $\Psi \in \mathrm{Form}_0(\mathcal{L})$, all $\mathcal{L}$-structures $\mathcal{M}$, and all index sets I,

$$\overline{\overline{I}} \geq f(\Psi) \Rightarrow (\mathcal{M}^I \models \Psi \Leftrightarrow \mathcal{M}^{f(\Psi)} \models \Psi).$$

We think of this as a stability theorem, implying in particular that $\langle \mathrm{Th}(\mathcal{M}^n)\rangle_{n<\omega}$ "converges recursively" to Th($\mathcal{M}^\omega$), in the usual topology for complete $\mathcal{L}$-theories. Another consequence is that the sets $\{n < \omega : \mathcal{M}^n \models \Psi\}$ are all finite or cofinite.

These results do not extend to direct powers with distinguished diagonal, as we will see later. However, there is a weakened version when $\mathcal{M}$ is finite.

4.1. Let $\mathcal{L}$ be a fixed logic, and k a fixed finite cardinal. In this subsection we are concerned exclusively with $\mathcal{L}$-structures $\mathcal{M}$ such that $\overline{\overline{\mathcal{M}}} = k$. Adjoin to $\mathcal{L}$ individual constants $c_0, \ldots, c_{k-1}$ to get a logic $\mathcal{L}^1$. Make each $\mathcal{M}$ into an $\mathcal{L}^1$-structure $\mathcal{M}^1 = (\mathcal{M}, a_0, \ldots, a_{k-1})$ by selecting some fixed enumeration $a_0, \ldots, a_{k-1}$ of $\mathcal{M}$.

LEMMA 2. There is a recursive map γ_k from Form($\mathcal{L}_\Delta$) to Form($\mathcal{L}^1$) such that

a) γ_k maps sentences to sentences, and

b) for any index set I, any $\mathcal{L}$-structure $\mathcal{M}$ with $\overline{\overline{\mathcal{M}}} = k$, and any $\mathcal{L}_\Delta$-sentence Φ, $\mathcal{M}^I_\Delta \models \Phi \Leftrightarrow (\mathcal{M}^1)^I \models \gamma_k(\Phi)$.

<u>Proof</u>. <u>Definition of</u> γ_k. If P doesn't occur in Φ, $\gamma_k(\Phi) = \Phi$. For a general Φ, $\gamma_k(\Phi)$ is got from Φ by replacing each subformula P(t) (t a term) by

$$(t = c_0) \vee (t = c_1) \vee \ldots \vee (t = c_{k-1}).$$

A simple induction now proves the lemma.

4.2. THEOREM 5. Let $\mathcal{L}$ be a fixed logic and k a fixed finite cardinal. There is a recursive map f_k from $\text{Form}_0(\mathcal{L}_\Delta)$ to ω such that for all $\mathcal{L}_\Delta$-sentences Ψ, all $\mathcal{L}$-structures $\mathcal{M}$ with $\overline{\overline{\mathcal{M}}} = k$, and all index sets I,

$$\overline{\overline{I}} \geq f_k(\Psi) \Rightarrow (\mathcal{M}^I_\Delta \models \Psi \leftrightarrow \mathcal{M}^{f_k(\Psi)}_\Delta \models \Psi).$$

<u>Proof</u>. We use the notation of 4.1. By the Feferman-Vaught result quoted earlier, there is a recursive map f from $\mathcal{L}^1$ sentences to ω, such that for all $\mathcal{L}^1$ sentences Φ, all $\mathcal{L}^1$ structures $\mathcal{N}$, and all index sets I,

$$\overline{\overline{I}} \geq f(\Phi) \Rightarrow (\mathcal{N}^I \models \Phi \leftrightarrow \mathcal{N}^{f(\Phi)} \models \Phi).$$

Now take f_k as $f \circ \gamma_k$, and use Lemma 2. The result is immediate.

COROLLARY 1. Suppose $\overline{\overline{\mathfrak{M}}}$ is finite. Then $\mathfrak{M}^{\omega}_{\Delta}$ is decidable.

Proof. Trivial from the theorem.

COROLLARY 2. Suppose $\overline{\overline{\mathfrak{M}}}$ is finite. Then $\mathrm{Th}(\{\mathfrak{M}^{n}_{\Delta} : n < \omega\})$ is decidable.

Proof. Trivial from the theorem.

This is the last of our positive results, and we now turn to counterexamples.

5. Some simple counterexamples

5.1. Let $\mathcal{L}$ be a logic with a single binary relation symbol $<$. Let $\mathfrak{M}$ be the $\mathcal{L}$-structure consisting of the rationals under the usual linear order. Let $\mathfrak{N}$ be $\mathfrak{M}^{I}/\mathfrak{D}$ where I is countable and $\mathfrak{D}$ is a non-principal ultrafilter on I. Then $\mathfrak{N} \equiv \mathfrak{M}$, and in $\mathfrak{N}$ every countable set of elements has an upper bound, by familiar saturation properties of ultrapowers. Let Φ be the $\mathcal{L}_{\Delta}$-sentence

$$(\forall x_0)(\exists x_1)(P(x_1) \wedge x_0 < x_1).$$

Then one easily checks that $\mathfrak{M}^{\omega}_{\Delta} \models \neg\Phi$ and $\mathfrak{N}^{\omega}_{\Delta} \models \Phi$, so $\mathfrak{M}^{\omega}_{\Delta} \not\equiv \mathfrak{N}^{\omega}_{\Delta}$. Thus we have a counterexample to a generalization of Theorem 5.1 of [3].

A significant defect of this example is that $\mathfrak{N}$ is uncountable. Further, if $\mathfrak{M}$ is as above, and $\mathfrak{M}_1$ is countable and $\mathfrak{M}_1 \equiv \mathfrak{M}$, then $(\mathfrak{M}_1)^{\omega}_{\Delta} \cong \mathfrak{M}^{\omega}_{\Delta}$, by the ω_0-categoricity of $\mathrm{Th}(\mathfrak{M})$.

Later we give another example avoiding this difficulty.

5.2. Let $\mathfrak{M}$, Φ be as in 5.1. Then clearly $\mathfrak{M}^{n}_{\Delta} \models \Phi$ for $n < \omega$, but $\mathfrak{M}^{\omega}_{\Delta} \models \neg\Phi$.

This gives a counterexample to a generalization of Corollary 6.7.1 of [3].

5.3. Let $\mathcal{L}$ have a single binary relation-symbol R. Let Ψ be the $\mathcal{L}_\Delta$-sentence

$$(\forall x_0)(\exists x_1)(P(x_1) \wedge R(x_0,x_1)).$$

Let $k < \omega$. Then, for any $\mathcal{L}$-structure $\mathfrak{M}$,

$$\mathfrak{M}^k_\Delta \models \Psi \Leftrightarrow \mathfrak{M} \models (\forall x_0)(\forall x_1)\ldots(\forall x_{k-1})(\exists x_k)(R(x_0,x_k) \wedge R(x_1,x_k) \wedge \ldots \wedge R(x_{k-1},x_k)).$$

We claim that for each $k < \omega$ there is a finite $\mathcal{L}$-structure $\mathfrak{M}_k$ such that $(\mathfrak{M}_k)^{k-1}_\Delta \models \Psi$ and $(\mathfrak{M}_k)^k_\Delta \models \neg\Psi$.

Take $\mathfrak{M}_k$ as a structure with k elements, in which $R(x,y)$ holds if and only if $x \neq y$. By the equivalence given above, it is trivial that $\mathfrak{M}_k$ has the required properties.

It follows that there is no function f from $\mathcal{L}_\Delta$-sentences to ω such that for all $\mathcal{L}_\Delta$-sentences Φ, all finite $\mathcal{L}$-structures $\mathfrak{M}$, and all n with $f(\Phi) \leq n < \omega$, $\mathfrak{M}^n_\Delta \models \Phi \Leftrightarrow \mathfrak{M}^{f(\Phi)}_\Delta \models \Phi$.

Thus we have a counterexample to a generalization of Theorem 6.6 of [3], and our Theorem 5 is the best that can be expected. Later we will obtain a deeper counterexample to a generalization of the stability theorem.

6. Counterexamples to Decidability

Throughout this section, until 6.3, $\mathcal{L}$ will be the logic for field theory. $\mathfrak{C}$ is the field of complex numbers.

By Tarski [16] $\mathbb{C}$ is decidable, so by Feferman-Vaught [3, Theorem 6.4] $\mathbb{C}^{\omega}$ is decidable, and the theory of the set $\{\mathbb{C}^n : n < \omega\}$ is decidable. We will show that these results do not generalize.

6.1. Each of the rings $\mathbb{C}^j$ $(j \leq \omega)$ is an algebra over $\Delta(\mathbb{C})$, so the notion of the spectrum of an element is applicable. We define this notion uniformly in the structures $\mathbb{C}^j_{\Delta}$ as follows:

$$y \in Sp(x) \leftrightarrow y \in \Delta(\mathbb{C}) \wedge (x - y) \text{ is not invertible.}$$

Clearly this informal definition can be formalized in $\mathcal{L}_{\Delta}$.

The ring $\Delta(\mathbb{C})$ has $\mathbb{Z}$ as a subring, and $\mathbb{N}$ as a subsemiring. We now give an $\mathcal{L}_{\Delta}$ definition of $\mathbb{N}$ in $\mathbb{C}^{\omega}_{\Delta}$.

DEFINITION.

$$N(t) \leftrightarrow t \in \Delta(\mathbb{C}) \wedge (\forall x)[(0 \in Sp(x)$$
$$\wedge (\forall y)(y \in Sp(x) \rightarrow y+1 \in Sp(x))) \rightarrow (t \in Sp(x)].$$

Obviously $t \in \mathbb{N} \rightarrow N(t)$. Conversely, suppose $t \in \Delta(\mathbb{C})$ and $t \notin \mathbb{N}$. Let $x \in \mathbb{C}^{\omega}_{\Delta}$ be the "identity" map from ω to $\mathbb{N}$. Then $Sp(x) = \mathbb{N}$, and $t \notin Sp(x)$. Thus $\neg N(t)$.

Thus the semiring of integers is first-order definable in $\mathbb{C}^{\omega}_{\Delta}$, so $Th(\mathbb{C}^{\omega}_{\Delta})$ is hereditarily undecidable.

<u>Remark</u>. $\mathbb{C}^{\omega}_{\Delta}$ is an algebra over $\mathbb{C}$. We get an appropriate logic for algebras by adjoining to $\mathcal{L}$ a new sort of variable ranging over scalars, and an operation corresponding to scalar multiplication. This gives a logic $\mathcal{L}_1$. We can interpret $\mathbb{C}^{\omega}_{\Delta}$ in the $\mathcal{L}_1$ theory of $\mathbb{C}^{\omega}$, by interpreting $\Delta(\mathbb{C})$ as the scalar multiples of the identity. Thus we can define $\mathbb{N}$ in the algebra $\mathbb{C}^{\omega}$, so the $\mathcal{L}_1$-theory of $\mathbb{C}^{\omega}$ is heredit-

arily undecidable. This gives an example of an algebra $\mathcal{C}^{\omega}$ over a decidable field $\mathcal{C}$, which is undecidable, but whose underlying ring is decidable.

6.2. It is more delicate to show that $Th(\{\mathcal{C}^n_\Delta : n < \omega\})$ is undecidable. By Theorem 2, each $\mathcal{C}^n_\Delta$ is decidable, so we cannot define $\mathbb{N}$ in $\mathcal{C}^n_\Delta$.

We make the following definition uniformly in the structures $\mathcal{C}^n_\Delta$.

$$\begin{aligned} N_1(t) \leftrightarrow\ & t \in \Delta(\mathcal{C}) \wedge \\ & (\exists x)(0 \in Sp(x) \wedge t \in Sp(x) \wedge \\ & (\forall y)(y \in Sp(x) \rightarrow (y+1 \in Sp(x) \vee y = t))). \end{aligned}$$

Clearly N_1 is $\mathcal{L}_\Delta$-definable, say by $\Phi(x_0)$.

It is easily verified that $\Phi(x_0)$ defines $\{0,1,\ldots,n\}$ in $\mathcal{C}^{n+1}_\Delta$, i.e.

$$\mathcal{C}^{n+1}_\Delta \models (\forall x_0)(\Phi(x_0) \leftrightarrow (x_0 = 0 \vee x_0 = 1 \vee \ldots \vee x_0 = n)).$$

Let $p(y_1,\ldots,y_k,z_1,\ldots,z_\ell)$ be a polynomial with coefficients in $\mathbb{Z}$. Then for $m_1, \ldots, m_k \in \mathbb{N}$,

$$\begin{aligned} & \mathbb{N} \models (\exists x_1)\ldots(\exists x_\ell)(p(m_1,\ldots,m_k,x_1,\ldots,x_\ell) = 0) \\ \Leftrightarrow\ & (\exists n{<}\omega)[\mathcal{C}^n_\Delta \models (\exists x_1)\ldots(\exists x_\ell)(p(m_1,\ldots,m_k,x_1,\ldots,x_\ell) = 0 \wedge \\ & \Phi(x_1) \wedge \ldots \wedge \Phi(x_\ell)]. \end{aligned}$$

Thus $\{\langle m_1,\ldots,m_k\rangle : \mathbb{N} \models \neg(\exists x_1)\ldots(\exists x_\ell)(p(m_1,\ldots,m_k,x_1,\ldots,x_\ell) = 0)\} = \{\langle m_1,\ldots,m_k\rangle : (\forall n{<}\omega)\ \mathcal{C}^n_\Delta \models \neg(\exists x_1)\ldots(\exists x_\ell)(p(m_1,\ldots,m_k,x_1,\ldots,x_\ell) = 0 \wedge \Phi(x_1) \wedge \ldots \wedge \Phi(x_\ell))\}$. Thus the predicate

$$\{<m_1,\dots,m_k>: \mathcal{N} \models \neg(\exists x_1)\dots(\exists x_\ell)(p(m_1,\dots,m_k,x_1,\dots,x_\ell) = 0)\}$$

is representable in $Th(\{\mathbb{C}^n_\Delta : n < \omega\})$. [For representability, see [15, Chapter III]].

Take k = 1, and apply Matijasevic's Theorem [10] to conclude that the complement of every recursively enumerable set is representable in $Th(\{\mathbb{C}^n_\Delta : n < \omega\})$. By [15, Chapter III, Theorem 9'], $Th(\{\mathbb{C}^n_\Delta : n < \omega\})$ is not recursively enumerable.

It is convenient for us at this point to sketch a proof that $Th(\{\mathbb{C}^n_\Delta : n < \omega\})$ is hereditarily undecidable. This result is quoted in [8], and applied to show that the theory of finite-dimensional semi-simple Banach algebras over $\mathbb{C}$ is hereditarily undecidable.

We refer to [2, page 90] for a statement of Cobham's Theorem and the theory R_0. In terms of N_1 defined above, it is easy to interpret R_0 in $Th(\{\mathbb{C}^n_\Delta : n < \omega\})$, whence by Cobham's Theorem $Th(\{\mathbb{C}^n_\Delta : n < \omega\})$ is hereditarily undecidable. (I am indebted to Gaifman for a remark during my talk at Bedford College, concerning coding Turing machines in $Th(\{\mathbb{C}^n_\Delta : n < \omega\})$. This led me to the present proof.)

It follows from the undecidability of $Th(\{\mathbb{C}^n_\Delta : n < \omega\})$ that there is no recursive function f from $\mathcal{L}_\Delta$-sentences to ω such that for each $\mathcal{L}_\Delta$-sentence Ψ, and $f(\Psi) \leq m < \omega$,

$$\mathbb{C}^m_\Delta \models \Psi \iff \mathbb{C}^{f(\Psi)}_\Delta \models \Psi.$$

Indeed we shall soon see that we can remove "recursive" from the above statement. But we conclude this section with a digression.

6.3. Now we improve Counterexample 5.1. Let $\mathcal{L}$ be the logic for ordered fields. Let $\mathcal{M}$ be the real-closed field of real algebraic numbers. $\mathbb{N}$ is a substructure of $\mathfrak{M}$, and $\mathbb{N}$ is cofinal in $\mathfrak{M}$. It is well-known that there exist $\mathfrak{N} \equiv \mathfrak{M}$ such that $\mathfrak{N}$ is countable and $\mathbb{N}$ is not cofinal in $\mathfrak{N}$. Now, the definition of $\mathbb{N}$ given in 6.1 works for $\mathfrak{M}^{\omega}_{\Delta}$ and $\mathfrak{N}^{\omega}_{\Delta}$. The proof is the same as before. But then

$$\mathfrak{M}^{\omega}_{\Delta} \models (\forall x_0)(P(x_0) \rightarrow (\exists x_1)(N(x_1) \wedge x_0 < x_1))$$

and

$$\mathfrak{N}^{\omega}_{\Delta} \models \neg(\forall x_0)(P(x_0) \rightarrow (\exists x_1)(N(x_1) \wedge x_0 < x_1)).$$

Thus $\mathfrak{M}^{\omega}_{\Delta} \not\equiv \mathfrak{N}^{\omega}_{\Delta}$.

7. Counterexamples to Stability

As we remarked before, a consequence of the classical results is that all sets

$$\{n : n < \omega \wedge \mathfrak{M}^{n} \models \Psi\}$$

are finite or cofinite. This result fails to generalize. We shall confine ourselves to the logic $\mathcal{L}$ for ring-theory, and look at two cases, namely when $\mathfrak{M}$ is $\mathbb{C}$, and when $\mathfrak{M}$ is $\mathbb{Z}$.

7.1. This continues 6.2. Define

$$D(t) \leftrightarrow N_1(t) \wedge \neg N_1(t+1).$$

Then clearly

$$\mathbb{C}^{n+1}_{\Delta} \models (\forall x_0)(D(x_0) \leftrightarrow x_0 = n).$$

We refer to [15] for a definition of the notion of constructive arithmetic predicate.

Let X be a constructive arithmetic subset of $\mathbb{N}$. Let $A(x_0)$ be a formula of number theory, involving only bounded quantifiers, defining X. Let $A^{(N_1)}(x_0)$ be the relativization of $A(x_0)$ to N_1. From the form of A it follows that if $0 \leq m \leq n$ then $\mathfrak{C}_\Delta^{n+1} \models A^{(N_1)}(m) \Longleftrightarrow \mathbb{N} \models A(m)$. Therefore

$$\begin{aligned} \mathfrak{C}_\Delta^{n+1} \models (\exists x_0)(A^{(N_1)}(x_0) \wedge D(x_0)) &\Longleftrightarrow \mathfrak{C}_\Delta^{n+1} \models A^{(N_1)}(n) \\ &\Longleftrightarrow \mathbb{N} \models A(n) \\ &\Longleftrightarrow n \in X. \end{aligned}$$

Take Ψ as $(\exists x_0)(A^{(N_1)}(x_0) \wedge D(x_0))$, and then

$$\{n : \mathfrak{C}_\Delta^{n+1} \models \Psi\} = X.$$

In particular, one can take X as the set of powers of 2 [15, Chapter IV, Prop. 3], so $\{n : \mathfrak{C}_\Delta^{n} \models \Psi\}$ need not be finite or cofinite.

7.2. Working with $\mathbb{Z}$ instead of $\mathfrak{C}$ we can get stronger instability results.

We define

$$y \in Sp^*(x) \leftrightarrow y \in \Delta(\mathbb{Z}) \wedge (\exists t)(t \neq 0 \wedge t.(x-y) = 0).$$

Now define $N_1^*(t)$ just as $N_1(t)$, except that each occurrence of Sp is replaced by Sp^*. Clearly N_1^* is $\mathcal{L}_\Delta$-definable, say by $\Phi^*(x_0)$.

It is easily verified that Φ^* defines $\{0,\ldots,n\}$ in $\mathbb{Z}_\Delta^{n+1}$. Define $D^*(t) \leftrightarrow N_1^*(t) \wedge \neg N_1^*(t+1)$. Then clearly

$$\mathbb{Z}_{\Delta}^{n+1} \models D^*(m) \Leftrightarrow m = n.$$

Let X be any arithmetical subset of $\mathbb{N}$. Since $\mathbb{N}$ is definable in $\mathbb{Z}$ (via Lagrange's Theorem), there is an $\mathcal{L}$-formula $A(x_0)$ such that $m \in \mathbb{N} \Rightarrow (\mathbb{Z} \models A(m) \leftrightarrow m \in X)$.

Let $A^{(P)}(x_0)$ be the relativization of A to P. Let Ψ be $(\exists x_0)(A^{(P)}(x_0) \wedge D^*(x_0))$. Then

$$\begin{aligned} \mathbb{Z}_{\Delta}^{n+1} \models \Psi &\Leftrightarrow \mathbb{Z}_{\Delta}^{n+1} \models A^{(P)}(n) \\ &\Leftrightarrow \mathbb{Z} \models A(n) \\ &\Leftrightarrow n \in X. \end{aligned}$$

Thus $\{n : \mathbb{Z}_{\Delta}^{n+1} \models \Psi\} = X$.

7.3. We conclude this section, and the main part of the paper, with some remarks and problems. Let $\mathcal{L}$ be a fixed logic, $\mathcal{M}$ an $\mathcal{L}$-structure, and Ψ an $\mathcal{L}_{\Delta}$-sentence. Define

$$\mathrm{Rep}(\mathcal{M},\Psi) = \{n < \omega : \mathcal{M}_{\Delta}^{n} \models \Psi\}.$$

PROBLEM 1. Must $\mathrm{Rep}(\mathcal{M},\Psi)$ be arithmetical?

For this problem, we may assume $\mathcal{L}$ is of finite signature. If $\mathcal{M}$ is elementarily equivalent to an arithmetical model, then we can show, via Lemma 1, that $\mathrm{Rep}(\mathcal{M},\Psi)$ is arithmetical. In particular, since $\mathbb{C}$ is elementarily equivalent to the arithmetical field of algebraic numbers, we can show that every set $\mathrm{Rep}(\mathbb{C},\Psi)$ is arithmetical. Similarly, every set $\mathrm{Rep}(\mathbb{Z},\Psi)$ is arithmetical.

Define

$$\mathcal{B}(\mathcal{M}) = \{\mathrm{Rep}(\mathcal{M},\Psi) : \Psi \text{ an } \mathcal{L}\text{-sentence}\}.$$

Clearly $\mathcal{B}(\mathcal{M})$ is closed under finite unions, and under complementation, so forms a Boolean algebra.

PROBLEM 2. Which Boolean algebras of subsets of ω are of the form $\mathcal{B}(\mathcal{M})$?

It follows from 7.2 and a remark above that $\mathcal{B}(\mathbb{Z})$ is the algebra of arithmetical sets. From 7.1 it follows that $\mathcal{B}(\mathbb{C})$ contains all constructive arithmetic sets, and is a subalgebra of the algebra of all arithmetic sets. So,

PROBLEM 3. What is $\mathcal{B}(\mathbb{C})$?

8. Possible extensions

We conclude by suggesting two ways in which the results of this paper might be extended.

8.1. Reduced powers. Let $\mathcal{D}$ be a proper filter on the index set I. Let $\mathcal{M}$ be an $\mathcal{L}$-structure. Form the reduced power $\mathcal{M}^I/\mathcal{D}$. As with the special case of direct powers, we have a canonical diagonal map $\Delta : \mathcal{M} \to \mathcal{M}^I/\mathcal{D}$. Let $(\mathcal{M}^I/\mathcal{D})_\Delta$ be the $\mathcal{L}_\Delta$-structure $(\mathcal{M}^I/\mathcal{D}, \Delta(\mathcal{M}))$.

Classical results are:

i) $$\mathcal{M} \equiv \mathcal{N} \Rightarrow \mathcal{M}^I/\mathcal{D} \equiv \mathcal{N}^I/\mathcal{D};$$

ii) $$\mathcal{M} \text{ decidable} \Rightarrow \mathcal{M}^I/\mathcal{D} \text{ decidable.}$$

(i) is direct from [3]. (ii) follows from [5, Theorem 4.10] and the fact that each Boolean algebra $2^I/\mathcal{D}$ is decidable. This follows from [1].

One should investigate analogues of (i) and (ii) for $(\mathcal{M}^I/\mathcal{D})_\Delta$, for various I and $\mathcal{D}$. We have not done this, but have a little information.

If $I = \omega$, and $\mathcal{D}$ is arbitrary, our example 5.1 generalizes to show elementary equivalence is not preserved. But we do not see how to avoid the defect that the $\mathcal{M}$ and $\mathcal{N}$ of this example have different cardinalities. We want an example in which $\mathcal{M}$ and $\mathcal{N}$ are countable.

As regards the analogue of (ii) we know little, even when $I = \omega$ and $\mathcal{D}$ is a non-principal ultrafilter. From [13] it follows that $\mathrm{Th}((\mathbb{C}^{\omega}/\mathcal{D})_{\Delta})$ is decidable. From [9] it follows that $\mathrm{Th}((\mathbb{R}^{\omega}/\mathcal{D})_{\Delta})$ is decidable, but $\neq \mathrm{Th}((K^{\omega}/\mathcal{D})_{\Delta})$, where K is the field of real algebraic numbers. $\mathrm{Th}((K^{\omega}/\mathcal{D})_{\Delta})$ is also decidable.

8.2. Permutations of I. Let I be an index set, and Perm(I) the group of permutations of I. Perm(I) acts on $\mathcal{M}^I$ by:

$$\gamma(f) = f \circ \gamma, \quad \text{for } \gamma \in \mathrm{Perm}(I),\ f \in \mathcal{M}^I.$$

It seems natural to extend $\mathcal{L}$ by adding operation-symbols corresponding to the members of Perm(I). More generally, let Γ be a subset of Perm(I). Form $\mathcal{L}_{I,\Gamma}$ by adding operation-symbols corresponding to the members of Γ. Then $(\mathcal{M}^I,\gamma)_{\gamma\in\Gamma}$ is an $\mathcal{L}_{I,\Gamma}$-structure.

If $I = n < \omega$, Perm(I) is the symmetric group S_n. We can define $\Delta(\mathcal{M})$ in $(\mathcal{M}^I,\gamma)_{\gamma\in S_n}$ by:

$$f \in \Delta(\mathcal{M}) \longleftrightarrow \gamma(f) = f \quad \forall \gamma \in S_n.$$

Thus, for finite I, we can interpret $\mathcal{M}^I_{\Delta}$ in $(\mathcal{M}^I,\gamma)_{\gamma\in \mathrm{Perm}(I)}$.

It is easily shown, along the lines of Lemma 1, that:

i) if $n < \omega$ and $\mathcal{M} \equiv \mathcal{N}$, and $\Gamma \subseteq S_n$, then $(\mathcal{M}^n,\gamma)_{\gamma\in\Gamma} \equiv (\mathcal{N}^n,\gamma)_{\gamma\in\Gamma}$;

ii) if $n < \omega$ and $\mathfrak{M}$ is decidable, and $\Gamma \subseteq S_n$, then $(\mathfrak{M}^n, \gamma)_{\gamma \in \Gamma}$ is decidable.

The obvious analogues of Theorems 3 and 4 go through.

We do not know what the possibilities are when I is infinite.

REFERENCES

[1] Yu. L. Ersov, Decidability of the elementary theory of relatively complemented distributive lattices and the theory of filters, Algebra i Logika Sem. 3 (1964), 17-38 (Russian).

[2] Yu. L. Ersov, I. A. Lavrov, A. D. Taimanov, and M. A. Taitslin, Elementary Theories, Russian Mathematical Surveys, vol. 20 (1965), 35-105.

[3] S. Feferman and R. L. Vaught, The first order properties of products of algebraic systems, Fund. Math. 47 (1959), 57-103.

[4] T. Frayne, A. C. Morel and D. S. Scott, Reduced direct products, Fund. Math. 51 (1962), 195-228.

[5] F. Galvin, Horn sentences, Ann. Math. Logic 1 (1970), 389-422.

[6] A. Grzegorczyk, Logical uniformity by decomposition and categoricity on $\aleph_0$, Bull. Acad. Polon. Sci., Sér. sci. math., astronom. et phys., 16 (1968), 687-692.

[7] A. Macintyre, ω_1-categorical theories of abelian groups, Fund. Math. 70 (1971), 253-270.

[7a] " " , ω_1-categorical theories of fields, Fund. Math. 71 (1971), 1-25.

[8] " " , On the elementary theory of Banach algebras, Ann. Math. Logic 3 (1971).

[9] " " , Ph.D. thesis, Stanford University, 1968, unpublished.

[10] Yu. V. Matijasevic, All recursively enumerable sets are diophantine, Dokl. Akad. Nauk CCCP (1970). CCCP 191, 279-282 (1970). (Russian).

[11] M. Morley, Categoricity in power, Trans. Amer. Math. Soc. 114 (1965), 514-538.

[12] A. Mostowski, On direct products of theories, J. Symb. Logic 17 (1952), 1-31.

[13] A. Robinson, Solution of a problem of Tarski, Fund. Math. 47 (1959), 179-204.

[14] C. Ryll-Nardzewski, On the categoricity in power $\aleph_0$, Bull. Acad. Polon. Sci., Sér. sci. math., astronom. et phys., 7 (1959), 545-548.

[15] R. M. Smullyan, *Theory of Formal Systems*, Annals of Mathematics Studies, No. 47, Princeton, (1961).

[16] A. Tarski and J. C. C. McKinsey, *A Decision Method for Elementary Algebra and Geometry*, Rand Corporation, Santa Monica (1948).

[17] J. Waszkiewicz and B. Węglorz, On ω_0-categoricity of powers, Bull. Acad. Polon. Sci., Ser. sci. math., astronom. et phys., 17 (1969), 195-199.

[18] J. M. Weinstein, *First order properties preserved by direct product*, Ph. D. thesis, University of Wisconsin, Madison, Wis., 1965.

SOLUTION OF PROBLEMS OF CHOQUET AND PURITZ

A. R. D. Mathias

Peterhouse, Cambridge

DEFINITIONS. A filter on $\omega = \{0,1,2,\ldots\}$ is a collection F of subsets of ω with the properties that $x \in F \;\&\; y \in F \rightarrow x \cap y \in F$ and that $\omega \supseteq y \supseteq x \;\&\; x \in F \rightarrow y \in F$. If $0 \in F$, F is improper; otherwise F is proper. F is principal or fixed if $F = \{y \mid \omega \supseteq y \supseteq x\}$ for some $x \subseteq \omega$; otherwise F is non-principal or free. If $\forall x{:}\subseteq\omega$ ($x \in F$ or $\omega - x \in F$), then F is an ultrafilter; that is equivalent to being a maximal proper filter. For $A \subseteq \mathcal{P}(\omega)$, write $\tilde{A} =_{df} \{x \mid x \subseteq \omega \;\&\; \omega - x \in A\}$. I is an ideal if $\tilde{I}$ is a filter, and I is further described as principal, proper, etc. accordingly. I is prime if $\tilde{I}$ is an ultrafilter.

Let $A \subseteq \mathcal{P}(\omega)$ and $f : \omega \longrightarrow \omega$. Write $f_* A =_{df} \{x \mid x \subseteq \omega \;\&\; f^{-1}``x \in A\}$. Then if A is an ideal, a prime ideal, a filter or an ultrafilter, so is $f_* A$. Further, write $f^{-1}A =_{df} \{x \mid \exists y \in A\; x \subseteq f^{-1}``y\}$. If I is an ideal, so is $f^{-1}I$, but it will only be prime if I is and if $\{n \mid f^{-1}``\{n\}$ has more than one element$\} \in I$.

A filter F is rare if F contains all cofinite sets and given any partition of ω into non-empty finite sets s_i ($i < \omega$) there is an $x \in F$ such that for all i, $\overline{\overline{x \cap s_i}} = 1$. An ultrafilter F on ω is a p-point if it is free and given any family $\{x_i\}_{i<\omega}$ of elements of F, there is a $y \in F$ such that $\forall i{:}<\omega\; y - x_i$ is finite. An ultrafilter is Ramsey if it is both a p-point and rare.

The term p-point arises from the following topological considerations: let βN be the set of all ultrafilters on ω, and take as a basis for a topology all sets of the form $\{F \mid x \in F\}$ where $x \subseteq \omega$. Then $F \in \beta N$ is a p-point in the sense defined above iff $\{F\}$ is not open and the intersection of countably many neighbourhoods of F is again a neighbourhood. Another formulation is that a free ultrafilter is a p-point iff given any partition of ω into non-empty pieces S_i $(i < \omega)$ there is an $x \in F$ such that for all but finitely many i $x \cap S_i$ is finite.

If $F \in \beta N$ and $f : \omega \longrightarrow \omega$, then $f_* F \in \beta N$. The <u>Rudin-Keisler ordering</u> of βN is defined by writing $G \preccurlyeq F$ iff $\exists f\ G = f_* F$. $F \preccurlyeq G$ & $G \preccurlyeq F$ iff there is a permutation h of ω with $F = h_* G$. If F is a p-point then $f_* F$ is fixed or a p-point; if F is Ramsey, $f_* F$ is fixed or Ramsey; but if $2^{\aleph_0} = \aleph_1$, then $\forall F{:}\in\beta N\ \exists G{:}\in\beta N$ G rare and $F \preccurlyeq G$. Kunen [5] has shown that $\exists F,G{:}\in\beta N\ \forall f\ f_* F \neq G$ & $f_* G \neq F$. Rudin [8] contains more information.

Choquet [2a, page 48] asked whether for every free ultrafilter F there is an f with $f_* F$ a p-point. He uses the term "ultrafiltre absolument 1-simple" for p-points. (Cf. [2b], where he also discusses Ramsey ultrafilters.) The present paper answers Choquet's question, subject to the continuum hypothesis, by proving the following

THEOREM. If $2^{\aleph_0} = \aleph_1$ then there is a free ultrafilter F such that for no f is $f_* F$ a p-point.

The proof may not be intelligible to persons unfamiliar with the foundational approach to the projective hierarchy, and it is hoped to publish a more lucid version in [6].

A subset A of $\mathcal{P}(\omega)$ is Σ^1_1 if there is an $a \subseteq \omega$ such that for all $x \subseteq \omega$,

$$x \in A \longleftrightarrow \exists y{:}\subseteq\omega\ R(a,x,y)$$

where R is arithmetical, that is, built up from a recursive matrix by quantifiers binding variables ranging over ω. By notorious tricks [9, page 174], if $R(a,x,y,z)$ and $S(a,x,y,n)$ are arithmetical, then $\{x \mid \exists y{:}\subseteq\omega\ \exists z{:}\subseteq\omega\ R(a,x,y,z)\}$ and $\{x \mid \forall n{:}\in\omega\ \exists y{:}\subseteq\omega\ S(a,x,y,n)\}$ are Σ^1_1; furthermore "there is a sequence $y_0, y_1, \ldots$ of subsets of ω" can be expressed in Σ^1_1 form by remembering that a sequence y_i can be coded by the single set $\{2^m 3^i \mid m \in y_i\}$.

<u>Examples</u> 1. For $g : \omega \longrightarrow \omega$ define $I_g =$ $\{x \mid \exists k{:}\in\omega\ \forall i{:}>k\ x \cap g^{-1}\{i\}$ is finite$\}$ and $I^r_g =$ $\{x \mid \exists k{:}\in\omega\ \exists \ell{:}\in\omega\ \forall i{:}>k\ \overline{x \cap g^{-1}\{i\}} < \ell\}$. I_g and I^r_g are both Σ^1_1 (indeed, arithmetical) sets and are possibly improper free ideals. A free ultrafilter $\mathcal{U}$ is Ramsey iff $\forall g\ \mathcal{U} \cap I^r_g \neq 0$, and is a p-point iff $\forall g\ \mathcal{U} \cap I_g \neq 0$.

2. If A is Σ^1_1 and $f : \omega \longrightarrow \omega$ then $f_* A$ and $f^{-1}A$ are Σ^1_1.

3. If α_i $(i < \omega)$ is a strictly decreasing divergent series of positive real numbers with limit 0, then $\{x \mid \Sigma_{i \in x}\, \alpha_i < \infty\}$ is a Σ^1_1 ideal.

4. If $\pi : [\omega]^2 \longrightarrow 2$, then $\{x \mid \exists y_0 \ldots \exists y_{k-1}\ \forall i{:}<k\ y_i$ is homogeneous for π and $x \subseteq y_0 \cup \ldots \cup y_{k-1}\}$ is a (possibly improper) Σ^1_1 ideal R_π. (Here $[A]^2 = \{\{i,j\} \mid i \in A,\ j \in A,\ i \neq j\}$; y is <u>homogeneous</u> for π if π is constant on $[y]^2$.) The term "Ramsey" stems from the fact (proved in [1]) that a free ultrafilter F is Ramsey iff $\forall \pi\ R_\pi \cap F \neq 0$: Ramsey's theorem asserts that each π possesses an infinite homogeneous y.

5. Put, for α a complex number of modulus 1 and ϵ a positive real number, $K(\alpha,\epsilon) = \{n \mid |1-\alpha^n| < \epsilon\}$. Then $K =_{df}$ $\{x \mid \exists \alpha_1 \ldots \alpha_n\ \exists \epsilon_1 \ldots \epsilon_n\ x \supseteq K(\alpha_1,\epsilon_1) \cap \ldots \cap K(\alpha_n,\epsilon_n)\}$ is a Σ^1_1 proper filter, by a theorem of Dirichlet. [4, Theorem 201.]

6. If A is Σ^1_1 then $\{y \mid \{x \mid x \in A\ \&\ x \cap y$ is infinite$\}$ is infinite$\}$ is Σ^1_1, and its complement in $\mathcal{P}(\omega)$ is an ideal which is proper

iff A contains infinitely many infinite subsets of ω.

DEFINITION. An ideal is <u>gaunt</u> if it is proper, contains all finite sets and is Σ^1_1. A filter is <u>gaunt</u> if its dual ideal is.

LEMMA 1. (Sierpiński) No gaunt ideal is prime.

<u>Proof</u>. A non-principal prime ideal would have to have Lebesgue outer measure 1 and inner measure 0, but every Σ^1_1 set is Lebesgue measurable. <u>Aliter</u>, the statement that a given Σ^1_1 set forms a free prime ideal is Π^1_2 and is therefore false by Shoenfield's absoluteness lemma and the fact, due to Feferman [3], that there is a Boolean extension of the universe in which the statement that there are no free prime ideals on ω has Boolean truth value $\underset{\sim}{1}$.

LEMMA 2. Let I be gaunt. There are subsets x_i $(i < \omega)$ of ω such that no $x_i \in I$, $i \neq j \rightarrow x_i \cap x_j = 0$ and $\bigcup_{i<\omega} x_i = \omega$.

<u>Proof</u>. By repeated application of Lemma 1. Say x is <u>undecided</u> by an ideal J if neither x nor $\omega - x$ is in J; and for $A \subseteq \mathcal{P}(\omega)$, write id(J,A) for the <u>ideal generated by</u> J <u>and</u> A, that is to say,

$$\mathrm{id}(J,A) = \{x \mid \exists y{:}\in J\ \exists z_0 \ldots z_k{:}\in A\ x \subseteq y \cup z_0 \cup \ldots \cup z_k\}.$$

If A is also an ideal, then $\mathrm{id}(J,A) = \{x \mid \exists y{:}\in J\ \exists z{:}\in A\ x = y \cup z\}$. id(J,A) is Σ^1_1 if both J and A are. Now for Lemma 2: by Lemma 1, there is an x_0 undecided by I. $\mathrm{id}(I,\{x_0\})$ is gaunt, and so some x_1' is undecided by it: put $x_1 = (x_1' \cup \{0\}) - x_0$. Then x_1 too is undecided by $\mathrm{id}(I,\{x_0\})$; $\mathrm{id}(I,\{x_0,x_1\})$ is gaunt and so fails to decide some x_2': put $x_2 = (x_2' \cup \{1\}) - (x_0 \cup x_1)$ ... Continuing in this manner we obtain a sequence x_i with $\bigcup_{i<\omega} x_i = \omega$ as required.

$f : \omega \rightarrow \omega$ is I-<u>infinite</u> if $\forall i{:}<\omega\ f^{-1}``\{i\} \in I$.

LEMMA 3. If I is gaunt, then there is a gaunt $I' \supseteq I$ and a $\psi : \omega \longrightarrow \omega$ such that $I_\psi \subseteq I'$; in particular ψ is I'-infinite and for each $h : \omega \longrightarrow \omega$, $\{n \mid h(\psi(n)) \geq n\} \in I'$.

<u>Proof</u>. Let $\{x_i\}_{i\in\omega}$ be a sequence as in Lemma 2. Define $\psi(n) = i$ for $n \in x_i$. Put $I' = id(I, I_\psi)$. I' is Σ^1_1 and contains all finite sets. Suppose that $\omega = x \cup y$ where $x \in I$ and $y \in I_\psi$. Let i be such that $x_i \cap y$ is finite: then $x_i - x$ is finite and so $x_i \in I$. ※ Hence I' is proper. Put $X = \{n \mid h(\psi(n)) \geq n\}$. $\psi(n) = i \rightarrow n \leq h(i)$ and so for each i, $X \cap \psi^{-1}\{i\}$ is finite, and so $X \in I_\psi$.

The last clause shows that, in the terminology introduced by Puritz [7], ψ will be in a lower sky than the identity with respect to any ultrafilter extending $\tilde{I}'$.

LEMMA 4. Let I be gaunt and f I-infinite. Then there is a gaunt $I' \supseteq I$ and a $\psi : \omega \longrightarrow \omega$ with $I_\psi \subseteq f_* I'$: in particular, for every $h : \omega \longrightarrow \omega$, $\{n \mid h(g(n)) \geq f(n)\} \in I'$, where g is the composition of f and ψ, viz. $\lambda n\ \psi(f(n))$.

<u>Proof</u>. $f_* I$ is gaunt: let ψ be as in Lemma 3, and put $I' = id(I, f^{-1} id(f_* I, I_\psi))$: in fact $I' = id(I, f^{-1} I_\psi)$.
$\{n \mid h(\psi(f(n))) \geq f(n)\} = f^{-1}``\{k \mid h(\psi(k)) \geq k\}$, whence the last part, as $\{k \mid h(\psi(k)) \geq k\} \in I_\psi$ by Lemma 3.

LEMMA 5. Let I_i $(i < \omega)$ be a sequence of gaunt ideals with $I_i \subseteq I_{i+1}$ for all i. Then $\bigcup_{i<\omega} I_i$ is gaunt.

<u>Proof</u>. $\omega \notin \bigcup_{i<\omega} I_i$; that $\bigcup_{i<\omega} I_i$ is an ideal containing all finite sets is trivial; that it is Σ^1_1 is immediate from the classical result that the union of countably many Σ^1_1 sets is Σ^1_1.

It is now easy using Lemmata 4 and 5 and the continuum hypothesis to construct a free prime ideal I such that for each I-infinite f there is an I-infinite g with $\{n \mid h(g(n)) \geq f(n)\} \in I$ for all $h : \omega \longrightarrow \omega$

and such that

for all $f : \omega \longrightarrow \omega$ there is a $\psi : \omega \longrightarrow \omega$ with $I_\psi \subseteq f_*I$.

From the second property, $\tilde{I}$ is a free ultrafilter such that $f_*\tilde{I}$ is never a p-point; from the first, there is no lowest sky in the ultrapower $\omega^\omega/\tilde{I}$, which answers a question of Puritz [7]. There is a connection between the two problems, for if f is in the lowest sky of $\omega^\omega/\mathcal{U}$, then $f_*\mathcal{U}$ is a p-point.

The theorem has been improved by Mr. R. A. Pitt of Leicester University, who has shown that

if $2^{\aleph_0} = \aleph_1$, there is a free ultrafilter F such that for no f is f_*F either rare or a p-point, and there is a p-point $\mathcal{U}$ such that for no f is $f_*\mathcal{U}$ Ramsey.

His proofs, which are presumably more "elementary" in that they do not use the notion of a Σ^1_1 ideal, will appear in his doctoral dissertation. The present author has proved both parts of Pitt's theorem using Σ^1_1 ideals (the first part after and the second part before hearing of Pitt's proofs); the key step in the proof of the first part being the following

THEOREM. No gaunt filter is rare.

The existent proof of that uses forcing: a direct proof would be welcome. It is intended that [6] shall contain a discussion of the properties of gaunt ideals and filters. Let us say that a filter F is tall if there is no infinite $x \subseteq \omega$ such that $\forall y{:}\epsilon F$ $x-y$ is finite. There are tall gaunt filters which can, assuming $2^{\aleph_0} = \aleph_1$, be extended to p-points, for instance, that dual to the ideal in Example 3, and there are tall gaunt filters which can be extended to rare filters, for example $\tilde{I}_g$ where $\forall i$ $g^{-1}\{i\}$ is infinite, but, and this is the essential fact in the author's proof of the second part of Pitt's theorem, no tall gaunt filter can be extended to a Ramsey ultrafilter; and indeed a free ultrafilter $\mathcal{U}$ is Ramsey iff it contains no tall gaunt filter. That is a corollary of the following theorem, which will be proved in [6]:

THEOREM. A free ultrafilter $\mathcal{U}$ is Ramsey iff for every Σ^1_1 set $A \subseteq \mathcal{P}(\omega)$ there is an $x \in \mathcal{U}$ such that for every infinite subset y of x,

$$x \in A \leftrightarrow y \in A.$$

REFERENCES

[1] D. P. Booth, Ultrafilters on a countable set, Annals of Math. Logic 2 (1970), 1-24.

[2a] G. Choquet, Construction d'ultrafiltres sur N, Bull. Sci. Math. 92 (1968), 41-48.

[2b] G. Choquet, Deux classes remarquables d'ultrafiltres, Bull. Sci. Math. 92 (1968), 143-153.

[3] S. Feferman, Some applications of the notions of forcing and generic sets, Fund. Math. 56 (1965), 325-345.

[4] G. H. Hardy and E. M. Wright, An Introduction to the Theory of Numbers, 4th ed., Oxford (1960).

[5] K. Kunen, On the compactification of the integers, Not. Amer. Math. Soc. 17 (1970), 299.

[6] A. R. D. Mathias, Lectures on ultrafilters (in preparation).

[7] C. Puritz, Skies and monads in non-standard analysis, Dissertation, University of Glasgow (1970).

[8] M. E. Rudin, Partial orders on the types in βN, Trans. Amer. Math. Soc. 155 (1971), 353-362.

[9] J. R. Shoenfield, Mathematical Logic, Addison-Wesley Publishing Company (1967).

SOME B. RUSSELL'S SPROUTS (1903 - 1908)

J. M. B. Moss

Mathematics Department, Manchester University, England

§0. Introduction

Between 1903, when The Principles of Mathematics [1] was first published, (the preface is dated December 1902), and 1908, in which year the first definitive presentation of the Theory of Types appeared, Russell (hereafter R) was an exceptionally prolific and creative writer, judged by the highest standards (of both). In this paper, I omit discussion not only of his nine publications during this period on the meaning of life, fiscal policy, ethics, history, pragmatism, women's suffrage, and free will, but also of those of his fundamental papers on the philosophy of logic which have little direct bearing upon the philosophy of mathematics, taking this last field admittedly in a somewhat narrow sense; I therefore exclude from present consideration the important and neglected Meinong articles, the equally important but by no means neglected On denoting, two papers on the nature of truth, and some further work on Leibniz, in order to concentrate upon R's work in the philosophy of mathematics during these years, some aspects of which have been either forgotten or developed by others in apparent ignorance of it.

In particular, I seek to develop some themes in the following papers, which were published between 1904 and 1906: The axiom of infinity [2], On some difficulties in the theory of transfinite numbers and order types [3], and Les paradoxes de la logique [4]. Each of these appeared in the course of controversies, with Keyser, Hobson and Poincaré respectively (see [8] to [15a]), and this fact may help to explain the fertility with which they explore various approaches to the formalisation of the foundations of mathematics, in comparison with the more rigid system developed later in [5] and [7].[1]

Russell's central philomathetic aim during the period under discussion was to establish in detail the identity of mathematics and logic, a thesis whose plausibility he and Frege had previously urged on a priori grounds. This work was stimulated by the discovery between 1896 and 1902 of the set-theoretic paradoxes due to Burali-Forti, Cantor and Russell himself, which show, in Gödel's words: "the amazing fact that our logical intuitions ... are self-contradictory" ([18] p 215-6). But no less important were the semantic paradoxes - some of ancient origin, though the Richard, Berry and Zermelo-König paradoxes were new, and the "heterological" paradox was first presented by Grelling and Nelson in 1908 - which R, unlike Peano (see §3 below), considered to be paradoxes belonging to logic, and they are so presented in [4]. In addition, the axiom of choice was much discussed following its formulation by Beppo Levi in 1902 and its use by Zermelo in 1904, and the problem of justifying this disputed principle of reasoning further emphasised the need for a re-examination of the foundations of logic. Consequently the 1906 articles [3] and [4] aim to explore the

---

[1]Bibliographical comment: I have found the works listed in the bibliography by Fraenkel, Ramsey, Gödel, Quine, and Wang helpful for Russellian exegesis, though only the first mentions the Keyser discussion, for which see also Church [25]. Of the contemporary reviews of [1], only Couturat's book, <u>Les Principes des Mathématiques</u>, has been useful, though Poincaré's papers, reproduced in part in his four volumes of essays, are important not only as a focus but also because of the influence of the earlier papers on Keyser (see below). Recent books by Bowne [26], Mooij [27], and Vuillemin [37] contain valuable summaries and bibliography; however there appears to be no adequate, let alone comprehensive, bibliography of work on the foundations of mathematics in this period, though one is needed.

principles lying behind the set-theoretic and semantic paradoxes and the axiom of choice, preparatory to a detailed formal presentation, as given in [7] for <u>one</u> of the approaches considered.

I wish here to urge an historical note of caution against the commonly held view that it was the discovery of the paradoxes that brought about the formalisation of the principles of mathematical reasoning. For the paradoxes were unknown until (about) 1896, whereas formalisation of logic and the foundations of mathematics had been in the air for some thirty years before this. The following points are relevant:

(i) Axiomatisation of logic and arithmetic, for its own sake, was developed in the eighteen-eighties by Frege, Peirce, Dedekind and Peano, and emphasis on the axiomatisation of particular branches of mathematics goes back at least to Grassmann (1861).

(ii) In addition to <u>specific</u> Kroneckerian doubts about some of Cantor's principles and arguments, there was felt to be a need to formulate precisely the <u>general</u> principles involved in Weierstrass's "arithmetisation of analysis". The need arose, partly to resolve such conflicts as that between du Bois-Reymond and Weierstrass as to the correctness of certain results in analysis, partly because the different naive approaches to the arithmetisation of analysis during the third quarter of the 19th century were not all obviously equivalent, and partly because of doubts connected with differing philosophical approaches to the infinite (Cantor, Kronecker and du Bois-Reymond are the important names here). Also, Dedekind's notorious argument for the existence of infinite sets points to a further motive for formalisation, in that advances in mathematics become possible before the foundational difficulties are resolved.

(iii) Further, in addition to the belief that axiomatisation was either a good thing or at least a valuable aid towards clarifying concepts and in resolving disputes about what had been established, there was a Zeitgeist in favour of effective procedures. Thus Peano, in 1890, wrote that "one cannot apply infinitely many times an arbitrary law by which one assigns to a class an individual of that class", and he therefore used instead a definite (i.e. effective) law. The notion of effectivity thenceforth became important in the Italian and French foundational work of the last decade of the 19th century (Burali-Forti, Padoa, Borel); its role can be seen very clearly in the famous Cinq lettres of 1904-5 between Borel, Baire, Hadamard and Lebesgue.

These three developments, all of which occurred before the discovery of the paradoxes, would have brought about the formalisation of the foundations of mathematics even, per impossibile, had there been no paradoxes to be discovered. However, the paradoxes and the axiom of choice were of importance during the period under discussion, although the earlier worries about different concepts of the infinite play a more central role in these developments, and in particular in R's work, than might at first appear.

§1. The Axiom of Infinity

The axiom was first introduced by Cassius J. Keyser of Columbia University in [8] and [9],[2] to the latter of which [2] is a reply; a rejoinder from Keyser appeared in the following year.

---

[2] [9] is reprinted, together with other essays, in Keyser's book [11]. Despite its somewhat Faulknerian style, I take this opportunity to commend this book to the non-mathematical reader, in view of its contemporary relevance, in Britain at least, and probably elsewhere.

Consider, for example: "the modern developments of mathematics constitute not only one of the most impressive but one of the most characteristic phenomena of our age. It is a phenomenon, however, of which the boasted intelligence of our "universalised" daily press seems strangely unaware; and there is no other great human interest ... regarding which the mind of the educated public is permitted to hold so many fallacious opinions and inferior estimates. The golden age of mathematics ... it is ours." ([11] p 274)

Keyser poses the question:

(i) "whether it is possible ... to demonstrate the existence of the infinite; whether, in other words, it can be proved that there are infinite systems"; ([11] p 157)

and he concludes, in opposition to Bolzano, Dedekind, Royce and Russell (of the Principles):

(ii) "The upshot, then, is this; that conception and logical inference alike presuppose absolute certainty that an act which the mind finds itself capable of performing is intrinsically performable endlessly, or, what is the same thing, that the assemblage of possible repetitions of a once mentally performable act is equivalent to some proper part of the assemblage. This certainty I name the Axiom of Infinity, and this axiom being, as seen, a necessary presupposition of both conception and deductive inference every attempt to "demonstrate" the existence of the infinite is a predestined begging of the issue ... But if we cannot deductively prove the existence of the infinite, what, then, is the probability of such existence? The highest yet attained. Why? Because the inductive test of the axiom, regarded now as a hypothesis, is trying to conceive and trying to infer, and this experiment, which has been world-wide for aeons, has seemed to succeed in countless cases, and to fail in none not explainable on grounds consistent with the retention of the hypothesis." ([11] p 161-2)

In his reply, R renounced the Bolzano-Dedekind type of justification espoused in [1] §339, and added:

(iii) "... there are no new axioms at all in the later part of mathematics, including ... ordinary arithmetic and the arithmetic of infinite numbers. Professor Keyser maintains, on the contrary, that a special axiom is covertly involved in all attempted demonstrations of the existence of the infinite."

R then outlines a proof, along Fregean lines, based on a class of classes definition of a natural number, of

"the existence of a number which is the number of finite numbers, [whence] it follows that this number is infinite. Hence, from the abstract principles of logic alone, the existence of infinite numbers is rigidly demonstrated. ... Accepting the five postulates enumerated by Professor Keyser ... as assumed by Dedekind, I deny wholly that any of the five presupposes the actual infinite. It is true that they together imply the actual infinite; it is indeed their purpose to do so. But it is too common ... to confound implications with presuppositions..." ([2] p 810).

To which Keyser, unconvinced, replied:

(iv) "to prove, no matter what, is to use infinitude. Hence to try to prove that there is infinitude involves trying to prove that proving is a possible thing. The nature of the circle is evident." ([10] p 382)

Later, of course, R postulated the existence of an infinite set as an axiom, which status it retains in systems of type theory and cumulative rank structure (c.r.s.) set theory, though not in those foundational approaches that aim to modify Frege's inconsistent system, such as those developed by Quine. Historically, therefore, Keyser has been substantially vindicated; but conceptually the matter is less clear, though it is plausible that there are principles, whether or not the axiom of infinity be one, which are true, evident, and unjustifiable in an absolute sense.

The following consideration might help to clarify what is at issue. Keyser's question (i) is not whether or not there are infinite sets or classes, but rather (cf. [1] §339) whether or not there is an infinite number of mathematical objects. The quotation (iv) and the remarks in [8] in support of Poincaré's thesis that the principle of mathematical induction is synthetic a priori (see [15] and also [14]) establish this beyond doubt. However, R, in (iii) above, claims to establish (what he knew to be) the stronger thesis that there exists an infinite class; he would surely not have disputed the metalinguistic claim that "proving is a possible thing", in the sense that there is no upper bound to the length of possible formal proofs. Poincaré in fact appears to have overstated his case in claiming that in arithmetic "the idea of mathematical infinity already plays a preponderating part"

([14] p 11), at least if a distinction is drawn between the (so-called) potential and actual infinite. For the equivalence (Zermelo 1908 and Ackermann 1937) between Peano arithmetic and the theory of finite sets[3] shows that the stronger claim, that arithmetic is committed to the existence of the 'actual' infinite, is false. However, the weaker claim that it is committed merely to the potential infinite, though vague, can be made unexceptionable. Indeed the fact that the metatheory of propositional logic is of the same strength as arithmetic can be seen to give a precise sense to the hitherto vague notion that the concept of the (potential) infinite is presupposed by all reasoning.

Since the Keyser-Russell debate, there has been a deep division between logicians as to whether or not foundational problems arise from the obscurity (or worse) of the concept of infinity, or from some other source. R has at different times been pulled both ways in this conflict, but in 1906 his general sympathies still lay with his remark quoted above in (iii); note in particular his comment "the contradictions have no essential connection with the infinite" ([4] p 633), in reply to Poincaré's "There is no actual infinity. The Cantorians forgot this, and so fell into contradiction." ([13] p 195)

The central task of the foundations of set theory is to determine which classes are, or correspond to, objects, and the common theme running through the papers [2], [3] and [4] is, not always explicitly, whether or not there is the logically relevant difference between finite and infinite sets that Poincaré, for example, claimed. In the final paragraph of this paper, I shall return to this theme in the light of some of the arguments suggested by Russell's two 1906 papers.

---

[3] Strictly speaking, there are two equivalences, of first and second order arithmetic to the first and second order theories of finite sets, respectively.

§2. On some difficulties in the theory of Transfinite Numbers and Order Types [3]

A. The set-theoretic paradoxes

The first two sections of [3] constitute a masterly discussion of the set-theoretic paradoxes, through the exploration of alternative foundations for set theory. R formulates a general result, quoted below, of which the three standard paradoxes (Burali-Forti, Cantor and Russell) are instances, and presents three approaches to their solution, which he calls the zigzag theory, the theory of limitation of size, and the no classes theory. In the text of the paper, he tentatively adopts the first, which can also be found in [1], and rejects the second,* though a note added in proof on 5 February 1906 ([3] p 53) comes down in favour of the no classes theory. As a plug for philosophy, he remarks that "the complete solution of our difficulties ... is more likely to come from clearer notions in logic than from the technical advance of mathematics."

According to the no classes theory, no sentential function determines a class because there are no classes; what there are instead is considered below in §3 B. However the general theme of [3] is that some but not all sentential functions determine classes, and that it is the proper business of logic to determine which of them do. Consequently, the alternative frameworks for set theory discussed in [3] are the zigzag theory and that of limitation of size. According to the former, sentential functions determine classes when they are "fairly simple" but may fail to do so "when they are complicated and recondite" ([3] p 33); it is not therefore bigness "that makes a class go wrong," but such complicated sentential functions "as might well be supposed to

---

*"[It] has, at first sight, a great plausibility and simplicity ... but [these qualities] tend rather to disappear upon examination." ([3] p 43)

have strange properties". Russell's further remarks show that Quine's system NF fits essentially what he had in mind, but this system is in certain ways now known to be unsatisfactory. It can however be replaced, once a distinction is drawn between sets and classes, by the system ML, also due to Quine, which because of its greater means of expression appears to be free from these objections. The zigzag theory "is that assumed in the definition of cardinal and ordinal numbers as classes of classes" ([3] p 39), for these classes of classes, except for zero (i.e. $\{\Lambda\}$), must contain universe-many members, and only systems resembling the zigzag theory in this respect can permit such a definition of number, based as it is on the Principle of Abstraction (discussed below). Also, the system ML, if it is consistent, is committed to the existence of denumerable proper classes,[5] which, as will be seen below, is of particular importance in connection with the semantic paradoxes.

The limitation of size theory, first formulated, unknown to R, by Cantor in his 1899 letter to Dedekind,[6] is the basis of the c.r.s. theories for which axioms were proposed by Zermelo and von Neumann. (An axiom of von Neumann for a strong version of the limitation of size theory states that a class is a set if and only if it is smaller than the class of all sets.) The theory therefore constitutes a development of the idea that sets are constructed "from below" ([3] p 44), and its basic idea is well conveyed by the following recent account: "We start off with certain objects which are not sets and do not involve sets in their construction. We call these objects <u>urelements</u>. We then form sets in successive stages. At each stage we have available the urelements and the sets formed at earlier stages; and we form into sets

---

[5] Specker told me of this result in 1966. I have never seen it stated in print.

[6] Translated in [28].

all collections of these objects. A collection is ... a set only if it is formed at some stage in this construction." ([29] p 238). The existence of a set is therefore dependent upon the existence of its members, each of which depends upon some previous construction - note that constructions are not assumed to be effective. A stage is determined by an ordinal number, and the only significant feature about the urelements is (usually) their cardinality.

The general formulation of the set-theoretic paradoxes runs ([3] p 35):

> "Given a property ϕ and a function f, such that if ϕ belongs to all members of u, f'u always exists, has the property ϕ, and is not a member of u; then the supposition that there is a class w of all terms having the property ϕ and that f'w exists leads to the conclusion that f'w both has and has not the property ϕ."

Applied therefore to Russell's paradox, for which f'u = u, the only possible conclusion is that not every sentential function determines a class, since there is no class of all classes that are not members of themselves. (R does not distinguish between sets and proper classes.)

For the other paradoxes (see Wang, [22] for further discussion), there is a choice between denying the existence of the function f or of the class w. However, in the case of the Burali-Forti paradox, for which ϕx is 'x is an ordinal' and f'u is 'the ordinal of u' whenever u is a von Neumann ordinal, i.e. a segment set of ordinals, R concludes that since the function f exists, there can be no class of all ordinals, according to both the zigzag and the limitation of size theories.[7] In

[7] The concepts of ordinal and cardinal are not definable in NF ([24]), which suggests that the intentions of the zigzag theory might be better realised by denying the existence of the function f, as defined above. It is, however, more natural to distinguish sets and classes, since the concept of ordinal can be defined without difficulty in ML.

[3], Cantor's paradox is not presented in the framework of the general result, though this can easily be done ([22] p 11 with a small modification), and the conclusion to which R is led is that there is no class of all cardinals. (See however footnote 7.)

In general, therefore, it is more natural to deny the existence of the class w than of the function f, unless a grammatical distinction is drawn between sets and classes, in which case both f and w exist as classes, although the latter, at least, is not a set. This observation, however, does not help to explain <u>why</u> certain classes are illegitimate, nor is any explanation offered by R's characterisation of the paradoxes as arising from "the fact that, according to current logical assumptions, there are what we may call <u>self-reproductive</u> processes and classes" ([3] p 36). It is appropriate therefore to look more closely at the competing systems of set theory in order to attempt to resolve which of them, if any, could be correct - a meaningful programme for non-formalist foundations.

I propose to consider below the zigzag and limitation of size theories, assuming that they are consistent, in connection with the following five questions, to each of which they give formally incompatible answers, in an attempt to formulate sharply the most important respects in which they are opposed:

(1) What account is to be given of cardinal and ordinal numbers?

(2) What justification can be given for the existence of countable and other infinite sets?

(3) Regardless of philosophical issues, which version of set theory, if any, is either factually or conceptually adequate for the foundations of mathematics?

(4) Does a class depend for its existence upon considerations

of constructibility (assuming that this is not true by stipulative definition)?

(5) Finally, how are the semantic paradoxes to be resolved?

(1), (2) and (3) are discussed in the present section, and (4) and (5) in B and C of §3 respectively.

(1) Cardinal and ordinal numbers and the principle of abstraction. That numbers, either cardinal or ordinal, are really sets, is, prima facie, not at all plausible. The notion of number appears to be both epistemologically and genetically more primitive than that of set. Also, if numbers are sets, there must subsist a unique answer to the question: which set is the number 3? However, no non-arbitrary answer to this question can actually be given (see Benacerraf [30] for an elaboration upon this theme). Consequently, neither the standard version of the limitation of size theory, according to which each cardinal or ordinal number is a particular set, e.g. the set of its predecessors, nor the zigzag theory, according to which every number is a class of classes, could be a correct account of the cardinal or ordinal numbers. Indeed, Frege's insight that cardinal numbers are essentially connected with quantifiers and that (cardinal) numerosity is a feature common to similar properties (and so in effect a property of classes), and Russell's further claim ([1] §231) that ordinal numerosity is a property of sequences, suggest that no extensional account of numbers, and therefore no account of numbers in terms of classes, could be fully correct. There remains, however, more to be said, since the limitation of size theory, as remarked earlier, presupposes the notions of cardinal and ordinal number to be antecedently understood, and to define numbers as sets is thus a merely technical device to simplify the vocabulary and presentation of the theory; surely no-one who 'believes in' sets believes that it is true that $17 \in 18$. (For further discussion, see [31].) Also, the classes of classes definition of

numbers of both [1] and the zigzag theory depends upon the principle of abstraction, which states that for any equivalence relation S, there is a set of just those objects equivalent under S. Hence, if S is an equivalence relation,

$$(\forall x)(\exists y)(\forall z)(z \in y \equiv xSz).$$

This is a natural principle to accept for classes considered as extensions of properties, but it is not generally true for limitation of size theories, since y might be too large. For example, if S is either the relation of cardinal equivalence or of ordinal similarity, y would with one exception have the cardinality of the universe. Remarks on p 39 of [3] suggest that R regarded the principle of abstraction as the most powerful objection to any limitation of size theory; its acceptance would in fact yield a formal refutation of most versions of that theory.

What arguments can be given in support of the principle? It can be said that _if_ a feature common to properties or classes is (to be construed as) a common property, and therefore represented extensionally by means of the abstraction class, and _if_ such a class is a possible _object_ of mathematical thought, then the abstraction principle follows. I have attempted to justify the second of these assumptions in the discussion of question (4) (on p 221) in §3, but justification of the first depends upon the most natural way to formalise Frege's insight presented above, a story I have unfolded elsewhere [32].

(2) _Infinite sets_. The existence of infinite sets (and stages) needs to be postulated for limitation of size theories. A stage, or rank, is (represented by) an ordinal, and to justify the existence either of sets of various infinite ranks or of these ranks themselves, more powerful axioms of infinity may be needed. The procedure is

currently familiar. However attempts to justify the introduction of these axioms are not easily to be found (Mostowski [33] is a distinguished exception) though for any logician who believes that there is a vast logical gulf between the finite and the infinite, such justification would appear to be appropriate. It may be felt that the existence of countable sets can be justified by the existence of a countable number of mathematical objects (e.g. numbers) or ideal objects (e.g. Dedekind's thoughts of thoughts of ...), but this is insufficient as such arguments omit to explain why a property with a countable number of instances can be objectified (see [1] above, and also §9 of [31]). However, since it is assumed by the theory as the fundamental principle of set construction that sets exist whenever previously constructed objects are collected together, it would appear to be more natural to restrict attention to predicative set theories, with predicatively definable ranks (of the constructible universe), though the standard c.r.s. theories entail the existence of sets that cannot be specified by predicative means. As for the higher orders of infinity, I know of no justification offered for e.g. the assumption of measurable cardinals other than: "if there are no obvious reasons why all sets should have the property P, we adjoin to the axioms an existential statement to the effect that there are sets without the property P" ([33] p 85).

The zigzag theories, as developed by Quine, contain, as R envisaged, theorems of the existence of certain infinite sets derived from axioms of quite general form - indeed, NF and ML each contain only one axiom schema of set existence, which asserts (roughly) the existence of sets whose defining condition is stratified. Rosser, however, proved that the class of natural numbers is not a set in ML, if it is consistent, and further axioms are needed for ordinal number theory (Orey) and for analysis. Orey has also proved that the axiom of counting, though evidently true, is not provable in NF, if it is consistent. Hence although an axiom of infinity is provable in NF and

ML, the situation is by no means satisfactory, and it is no longer easy to believe that "there are no new axioms at all in the later part of mathematics".

In connection with the relationship between the finite and the infinite, which appears to be difficult to tackle directly, a remark by Wang is relevant: "Russell's reluctance to treat finite classes and infinite classes in basically different manners is closely tied up with the wish to identify logic with mathematics. Since infinity is central to mathematics, a theory which blames the paradoxes on the peculiarity of infinite classes would cast doubt on the claim that logic contains the full richness of mathematics." ([22] p 22 - see also pp 25 and 27) This however appears to be the wrong way round; what is emphasised above is that <u>because</u> the concepts of the finite and the infinite are equally in need of clarification, and <u>because</u> of what is shown by the paradoxes and the considerations relating to the principle of abstraction, one is thereby led to develop a theory in which the existence of an infinite set is a non-trivial theorem. Further, Wang's doubts, with respect to the identity thesis (of logic and mathematics), as to whether: "the logic which is true for finite sets automatically applies also to infinite sets" ([22] p 25) suggest a difference between e.g. languages with sentences of arbitrary finite length and those with sentences of infinite length not easily reconciled with the need to use the latter to give a first-order characterisation of arithmetic notions.

(3) <u>The adequacy of set theories</u>. With regard to limitation of size theories, Zermelo's aim, as presented in his fundamental paper of 1908, was to provide a system of axioms sufficient to establish set theory "as it is historically given" (see [27] p 200), so as "to develop thereby the logical foundations of all of arithmetic and analysis," - in short, to provide foundations for the working mathematician. At the present time, the question of adequacy vis-à-vis mathematics arises

most naturally for category theory, in connection with which Feferman [34] has proposed the Hypothesis: "Every theorem of current category theory is provable, or has an adequate version which is provable in" a conservative extension of ZFC,[8] whose language contains an additional primitive symbol designating the universe of small sets. However, this hypothesis can only be established if it is known what modifications in the statement of a theorem are permissible in the "adequate version which is provable". The difficulty arises from what is from a naive point of view the possibility of ascribing properties, and therefore membership in other classes, to some Chang classes, i.e. classes which contain universe-many members; and even if the difficulty in the last sentence be resolved, it seems unnatural to have to claim that sentences apparently about Chang classes are not really about them at all.

Zigzag theories have not been shown to be vulnerable to the above objection, though this may be because little attention has been paid to them in recent years. As mentioned above, the systems NF and ML proposed by Quine are not sufficiently strong to serve as foundations for the whole of mathematics, and no very natural way to strengthen them appears to have yet been proposed. More generally, the semantics of zigzag theories is not well understood at the present time, so despite the attractive features of ML of containing (i) some Chang sets and some countable proper classes; (ii) an axiom of infinity provable as a theorem; and (iii) the principle of abstraction (cf [35] p 340 f), it has not so far proved to be the most attractive set-theoretic foundation for mathematics.

One general note of caution about set theory should be urged. Kreisel has remarked, in an appendix to [34], that such notions as constructive rule, abstract structure, and abstract property "present

[8] ZFC = ZF + the axiom of choice.

serious problems for set-theoretical foundations if we seriously wanted to reduce these notions to set theory" (p 243). Indeed, behind any difficulties that arise in connection with particular notions, it can be shown generally (see [31] and [36]), that not all mathematical objects are sets. The formal implications of this do not appear at present to be well understood.

B. The Axiom of Choice

The third section of [3] is concerned with the Axiom of Choice, called Zermelo's axiom. Though still a valuable exposition, it lacks the importance of the earlier parts of the paper, in part because the equivalence between the multiplicative axiom that Russell formulated, and the "generalised form" of the axiom, which R believed to be doubtful, was in fact shown by Zermelo in his two 1908 papers (see [28]). One point, however, which might merit further analysis is that R expresses doubts as to the existence of the uncountable infinite; there is, he argues, "no ground for thinking that there are classes of finite numbers which are not definable by a formula" ([3] p 52).[9] Although the emphasis upon definability is much less pronounced than in the whole-hearted constructivism of Borel ("such reasonings lie beyond mathematics") and the French school, R believed that neither form of the axiom of choice is provable, though either may be disprovable. The axiom is, he supposed, unlikely to be true without some restriction, but "the more we restrict the notion of class, the more likely ... Zermelo's axiom is to be true" ([3] p 52). Together with the emphasis on definability, this appears to suggest that R thought the axiom might

[9] Compare also the much neglected thesis in [1] §141, referred to in [7]: "Whatever we can apprehend must be of finite complexity." ([7] p 50) It appears likely that this thesis could have an important role to play within the epistemology of mathematics, particularly in connection with the significance of the Löwenheim-Skolem theorem.

hold within a predicative version of set theory, though he would presumably have welcomed Specker's 1953 result that it is refutable in NF - this depends, however, on the inability of NF to express intuitive notions, and the result does not extend to the more satisfactory ML.

Historical note. The axiom was implicitly formulated in 1890 by Peano, who rejected it, (see the quotation on p 214 above), reformulated in 1901 and 1902 by Beppo Levi, and was suggested by Erhard Schmidt to Zermelo for the proof of the well-ordering principle. However, (see [3] p 49 n 1), Levi's proposal arose from an attempt to prove a version of an axiom proposed by Burali-Forti in 1896 to establish the coextensionality of the finite and the Dedekind-finite. Moreover, Burali-Forti's axiom is only one of a number of alternative axioms proposed before 1904 to legitimise, or at least clarify, dubious principles of reasoning. Russell's axiom 4.3 in [6] is another axiom apparently formulated before the axiom of choice; the question of its equivalence with the axiom of choice may still not have been decided.

§3. Les paradoxes de la logique [4]

The main topics discussed in [4] are the semantic paradoxes, the no classes theory (also mentioned in [3]), and the Vicious Circle Principle (V.C.P.). The paper also contains the first formulation of the axiom of reducibility, and some stimulating philosophical asides about the existence of propositions and their identity with facts. 'Ranges of significance' are mentioned but discussed only in connection with the illegitimacy of a genuine universal quantifier, which is a recurrent theme in [4] and [5], though it barely survives in [7]. The paper begins with a useful methodological discussion in support of an inductive justification of principles of logic. I have no space to discuss this, and proceed therefore to consider the three central topics of the paper.

A. <u>The V.C.P., predicative set theories, and the axiom of reducibility</u>

In a sequence of three papers in the Revue de Métaphysique et de Morale, Poincaré had proposed that "the definitions that must be regarded as non-predicative are those which contain a vicious circle" ([13] p 190-1), though he does not there demarcate the class of definitions to be excluded; the expression "non-predicative" had previously been used by R in [3] to apply to one-place predicates that fail to determine a class, without reference to any prohibition on vicious circles. In the above papers, Poincaré conflates his objections to the use of the 'completed' infinite and to non-predicative definitions, but in [4] R discusses them separately, claiming that neither the paradoxes nor impredicative definitions essentially involve the notion of infinity (see p 217 above).

He agrees, however, with Poincaré as to the importance of the V.C.P., which he reformulates "what contains an apparent variable cannot be a possible value of that variable" ([4] p 643-4), and about which he makes the following two claims: firstly, that only some version of the no classes theory can prevent vicious circles; secondly, that "the vicious circle principle is not itself the solution of the paradoxes, but only a consequence that a theory must supply to lead to a solution. In other words, one must construct a theory about expressions containing apparent [bound] variables which implies the vicious circle principle as a consequence. A reconstruction of the most basic logical principles is therefore needed, and we cannot remain content with the simple fact that the paradoxes arise from vicious circles" ([4] p 640-1).

The new theory of first-order logic thus required would, however, apparently depose the part of logic which is most secure. This will not be pursued here, as the systems of predicative analysis and set theory mentioned below have usually been formulated within classical logic, though more attention should perhaps be directed to the non-standard systems proposed e.g. by Church in the early thirties, in which there is no proper universal quantifier, and by Fitch after the war, in which the law of excluded middle is rejected, thought not on intuitionist grounds. A more fundamental difficulty is that the V.C.P. is a restrictive principle, and any theory which entails it must be at least as restrictive. But, as was soon discovered, standard mathematical theorems such as the least upper bound theorem, the Cantor-Bendixson theorem, and, in a certain sense, even the principle of mathematical induction, are essentially impredicative, and hence could not be theorems of any theory that entails the V.C.P.. This is a crucial difficulty for predicative foundations, if extant mathematics is taken as given and not as possible nonsense in need of Procrustean trimming, although predicative systems continue to play an essential role in determining the nature and extent of impredicative reasoning within classical mathematics.

R's immediate reaction, in [4], to the difficulty over mathematical induction caused by the prohibition on bound variables whose values include properties definable by quantification over all properties of numbers, was to postulate the principle now known as the axiom of reducibility, viz., that any property is equivalent to a first-order (so-called predicative) property, specifiable without the use of bound variables ranging over totalities not yet constructed. He is therefore led to claim (see C below) that the semantic paradoxes are due to intensional features ([4] p 648); but more important for the present discussion is that the axiom of reducibility, as R admitted ([5] §5 and [7] pp 68ff) is an axiom of class (i.e. set) existence, which allows for the existence both of sets which lack defining conditions and of sets which cannot be defined without violating the V.C.P.. Hence from a predicative point of view, or that of the no classes theory, the axiom of reducibility is clearly false, unless the notion of definability is extended to allow for the possibility of infinitely long definitions (cf. [18] p 225).

Even within a predicative framework, the ramified theory of types, based as it is upon a hierarchy of orders determined by syntactic complexity,[10] has been felt to be too restrictive, but an important liberalisation, due to Wang ([23] Chs. 23 and 24) constructs a sequence Σ of cumulative extensions of the original theory, extended to transfinite levels. Also, predicative formulations of the limitation of size theory have been given by Wang ([23] Ch. 25, and references given there - see also [22]), and more recently by Feferman; I do not know if these two systems are equivalent, nor how either of them relates to the systems Σ. A predicative version of zigzag theory is mentioned by Quine in [20] p 125. In addition to these predicative versions of set theory, predicative analysis, originated by Weyl and developed recently by Kreisel, Feferman, and Schütte, has led to a fuller understanding of the notion of predicatively provable sentence of analysis, and to partial clarification of that of predicatively definable set of numbers.

This technical progress, however, notwithstanding its foundational importance, has not yet succeeded in clarifying some general issues. Firstly, there is the question as to which inductive definitions are legitimate, when considered from a constructive point of view. Secondly, the recent work, in which the existence of the set of natural numbers and the principle of mathematical induction are both "presup-

---

[10] Reference is sometimes made to the double hierarchy of types and orders to which the ramified theory is said to be committed. However, the assumption of ranges of significance in [4] is weaker than that of types, since such ranges need not be exclusive. Moreover, though the theory of orders naturally suggests the use of ranges of significance, it is not committed to them (see Wang [22] pp 9-10 and Quine [21] last paragraph of §34). Essentially therefore the ramified theory is a theory of orders and not of types.

posed", yields more powerful results than ramified type theory as first proposed. Since the set of natural numbers is not definable without quantification over all properties, the legitimacy of this assumption may be questioned, if predicative <u>definitions</u> are emphasised. More fundamentally, the problem remains of giving an exact formulation of the V.C.P.. Recently, use has been made of Poincaré's semantic adequacy criterion of 1909 that a predicatively defined set is one that is not "disordered by the introduction of new elements" ([15a] p 47), i.e. by enlarging the universe, an idea also employed in connection with the so-called <u>basis theorems</u>. However, R's original proposal was that a predicative classification be specified through a syntactic condition, and insofar as predicative <u>definition</u> is concerned, such a syntactic account would appear to be essential. Later, in [7] pp 37-9, R gave (implicitly) three seemingly different accounts, of predicative definition, presupposition, and involvement, of which the two last, which may turn out to be identical, alone seem close in spirit to Poincaré's criterion (see [18] for a full discussion).

To illustrate the difference between these various accounts of predicativity, consider the confusion in the literature as to whether the existence of the union set or of the power set is responsible for impredicative set formation in limitation of size and type theories. It is easily seen that given a set x the existence of $\cup x$ is syntactically impredicative for non-cumulative theories, but not for cumulative ones, and it raises no semantic problems. On the other hand, the definition of $\mathcal{P}x$ is syntactically unobjectionable, since

$$(\forall x)(\exists y)(\forall z)(z \in y \equiv (\forall w)(w \in z \supset w \in x))$$

is a correct definition, as can be seen by inserting type indices. However, the existence of $\mathcal{P}x$ cannot be assumed for the purpose of defining subsets of x, and it is in practice more useful to work with the set of predicatively definable subsets of x. The real culprit, of course, is the comprehension axiom.

The system of the ramified theory of types, with the axiom of reducibility added, is equivalent in extensional contexts to what is now called the simple theory of types, as Ramsey, Chwistek, and Quine first observed. In this, as currently formulated, there is no distinction of orders, which in the ramified theory are based upon the syntactic quantifier complexity of predicates, but instead a division of all things in the universe (on one interpretation) or of all linguistic expressions (on another interpretation)[11] into exclusive ranges of significance. However, since the axiom of reducibility, as mentioned above, is an axiom of set existence, it follows that the simple theory of types is committed to the existence of sets in just the same way, though not to the same extent since type theory is weaker than Zermelo set theory, as any other theory of sets. There is therefore no basis for characterising it as a no classes theory.

Any version of type theory, considered in other than a purely formal way, must have either an ontological or a linguistic basis. Taken in the former sense, the supposed ontology, which the theory reflects, has no bearing upon the meaningfulness, as ordinarily understood, of sentences, and it is in fact plausible to suppose, as Fitch in particular has argued, that some classes belong to themselves and some properties apply to themselves. Also, it is difficult to believe in the infinite reduplication of the natural numbers, and everything constructed from them, to which the theory is committed. Finally, if the theory were true, it could not, in a precise sense, be coherently stated, because the metatheory would also reflect the background ontology in a similar way to the theory.

---

[11] Suggested in [7] p 161.

Considered as a grammatical metalinguistic thesis, certain intuitively meaningful expressions, as is well-known, are adjudged to be not well-formed. This limitation on what can be expressed yields good reasons for not adopting this version of the theory, since other meaning stipulations are formally simpler, or, by realists, judged to be more in accord with the way things are.

The theory of types is consequently either false or misleading.

B. The no classes theory and the meaning of class existence

The no classes theory, which R invented in order to resolve the paradoxes, is summarised by him: "classes are simply linguistic or symbolic abbreviations" ([4] p 636). Classes, like definite descriptions, are therefore non-denoting symbols,[12] and the technical problem immediately arose of justifying those results which had previously been obtained using substantial class-existence assumptions; in [4], R aimed to solve this by a rather dicey principle, to the effect that there is always something else (p 638), of which he claims that it implies the existence of all cardinal numbers less than $\aleph_\omega$. There follows a polemic against quantifiers ranging over all entities, in which he remarks that $(\forall x)\phi x$ does not imply $\phi((\forall x)\phi x)$, and that there are no general propositions; in particular the law of excluded middle is not a proposition, since its expression contains a bound variable purporting to range over all propositions. The word "true", he claims, changes its meaning with every change in the number of bound variables in a sentence, a suggestion developed by Ramsey in the latter's solution of the semantic paradoxes (see C below).

---

[12]However, class terms fail to denote in a somewhat different way from definite descriptions, since the latter, unlike the former, sometimes have a reference.

The important interpretational problem is to discover what, according to the no classes theory, are intended to replace classes. Two apparently quite different answers to this are to be found in the literature: there is a nominalist interpretation, presented by Gödel in [18], though apparently formulated by Hahn (in [38] which I have not seen), which interprets R as denying the existence of both classes and properties in favour of linguistic predicates. And there is also a realist view, presented (critically) by Quine, which rejects the interpretation just given ([19] p 21 and [20] p 122), and which takes issue with R for replacing classes by the nebulous slum of properties. There would, however, be no incompatibility between these two interpretations if quantifiers over predicate variables, the determining factor in Quine's view, could be read substitutionally; on this account the no classes theory would be nominalist, in accord with the quotation from [4] on p 25 above, since predicate quantifiers interpreted substitutionally yield no concessions to realism - for a fuller discussion of the role of substitutional quantification for a predicative theory of classes, see Parsons [39]. Moreover, the natural development of the no classes theory must be along the lines suggested by Chwistek and Myhill, of systems in which all classes are nameable and in which, therefore, unless infinite names are allowed, the universe must be denumerable; the simple theory of types, however, assumes, as mentioned above (p 22), the existence of uncountable and undefinable classes.

A further problem for the no classes theory is the status of its lowest level objects and primitive predicates, since R's assumption in [7] of an infinite collection of empirical <u>urelemente</u> is clearly not a satisfactory foundation for mathematics. This will not be further considered here, though clarification might be expected to come through the development of modalised versions of set theory; alternatively, Parsons has an interesting proposal in [40] of a modal foundation for arithmetic, deriving (apparently) from the study of Kant.

There remains the problem of the nature of class existence assumptions, mentioned (as (4)) on p 221 above. On the no classes theory, there is no problem. Sentences containing either purported class names or purported class descriptions alike mislead as to their logical form. On the limitation of size theory, finite classes only can be named, though many infinite classes can be described; it is sometimes consistent to assume that all classes can be (Myhill [41]). However the intended von Neumann model, referred to on p 219 above, assumes unexpressible sets and classes, with the distinction between sets and proper classes drawn solely on grounds of size, regardless of expressibility. Consequently, classes describable by apparently intelligible conditions may turn out to be proper classes, e.g. the class of one-membered sets - call it U - and therefore lack properties themselves, since they cannot belong to other classes. This is counter-intuitive, as the meaningfulness of predications about a class such as U should not depend upon its extension, which could have been difficult to discover. This extensional view of classes appears therefore to lead to the conclusion that the intelligibility of the description of an infinite class has no bearing upon the question of its set existence.

On the other hand, according to the zigzag theory, the existence of a set depends upon the syntactic structure of some expression that describes it. Though this may at first sight appear unnatural - it goes against the deeply held belief that ontology is prior to language - it is in fact a highly desirable feature. For, assuming the usual (ω,ω) language, the existence of sets is made dependent on there being some satisfactory means of characterising them, as by a stratified condition (see Quine [20]). And since our only purchase on infinite sets is through the limited range of procedures obtainable by reflection upon a small finite number of fundamental notions, it is to be expected that there should be countable classes which, because we cannot describe them without circularity or regress, cannot be talked about. Hence this intensional feature of the zigzag theory yields the basic clue to the problem of class existence; it is an insight partially anticipated by R in [4] in connection with the no classes theory - that it is how one talks about what classes would be if they existed that is important - ironically after he had renounced the zigzag theory.

C. The semantic paradoxes

A common view of the semantic paradoxes is that expressed in Peano's much quoted Interlingua remark of 1906: "Exemplo de Richard non pertine ad Mathematica, sed ad linguistica" (quoted from [17]; the use of capitals is notable). On Peano's view, therefore, such paradoxes as the Liar, Richard's, and the heterological one, are not a proper concern of the logician, but belong to linguistics, or, to use Ramsey's word, to epistemology. However the reservation that Ramsey, here following Chwistek, expressed about this approach has often been forgotten: "The only solution which has ever been given, that in Principia Mathematica, definitely attributed the contradictions to bad logic, and it is up to opponents of this view to show clearly the fault ... to which [they] are due" ([17] p 21). (The phrase "the only solution" is surprising in view of Chwistek's discussion, with which Ramsey was familiar.)

The consequence has been that discussions of the semantic and set-theoretic paradoxes have usually been separated in the recent literature, to the disadvantage of the former. A remark by Tarski in 1931 characterises the situation perfectly: "Mathematicians, in general, do not like to operate with the notion of definability; their attitude towards this notion is one of distrust and reserve" ([42] p 110), and in spite of Tarski's own work, the situation has hardly changed. The historical and cultural background to this separation between mathematics and (what was seen as) philosophy is a fascinating story which cannot be told here; it goes much further back than the attitude of Hilbert and his school, important though that was. But rationally, at least, the following <u>three</u> arguments should be decisive:

(i) From a naive, man-on-a-Belsize-Park-omnibus point of view, logic is essentially the study of such paradoxes as Berry's and the Epimenides (see, for example, de Quincey's discussion of the latter in his article on Sir William Hamilton), and the arguments that lead to paradox are ones that the layman expects a mathematical logician to be able to clarify for him. Russell's paradox, however, and those of Burali-Forti and Cantor, involve the unfamiliar mathematical notion of class, and can therefore be left to set-theorists. Though such arguments about Anglo-Saxon usage can prove too much, the concepts of truth and falsity, meaning, and definability, must surely be held to fall within the province of logic, and semantic notions deserve to be, and are capable of being, studied with the same precision as set-theoretic ones.

(ii) The difficulties of resolving the set-theoretic paradoxes (see §2) suggest that to follow R in considering the two sorts of paradox together is likely to lead to more insight than treating them separately. The conclusions below support this conjecture, as even if there is no unique solution to the paradoxes, the partial insights (formal systems) developed for one sort of paradox correspond with and illuminate those developed for the other.

(iii) The two kinds of paradox are structurally similar; compare the following development of the heterological paradox with the well-known presentation of Russell's paradox by Rosser ([35] p 202):

<u>Additional logical symbol</u>: des
("des" is a two-place relational symbol intended to hold between predicates and the properties that they designate.)

<u>Definition D</u>: N is het(erological) $=_{df}$ $(\exists\phi)(N \text{ des } \phi \,\&\, \neg\phi(N))$.

<u>Assumption 1</u>: "het(N)" is a well-formed sentential function

<u>Assumption 2</u>: "het(N)" contains free occurrences of N (only)

hence <u>3</u>: "het" is a one-place predicate (corresponding to what Rosser calls a "condition")

<u>Assumption 7</u>: Every predicate designates a unique property

<u>Assumption 4</u>: $(\forall\phi)(\text{"het" des } \phi \equiv \phi = \text{het})$ - thus the predicate "het" uniquely designates the property of heterologicality

hence <u>5</u>: $(\forall\phi)(\text{"het" des } \phi \;.\!\supset\!.\; \text{het("het")} \equiv \neg\phi(\text{"het"}))$

<u>Assumption 6</u>: $P \equiv \neg P \;.\!\supset\!.\; Q \,\&\, \neg Q$.

Note that <u>5</u> follows from <u>3</u> and <u>7</u> with the help of <u>D</u>. It is easily seen that a contradiction is derivable from <u>4</u>, <u>5</u>, and <u>6</u>.

Assumption <u>2</u> cannot reasonably be denied, and if classical logic is not considered to be in question, neither can assumption <u>6</u>. There remain only assumptions <u>1</u>, <u>4</u>, and <u>7</u>, one of which must be rejected, on pain of contradiction. To deny <u>1</u> is to deny that the predicate "het" is well-formed, perhaps on the lines of R's theory of orders, for which "het" is meaningless because of the quantifier on the right-hand side of <u>D</u>; this corresponds to the simple theory of types, according to which $\neg(x \in x)$ is not well-formed. To deny <u>7</u> is to suppose that there is no property of heterologicality; this corresponds in set theory to the non-existence of the Russell class. Finally, to deny <u>4</u>, supposing that heterologicality is a property, implies that no predicate designates this property uniquely. Various reasons might be given for this (see below), as, for example, that though there is a property of heterologicality, it is not an object and can therefore be neither named nor designated; this corresponds to the non-sethood of the Russell class (see [44]).[13] Just as for Russell's paradox, each of the three assumptions has been denied, some for more than one reason, bringing about a spectrum of semantic theories.

I proceed to consider in detail the various approaches to the heterological-cum-Richard paradox just mentioned, but it should by now be clear that the reasons (i) to (iii) above suffice to establish that the semantic paradoxes belong properly within the field of logic, and that the "clearer notions in logic" that R sought should be equally applicable to both types of paradox.

It is convenient, in considering the alternative possibilities of solution, to make the following linguistic stipulation, which is intended not to be controversial, namely, that 7 is trivially true, but that some properties are not objects and do not themselves have properties. This corresponds to positing the set-theoretic distinction between sets and proper classes, which is likewise assumed to be uncontroversial, except in the case when strong class-existence assumptions are made, as with the system MK (Morse-Kelley). Although doubts about the existence of the property of heterologicality have sometimes been expressed, as (implicitly) by R (see below), it is surely sufficiently perspicuous for it to be evident, for example, that "long" has it and "fifteen-lettered" lacks it.

Some previous discussions of the semantic paradoxes, by R ([4] and [7]), Ramsey [17], Chwistek [43], and Tarski [42] will now be related to the above classificatory scheme.

Russell. The solution given in [4] to the original (functional)

[13]Exactly the same development can be given of the relational version of the Richard paradox given by Chwistek [43]. In this, "het(N)" would be read as "N is the (gödel) number of a definable one-place predicate false of N". The liar paradox is somewhat different (see [23] Ch. 22).

version of the Richard paradox is that the Richard class E (see [28] p 143) is "an ill-defined notion. The reason is that ... the notion of definition is itself ... not definable, and is not even a definite notion ... There is therefore no such collection as E, not just in the sense that all classes are non-entities, but in the sense that there is no property common and proper to the members of E" ([4] p 645). This brilliant but not very happily expressed passage conceals how the solution is to be classified: is there, despite one phrase, a collection E, some of whose members can be exhibited, but which cannot be defined, thus denying 1 on p 239? Or is there no such collection, leading to the denial of 7? The first answer seems to be more in accord with the account in [7],[14] since if the second were correct, one could not understand the formulation of the paradox.

Ramsey's account of the heterological paradox, according to the principles of [7], confirms the above interpretation. Using the notation above, the variable ϕ in the definition D, he says, must, because of the hierarchy of orders, "have a definite range of values, of which [heterologicality] cannot itself be a member. So that 'heterological' ... is not an adjective in the sense in question, and is neither heterological nor autological." ([17] p 27) The objection that Ramsey makes to this approach is that classical mathematics cannot be developed within the theory of orders unless an axiom is added such as that of reducibility (see subsection A above) which there is no reason to suppose true, let alone necessary, and he therefore proceeds to develop his own alternative solution, presented below. It should perhaps be emphasised that if the axiom of reducibility is not assumed, the designation relation in the solution just given is not required to be intensional.

---

[14] In which the solution given is that though there are names of different orders, there can be "no such thing as a totality of names", ([7] p 63), and consequently "any name in which the phrase "nameable by names of order n" occurs is necessarily of a higher order than the nth."

<u>Chwistek and Ramsey</u>. In 1921, Chwistek put forward a purported proof that Richard's paradox was derivable in the theory of types, assuming the axiom of reducibility, and he concluded that the axiom was thereby shown to be false. He tacitly assumed, however, that all contexts were extensional; and the conclusion properly to be drawn from his argument is that in the simple theory of types, the semantic paradoxes can only be resolved, as R had suggested in [4] p 648, by appeal to the intensional features of semantical notions, a view that he rightly felt to be unsatisfactory. This intensional solution is presented more fully in [17] pp 42-46, in connection with the heterological paradox, where Ramsey claims to establish that "the contradiction is simply due to an ambiguity in the word "meaning" and has no relevance to mathematics whatever" ([17] p 43), - surprisingly, he appears to overlook the reservation he had expressed 22 pages earlier (quoted above p 237). There is, according to Ramsey, a different designation relation, indeed more than one, for each order, and "the meanings of meaning form an illegitimate totality", heterologicality not being meant by any of them. Thus unlike Russell (of the ramified theory) who denied either <u>1</u> or <u>7</u> above, Ramsey denies <u>4</u>. Indeed, given both the predicate "het" and the property het, the only possibility is that the predicate does not designate the property, and the important question is: why not? Ramsey locates the source of the difficulty in the (intensional) designation relation, but his solution, though suggestive for one mode of meaning, fails to show that there is no (quasi-)referential notion of meaning which is correctly designated by "designates". Tarski later developed the formal results and methods that helped to make this question more tractable, but it should be emphasised that nothing in Tarski's work precludes the possibility of an alternative approach, whereby the paradoxes arise <u>not</u> through any intensional feature of the meaning relation, but because the extension of e.g. the property of heterologicality is a proper class.

<u>Tarski</u>. Tarski deployed arguments similar to those that yield the semantic paradoxes in order to state and prove formal undefinability results for semantic notions; most of his work can be found in [42] and [45], in the latter of which he uses the argument of Richard's paradox to prove partial results concerning the undefinability of definability, which R had conjectured in [4] (cited above p 32). Tarski works throughout within a nominalist framework, eschewing properties and propositions in favour of their linguistic counterparts; this has no bearing upon the formal results, but may have misled some people. Consider, for example, his best-known theorem on the undefinability of truth, in which it is shown that a language containing formal arithmetic and the standard logical connectives (including negation) cannot express a predicate true of just the gödel numbers of the true sentences of the language. For, if a truth predicate could be expressed, so could a falsity predicate $\mathcal{F}$, and a sentence **g** in the language could then be constructed provably equivalent in any extension of a weak sub-system of arithmetic to $\mathcal{F}[g]$, where [g] is a gödel number of **g** (see Montague [46]). Hence a truth-predicate for such a language can only be obtained in a more expressive metalanguage, either by means of a new semantic primitive or by conversion of an inductive definition. Since the metalanguage can again be formalised, a hierarchy of increasingly expressive metalanguages can thus be constructed, and this regress is not stopped because it is possible to construct so-called non-translational metalanguages (Martin [47]) which contain, in a weak sense, truth definitions for themselves.

To disentangle what is established by these formal results from what is genuinely disputable and arises from the nominalist manner in which they were first presented, consider the question of whether they establish that there is what might be called an intensional hierarchy of semantic predicates. In other words, to what extent, for example, are the truth predicates of different languages really different, since from a platonist viewpoint they can all be said to designate the same property, viz., truth? It might therefore be thought more natural to say that the same predicate has a varying extension in different languages. Of course, for a given formal language with a determinate set of sentences, the truth predicate can be formally defined only in a richer metalanguage. But these specific truth predicates differ only insofar as their extensional definitions are language-dependent; and since to identify these predicates would result in a new language, requiring yet another truth definition, it is linguistically simpler to say that different predicates designate the same property.

Consider, for example, the sentence (cf. Kneale [48] p 666):

"Designation in $English_1$" designates in $English_2$, designation in $English_1$.

Despite its syntactic appearance, the occurrences of "designates in $English_2$" and "designation in $English_1$" have the same designation, and the proposition it expresses can just as well be conveyed by

"Designation in $English_1$" designates in $English_2$ designation in $English_2$.

It is of course in no way being suggested that the hierarchy of languages fails to provide a resolution of the semantic paradoxes; the question is only whether this solves the paradoxes or avoids them (as I have claimed that the limitation of size theory does with regard to the set-theoretic paradoxes). However, once the semantic notions, though undefinable, are recognised to be clear, there is less reason to believe that the semantic paradoxes arise from linguistic or intensional considerations, and the way becomes open to seeing them as posing, in a somewhat different form, the same logical puzzles as the set-theoretic ones.

The argument above is intended to establish not that the designation relation is extensional, but that neither its status nor the undefinability results determine how the semantic paradoxes are to be resolved. Consequently, it is appropriate to look at the alternative approach suggested two pages back, according to which the extension of the property of heterologicality is a proper class. The property of heterologicality, therefore, is not an object, and so cannot be named, or designated in the required manner,[15] since only objects can be named. The objection that might be raised against this suggestion is that, as R claimed in [4], the semantic paradoxes have nothing to do with infinity. However it is easily seen that the number of heterological predicates in any (ω,ω) language is $\aleph_0$, and the least undefinable ordinal, assuming that there is one, is certainly countable; thus the real burden of the objection just made is that there are no 'small' infinite proper classes. As shown above, however, though this is an insuperable difficulty for the limitation of size theory, the zigzag theory, assuming the unpublished result communicated by Specker in §2 above, requires countable proper classes. Indeed the definition of "heterological" above would be unstratified if, whenever N denotes ϕ, N and ϕ are to be of the same potential type. It is to be expected that the arguments which lead to the semantic paradoxes could be mirrored to obtain a countable proper class in Quine's system ML, but neither this nor the more general programme of characterising the formal relations between semantic theories and set theories has yet been carried out.

§4. Conclusion

The discovery of the set-theoretic paradoxes around the turn of

---

[15] More precisely, the word "heterological" can be said to designate when it appears as a predicate, but not when it appears as a subject. Compare Frege on the concept horse.

the century gave added impetus to the study of the foundations of mathematics; a subject previously concerned largely with problems concerning the nature and existence of the infinite, and the bounds of constructivity, was thenceforth plunged into a reexamination of the basic principles of logic. At the centre of this development was Russell, who in 1903 published a major philosophical work in support of the identity of logic and mathematics, and in the years following he suggested, in outline, a number of systems of what are now known as set theory and type theory. His aim throughout was to formulate the correct though previously undiscovered logical principles required to resolve the set-theoretic and semantic paradoxes; in a well-known passage about "the advantages of theft over honest toil", he later rejected with scorn the thesis that sufficient justification for a foundational theory is that it yields implicit definitions of its primitive notions.

The three theories he proposed between 1903 and 1908 were the limitation of size theory, later developed by Zermelo, the zigzag theory, later developed by Quine, and the no classes theory, later known as the theory of ramified types and developed by Russell himself and by Chwistek; an early version of the theory of <u>Principia Mathematica</u>, later known as the simple theory of types, was also propounded at that time. Of these theories, the simple theory of types, as a realist theory, is easily shown to be implausible, and the ramified theory is inadequate to serve as a foundation for classical mathematics. Though reasons can be given for holding that no system of set theory can be an adequate foundation for mathematics, it is important for the philosophy of mathematics to appraise the partial adequacy of the other two theories, that of the limitation of size and the zigzag theory.

Comparison between these two theories suggests several grounds on which the latter is to be preferred. The zigzag theory allows for the existence of numbers defined as classes of equivalent classes, and of other large classes similarly defined by an equivalence relation. It allows for the reference of class descriptions whenever the given description is not circular. It further enables the structural similarity that holds between the set-theoretic and the semantic paradoxes to be explained. The limitation of size theory, however, has none of these advantages, and leads instead to the exclusion of semantic notions from the province of logic. A final advantage of the zigzag theory is that it allows for the existence of large classes, and it may therefore be possible to embed within it certain branches of mathematics (e.g. category theory) that cannot without distortion be developed within a limitation of size theory.

If, on these grounds, the zigzag theory is to be preferred to the limitation of size theory, the dispute between Russell and Keyser in 1903-4 about the status of the axiom of infinity is resolved in favour of the former; the existence of an infinite set becomes a non-trivial consequence of a very general cardinal-free assumption about set existence.

REFERENCES

[1] Russell, B., The Principles of Mathematics, Cambridge (1903), 2nd edition London, 1937.

[2] ibid., The axiom of infinity, Hibbert J. 2 (1903-4).

[3] ibid., On some difficulties in the theory of transfinite numbers and order types, Proc. Lond. Math. Soc. 4 (1906).

[4] ibid., Les paradoxes de la logique, Rev. mét. mor. 14 (1906).

[5] ibid., Mathematical logic as based on the theory of types, A.J.M. 30 (1908), reprinted in [28].

[6] ibid., Section III of Whitehead, A. N., On cardinal numbers, A.J.M. 24 (1902).

[7] Whitehead, A. N. and Russell, B., Principia Mathematica, Cambridge (1910-13). Page references are to Volume I.

[8] Keyser, C. J., Concerning the axiom of infinity and mathematical induction, Bull.A.M.S. 9 (1902-3).

[9] ibid., The axiom of infinity: A new presupposition of thought, Hibbert J. 2 (1903-4), reprinted in [11].

[10] ibid., The axiom of infinity, Hibbert J. 3 (1904-5).

[11] ibid., The Human Worth of Rigorous Thinking, Essays and Addresses, New York (1916).

[12] Hobson, E. W., On the general theory of transfinite numbers and order types, Proc. Lond. Math. Soc. 3 (1905).

[13] Poincaré, H., Les mathématiques et la logique (3 papers), Rev. mét. mor. 13 and 14 (1905-6), as translated in Science and Method, New York, n.d., originally Paris, 1908.

[14] ibid., Sur la nature du raisonnement mathématique, Rev. mét. mor. 2 (1894), as translated in Science and Hypothesis, New York, n.d., originally Paris, 1902.

[15] ibid., Du rôle de l'intuition et de la logique en mathématiques, 2nd Int. Cong. Math (1900), as translated in The Value of Science, New York, n.d., originally Paris, 1905.

[15a] ibid., La logique de l'infini, Rev. mét. mor. 17 (1909), as translated in Mathematics and Science: Last Essays, New York, 1963, originally Paris, 1913.

[16] Fraenkel, A., Abstract Set Theory, Amsterdam (1953), based upon Einleitung in die Mengenlehre, 2nd edn., Berlin, 1923, 3rd edn., Berlin, 1928. Cf. also Fraenkel, A. and Bar-Hillel, Y., Foundations of Set Theory, Amsterdam (1958).

[17] Ramsey, F. P., The Foundations of Mathematics and other Logical Essays, ed. R. B. Braithwaite, London (1931).

[18] Gödel, K., Russell's mathematical logic, as reprinted in Philosophy of Mathematics, Selected Readings, eds. Benacerraf, P. and Putnam, H., Oxford (1964). Originally published in The Philosophy of Bertrand Russell, ed. P. A. Schilpp, New York (1944).

[19] Quine, W. V., Selected Logic Papers, New York (1966).

[20] ibid., From a Logical Point of View, Cambridge, U.S.A. (1953).

[21] ibid., Set Theory and its Logic, Cambridge, U.S.A. (1963 and 1969).

[22] Wang, H., Russell and his logic, Ratio 7 (1965).

[23] ibid., A Survey of Mathematical Logic, Peking and Amsterdam (1963).

[24] Rosser, J. B. and Wang, H., Non-standard models for formal logic, J.S.L. 15 (1950).

[25] Church, A., Mathematics and logic, in Contemporary Philosophy, Vol. I, ed. R. Klibansky, Florence (1968). Originally published without bibliography in Logic, Methodology, and Philosophy of Science, eds. Nagel, Suppes, and Tarski, Stanford, U.S.A. (1962).

[26] Bowne, G. D., The Philosophy of Logic 1880-1908, The Hague (1966).

[27] Mooij, J. J. A., La philosophie des mathématiques de Henri Poincaré, Paris and Louvain (1966).

[28] van Heijenoort, J., ed., From Frege to Gödel, A Source Book in Mathematical Logic, Cambridge, U.S.A. (1967).

[29] Shoenfield, J., Mathematical Logic, Reading, U.S.A. (1967).

[30] Benacerraf, P., What numbers could not be, Phil. Rev. 74 (1965).

[31] Moss, J. M. B., Kreisel's work on the philosophy of mathematics, I. Realism, in Logic Colloquium '69, eds. Gandy and Yates, Amsterdam and London (1971).

[32] ibid., Quantifiers, numbers, and the bounds of logic, A.S.L. meeting, Cambridge, August 1971. Abstract to appear J.S.L. 37 (1972).

[33] Mostowski, A., Recent results in set theory, in Problems in the Philosophy of Mathematics, ed. I. Lakatos, Amsterdam (1967).

[34] Feferman, S. (with appendix by Kreisel, G.), Set-theoretical foundations of category theory, in Reports of the Midwest Category Seminar III, ed. S. MacLane, Springer Lecture Notes in Mathematics 106 (1969).

[35] Rosser, J. B., Logic for Mathematicians, New York (1953).

[36] Pollock, J. L., On logicism, in Essays on Bertrand Russell, ed. E. D. Klemke, Urbana, U.S.A. (1970).

[37] Vuillemin, J., Leçons sur la première philosophie de Russell, Paris (1968).

[38] Hahn, H., Überflüssige Wesenheiten, Vienna (1930).

[39] Parsons, C. D., A plea for substitutional quantification, J. Phil. 68 (1971).

[40] ibid., Ontology and mathematics, Phil. Rev. 80 (1971).

[41] Myhill, J. R., The hypothesis that all classes are nameable, Proc. Nat. Acad. Sci. 38 (1952).

[42] Tarski, A., Logic, Semantics, Metamathematics, Papers from 1923 to 1938, Oxford (1956).

[43] Chwistek, L., Antynomje logiki formalnej, Przegląd Filozoficzny 24 (1921), as translated in Polish Logic, 1920-1939, ed. S. McCall, Oxford (1967).

[44] Moss, J. M. B., Syntactic and semantic paradoxes (abstract), J.S.L. 31 (1966).

[45] Tarski, A., A problem concerning the notion of definability, J.S.L. 13 (1948).

[46] Montague, R. M., Theories incomparable with respect to relative interpretability, J.S.L. 27 (1962), publ. 1963.

[47] Martin, R. M., *Truth and Denotation*, London (1958).

[48] Kneale, W. C. and M., *The Development of Logic*, Oxford (1962).

ON MODELS OF ARITHMETIC

J. B. Paris

Manchester, England

Introduction

Let T be a complete consistent theory in some countable language $\mathbb{L}$ such that T extends Peano arithmetic, has the induction axiom for each formula of $\mathbb{L}$, and if $\theta(x_1,\ldots,x_n,y)$ is a formula of $\mathbb{L}$ and

$$(\forall x_1\ldots x_n)(\exists! y)\theta(x_1,\ldots,x_n,y) \in T$$

then for some n-place function symbol F in $\mathbb{L}$,

$$(\forall x_1\ldots x_n)\theta(x_1,\ldots,x_n,F(x_1,\ldots,x_n)) \in T.$$

Suppose B is a model of T and d a subset of the universe of B (- which we shall write as $|B|$). Let $[d]_B$ be the substructure of B generated by d. Then the conditions on T ensure that $[d]_B$ is an elementary substructure of B. Let $\mathfrak{B}(B)$ be the set of (elementary) substructures of B. Then $\langle\mathfrak{B}(B),\subset\rangle$ is a lattice, join and meet being given by

$$A \vee C = [\,|A| \cup |C|\,]_B \qquad (A,\ C \in \mathfrak{B}(B))$$
$$A \wedge C = [\,|A| \cap |C|\,]_B,$$

and furthermore this lattice has a least element, namely $[\emptyset]_B$.

This naturally raises the following question: For which lattices

L are there models B of T such that $L \cong \langle \mathcal{B}(B), \subset \rangle$?

In this paper we shall show the following results.

THEOREM 1. (Gaifman) There is a model B of T such that

$$\langle \mathcal{B}(B), \subset \rangle \cong \langle \omega_1, < \rangle.$$

THEOREM 2. (Gaifman) For every set A there is a model B of T such that

$$\langle \mathcal{B}(B), \subset \rangle = \langle \mathcal{P}(A), \subset \rangle.$$

THEOREM 3. Let L be a complete distributive ω_0-compactly generated lattice. Then there is a countable model B of T such that

$$\langle \mathcal{B}(B), \subset \rangle = L.$$

The paper is divided into three sections. In section 1 we prove theorems 1 and 2. Section 2 is devoted to a proof of theorem 3 and in section 3 we state some results for non-distributive lattices. We shall assume the Axiom of Choice throughout. I wish to thank H. Gaifman and W. Hodges for their help in preparing this paper.

Apparatus

Let $\mathcal{U}$ be a non-principal ultrafilter on ω_0 and let M be the minimal model of T. Set N to be the ultrapower of M with respect to $\mathcal{U}$ and for $f \in M^{\omega_0}$ let $\bar{f}$ be the class in N containing f.

Let $\mathcal{S}_i = \{F^i_j \mid j < \omega_0\}$ $(i > 0)$ enumerate the i-place functions in $\mathbb{L}$ and for $F \in \mathcal{S}_i$, $f_1, \ldots, f_i \in |M|^{\omega_0}$ let $F(f_1,\ldots,f_i)$ be the map

$$\lambda n \in \omega_0 : F(f_1(n),\ldots,f_i(n)) \quad (\in |M|^{\omega_0})$$

Let $\underline{0}$ be the map from ω_0 to $|M|$ with constant value 0 ($\in |M|$).

For f, g $\in |M|^{\omega_0}$ define

$$f \trianglelefteq g \leftrightarrow (\exists F \in \mathscr{S}_1)(\overline{F(g)} = \overline{f})$$
$$f \equiv g \leftrightarrow f \trianglelefteq g \text{ and } g \trianglelefteq f$$
$$f \triangleleft g \leftrightarrow f \trianglelefteq g \text{ and } \neg(g \trianglelefteq f)$$

Clearly $\equiv$ is an equivalence relation on $|M|^{\omega_0}$. For f $\in |M|^{\omega_0}$ let $\tilde{f}$ be the equivalence class mod $\equiv$ containing f and define

$$\tilde{f} \trianglelefteq \tilde{g} \leftrightarrow f \trianglelefteq g$$
$$\tilde{f} \triangleleft \tilde{g} \leftrightarrow f \triangleleft g.$$

It is easy to check that $\mathcal{L} = \langle\{\tilde{f} \mid f \in |M|^{\omega_0}\}, \triangleleft\rangle$ is an upper semi-lattice. (For $\tilde{f}, \tilde{g} \in \mathcal{L}$, $\tilde{f} \vee \tilde{g} = \widetilde{2^f.3^g}$.) Also $\tilde{\underline{0}}$ is the least element of $\mathcal{L}$.

There is a very close connection between $\mathcal{L}$ and $\langle\mathfrak{P}(N),\subset\rangle$ which we state after the following definition.

DEFINITION. Let L be an upper semi-lattice. I $\subseteq |L|$ is an _ideal_ if (1) I $\neq \emptyset$, (2) a $\in$ I and b < a in L $\rightarrow$ b $\in$ I, and (3) a, b $\in$ I $\rightarrow$ a $\vee$ b $\in$ I. Set Id(L) to be the set of ideals on L. Then $\langle Id(L),\subset\rangle$ is a lattice.

LEMMA 0. (P. Aczel) $\langle Id(\mathcal{L}),\subset\rangle \cong \langle\mathfrak{P}(N),\subset\rangle$.

Proof. For I $\in$ Id($\mathcal{L}$) set

$$\theta(I) = \{\overline{f} \mid \tilde{f} \in I\}.$$

θ is the required isomorphism. □

We shall show theorems 1 - 3 by showing corresponding results for $\mathcal{L}$ and then applying lemma 0.

Section 1

Notation. All formulas we mention are assumed to be in $\mathbb{L}$. For B a subset of $|M|$, B is M-inf if B is definable and unbounded in M; $[B]^n = \{\langle a_1,\dots,a_n\rangle \mid a_1,\dots,a_n \in |M| \ \& \models_M a_1 < a_2 < \dots < a_n\}$. For $B(x_1,\dots,x_n,y)$ a formula of $\mathbb{L}$ and $a_1, \dots ,a_n \in |M|$ $Ba_1\dots a_n$ is the set of $b \in |M|$ such that $\models_M B(a_1,\dots,a_n,b)$.

For each $n > 0$ let $<_n$ be the M-definable ordering of n-tuples of elements of $|M|$ given by

$$\langle a_1,\dots,a_n\rangle <_n \langle b_1,\dots,b_n\rangle \leftrightarrow \max(a_1,\dots,a_n) < \max(b_1,\dots,b_n),$$

or $\max(a_1,\dots,a_n) = \max(b_1,\dots,b_n)$ and $\langle a_1,\dots,a_n\rangle$ is less than $\langle b_1,\dots,b_n\rangle$ in the usual lexicographic ordering of n-tuples of elements of $|M|$.

Clearly from this ordering we can find an M-definable order preserving map from n-tuples of elements of $|M|$ onto $|M|$. Thus we shall speak of the $\langle a_1,\dots,a_n\rangle$'th element of $|M|$ etc.

LEMMA 1.1. Let $P = \{a \in |M| \mid \models_M B(a)\}$ be M-inf and $C(x_1,\dots,x_n,w)$ a formula in $\mathbb{L}$. Then there is a formula $A(w,z)$, defined uniformly from B and C such that for all a, b $\in |M|$,

i) Aa is an M-inf subset of P and is a set of indiscernibles for $C(x_1,\dots,x_n,a)$ in M.

ii) $a < b$ (in M) $\rightarrow Aa \supseteq Ab$.

Proof. The proof is by induction on n. For the sake of hygiene we only give an informal proof.

Case 1. $n = 1$. For each $a \in P$, $b \in |M|$ we can associate an element of $|M|$ which codes a map h_a^b from $\{u \mid u \leq b\}$ into $\{0, 1\}$ such that for $u \leq b$,

$$h_a^b(u) = 0 \leftrightarrow C(u,a).$$

Let $\preceq$ be the usual lexicographic ordering of these maps. The required formula A is now defined by

$$\begin{aligned} A(w,z) \leftrightarrow\ & z \in P \wedge \exists \text{ unboundedly many } v \in P \text{ such that } h_v^w = h_z^w \\ & \wedge \text{ if } u \in P \text{ and } \exists \text{ unboundedly many } v \in P \text{ such that} \\ & h_v^w = h_u^w \text{ then } h_z^w \preceq h_u^w. \end{aligned}$$

Case 2. $n > 1$ and the result holds for $n - 1$. Then we may assume we have found a formula $A'(w,y,z)$ such that for all $a_1, a_2, b_1, b_2 \in |M|$,

i) $A'a_1b_1$ is an M-inf subset of P and is a set of indiscernibles for $C(b_1,x_2,\ldots,x_n,a_1)$,

ii) $2^{a_1}3^{b_1} < 2^{a_2}3^{b_2} \rightarrow A'a_1b_1 \supseteq A'a_2b_2$.

Let F be the M-definable function such that

$$F(0) = \mu y\colon y \in A'00$$
$$F(w+1) = \mu y\colon y > F(w) \wedge y \in A'wF(w), \quad (w \in |M|)$$

and let Q be the range of F. Then Q is an M-inf subset of P and for $w \in |M|$, $\{z \in Q \mid z > F(w)\}$ is a set of indiscernibles for

$C(F(w),x_2,\ldots,x_n,w)$.

Define

$$E(w,z) \leftrightarrow z > F(w) \wedge z \in Q \wedge (\exists z_2,\ldots,z_n \in Q)(z < z_2 < \ldots < z_n \wedge C(z,z_2,\ldots,z_n,w))$$

and by case 1 pick $A(w,z)$ such that for $a, b \in |M|$,

i) Aa is an M-inf subset of Q and is a set of indiscernibles for $E(a,x_1)$,

ii) $a < b \rightarrow Aa \supseteq Ab$.

We may also suppose that $A(a,b) \rightarrow b > F(a)$. Then for $a \in |M|$, $b_1, \ldots, b_n, c_1, \ldots, c_n \in Aa$ and $b_1 < b_2 < \ldots < b_n$, $c_1 < c_2 < \ldots < c_n$,

$$E(a,b_1) \leftrightarrow E(a,c_1) \text{ and } b_2,\ldots,b_n,c_2,\ldots,c_n \in Q$$

$\therefore$ $C(b_1,\ldots,b_n,a) \leftrightarrow C(c_1,\ldots,c_n,a)$; so A is the required formula.

□

COROLLARY 1.2. Let $C(x_1,\ldots,x_n)$ be a formula of $\mathcal{L}$ and $P = \{a \in |M| \mid \models_M B(a)\}$ M-inf. Then there is a formula $A(x)$, defined uniformly from B, C such that $\{a \in |M| \mid \models_M A(a)\}$ is an M-inf subset of P and is a set of indiscernibles for $C(x_1,\ldots,x_n)$. □

LEMMA 1.3. Let $F \in \mathcal{S}_n$ and $P = \{a \in |M| \mid \models_M B(a)\}$ be M-inf. Then we can find a formula $A(x_1)$, uniformly from F and B, such that $Q = \{a \in |M| \mid \models_M A(a)\}$ is an M-inf subset of P and $\exists\, j_1, \ldots, j_m \leq n$ such that for any $\langle a_1,\ldots,a_n\rangle, \langle b_1,\ldots,b_n\rangle \in [Q]^n$,

$$* \qquad F(a_1,\ldots,a_n) = F(b_1,\ldots,b_n) \leftrightarrow \bigwedge_{k=1}^{m} a_{j_k} = b_{j_k}.$$

<u>Proof</u>. By repeated use of 1.2 we can pick $A(x_1)$ uniformly from F and B such that if $Q = \{a \in |M| \mid \models_M A(a)\}$ then Q is an M-inf subset of P and Q is a set of indiscernibles for all formulas of the form

$$F(x_{i_1},\ldots,x_{i_n}) = F(x_{k_1},\ldots,x_{k_n})$$

where $i_1, \ldots, i_n, k_1, \ldots, k_n \leq n$. Let $H = \{i \leq n \mid (\exists <a_1,\ldots,a_{n+1}> \in [Q]^{n+1})(F(a_1,\ldots,a_i,a_{i+2},\ldots,a_{n+1}) = F(a_1,\ldots,a_{i-1},a_{i+1},\ldots,a_{n+1})\}$ and let $\{j_1,\ldots,j_m\} = \{1,\ldots,n\} - H$. It is straightforward to show * holds from right to left for this $j_1, \ldots, j_m$. To show the converse suppose $<a_1,\ldots,a_n>, <b_1,\ldots,b_n> \in [Q]^n$, $F(a_1,\ldots,a_n) = F(b_1,\ldots,b_n)$ and $\neg(\bigwedge_{k=1}^m a_{j_k} = b_{j_k})$. Let j_i be minimal such that $a_{j_i} \neq b_{j_i}$. Then for $k < j_i$ either $k \in H$ or $a_k = b_k$ so, by suitable sliding if necessary, we may assume $a_k = b_k$ for all $k < j_i$. Suppose $a_{j_i} < b_{j_i}$. Then, by suitable sliding if necessary, we may assume there is a $y \in Q$ such that $a_{j_i} > y > \max_{k<j_i} a_k$. But then, since $F(a_1,\ldots,a_n) = F(b_1,\ldots,b_n)$,

$$\begin{aligned} F(a_1,\ldots,a_{j_i-1},y,a_{j_i+1},\ldots,a_n) &= F(b_1,\ldots,b_n) \\ &= F(a_1,\ldots,a_n) \end{aligned}$$

contradicting $j_i \notin H$. $\therefore$ * also holds from left to right. □

Since $A(x_1)$ is obtained uniformly from B and F we can obtain the following generalization of 1.3.

LEMMA 1.4. Let $F \in \mathcal{S}_n$ and let $B(x_1,\ldots,x_p,y)$ be a formula of $\mathcal{L}$ such that for all $a_1, \ldots, a_p \in |M|$, $Ba_1\ldots a_p$ is M-inf. Then there is a formula $A(x_1,\ldots,x_p,y)$ of $\mathcal{L}$ such that for all $a_1, \ldots, a_p \in |M|$,

i) $Aa_1\ldots a_p$ is an M-inf subset of $Ba_1\ldots a_p$,

ii) $\exists\, j_1, \ldots, j_m \leq n$ such that for all $<b_1,\ldots,b_n>, <c_1,\ldots,c_n> \in [Aa_1\ldots a_p]^n$,

$$F(b_1,\ldots,b_n) = F(c_1,\ldots,c_n) \leftrightarrow \bigwedge_{k=1}^{m} b_{j_k} = c_{j_k}.$$

□

We now give a lemma from which theorem 1 follows.

LEMMA 1.5. Let $\{f_i \mid i \in \mathbb{N}\} \subseteq |M|^{\omega_0}$. Then there is a $g \in |M|^{\omega_0}$ such that for all $h \in |M|^{\omega_0}$,

$$\tilde{h} \triangleleft \tilde{g} \leftrightarrow \tilde{h} \trianglelefteq \tilde{f}_1 \vee \ldots \vee \tilde{f}_n \quad \text{some } n \in \omega_0.$$

$(\mathbb{N} = \omega_0 - \{0\}.)$

Proof. We pick a sequence of formulas $B^p(x_1,\ldots,x_p,y)$, $p \in \omega_0$ as follows. First set $B^o(y)$ to be, say, $y = y$. Suppose $B^p(x_1,\ldots,x_p,y)$ has been defined so that for $a_1, \ldots ,a_p \in |M|$, $B^p a_1 \ldots a_p$ is M-inf. Now by 1.4 pick $A(x_1,\ldots,x_p,y)$ such that for all $a_1, \ldots ,a_p \in |M|$, $Aa_1\ldots a_p$ is an M-inf subset of $B^p a_1 \ldots a_p$ and F_p^1 ($\in \mathscr{S}_1$) is either 1-1 or constant on $Aa_1\ldots a_p$.

Now define

$B^{p+1}(x_1,\ldots,x_{p+1},y) \leftrightarrow \exists u$, w such that u codes a map f with domain $\{\langle y_1,\ldots,y_{p+2}\rangle \mid \langle y_1,\ldots,y_{p+2}\rangle \leqslant_{p+2} \langle x_1,\ldots,x_{p+1},w\rangle\}$ such that

i) for all $\langle y_1,\ldots,y_{p+2}\rangle \in \mathrm{dom}(f)$, $f(y_1,\ldots,y_{p+2}) \in Ay_1\ldots y_p$, and if F_p^1 is 1-1 on $Ay_1\ldots y_p$ then $f(y_1,\ldots,y_{p+2}) = \mu z\colon z \in Ay_1\ldots y_p \;\&\; F_p^1(z) \notin F_p^1\text{`}f\text{`}t$, where $t = \{\langle z_1,\ldots,z_{p+2}\rangle \mid \langle z_1,\ldots,z_{p+2}\rangle <_{p+2} \langle y_1,\ldots,y_{p+2}\rangle\}$;

ii) $f(x_1,\ldots,x_{p+1},w) = y$.

Then for any $a_1, \ldots, a_{p+1}, b_1, \ldots, b_{p+1} \in |M|$, if $\langle a_1, \ldots, a_{p+1}\rangle \neq \langle b_1, \ldots, b_{p+1}\rangle$ then

i) $B^{p+1}a_1\ldots a_{p+1}$ is an M-inf subset of $B^p a_1 \ldots a_p$,

ii) F_p^1 is either 1-1 or constant on $B^{p+1}a_1\ldots a_{p+1}$,

iii) if F_p^1 is 1-1 on both $B^{p+1}a_1\ldots a_{p+1}$ and $B^{p+1}b_1\ldots b_{p+1}$ then $F_p^1 {}^{\backprime\backprime} B^{p+1}a_1\ldots a_{p+1} \cap F_p^1 {}^{\backprime\backprime} B^{p+1}b_1\ldots b_{p+1} = \emptyset$,

iv) $B^{p+1}a_1\ldots a_{p+1} \cap B^{p+1}b_1\ldots b_{p+1} = \emptyset$.

Now given $\{f_i \mid 1 \leq i < \omega_0\}$ pick $g \in |M|^{\omega_0}$ such that for all $m \in \omega_0$, $g(m) \in B^m f_1(m)\ldots f_m(m) - \bigcup_{i,j\leq m} \{F_j^i(f_1(m),\ldots,f_i(m))\}$. It remains to show that g satisfies the required conditions. Notice that for $1 \leq n < \omega_0$, $j < \omega_0$,

$$\bar{g} \neq \overline{F_j^n(f_1,\ldots,f_n)}$$

so $\neg(\tilde{g} \trianglelefteq \tilde{f}_1 \vee \ldots \vee \tilde{f}_n)$ all $1 \leq n < \omega_0$.

Now suppose $h \in |M|^{\omega_0}$ and $\tilde{h} \triangleleft \tilde{g}$. Then for some p, $\overline{F_p^1(g)} = \bar{h}$ and for all $n \in \omega_0$, F_p^1 is either 1-1 or constant on $B^{p+1}f_1(n)\ldots f_{p+1}(n)$. Suppose $\{m \in \omega_0 \mid F_p^1$ is 1-1 on $B^{p+1}f_1(m)\ldots f_{p+1}(m)\} \in \mathcal{U}$. Then since $g(m) \in B^m f_1(m)\ldots f_m(m) \subseteq B^{p+1}f_1(m)\ldots f_{p+1}(m)$ all $m > p$, by iii), $\{m \in \omega_0 \mid g(m) = \mu z{:}\ (\exists z_1,\ldots,z_{p+1})(F_p^1$ is 1-1 on $B^{p+1}z_1\ldots z_{p+1}$ & $z \in B^{p+1}z_1\ldots z_{p+1}$ & $F_p^1(z) = h(m))\} \in \mathcal{U}$. $\therefore\ \tilde{g} \trianglelefteq \tilde{h}$ – contradiction. Thus we must have

$$\{m \in \omega_0 \mid F_p^1 \text{ is constant on } B^{p+1}f_1(m)\ldots f_{p+1}(m)\} \in \mathcal{U}.$$

$\therefore\ \{m \in \omega_0 \mid h(m) = \mu z{:}\ (\exists y \in B^{p+1}f_1(m)\ldots f_{p+1}(m))(z = F_p^1(y)\} \in \mathcal{U}$.
$\therefore\ \tilde{h} \trianglelefteq \tilde{f}_1 \vee \ldots \vee \tilde{f}_{p+1}$.

Conversely suppose $h \in |M|^{\omega_0}$ and $\tilde{h} \trianglelefteq \tilde{f}_1 \vee \ldots \vee \tilde{f}_n$. By iv), for $1 \leq k < \omega_0$, $\{m \in \omega_0 \mid f_k(m) = \mu z_k : (\exists z_1, \ldots, z_{k-1})(g(m) \in B^k z_1 \ldots z_k)\} \in \mathcal{U}$. $\therefore \tilde{f}_k \trianglelefteq \tilde{g}$. $\therefore \tilde{h} \trianglelefteq \tilde{g}$ and so, since $\neg(\tilde{g} \trianglelefteq \tilde{f}_1 \vee \ldots \vee \tilde{f}_n)$, $\tilde{h} \triangleleft \tilde{g}$.

□

COROLLARY 1.6. There is a sequence a_α, $\alpha < \omega_1$ of elements of $|M|^{\omega_0}$ such that for all $\lambda \in \omega_1$, $h \in |M|^{\omega_0}$,

$$\tilde{h} \triangleleft \tilde{a}_\lambda \leftrightarrow \tilde{h} = \tilde{a}_\gamma \quad \text{some } \gamma < \lambda.$$

<u>Proof</u>. Set $a_0 = \underline{0}$ and suppose a_α have been picked for $\alpha < \lambda < \omega_1$. By lemma 1.5 pick a_λ such that for $h \in |M|^{\omega_0}$,

$$\tilde{h} \triangleleft \tilde{a}_\lambda \leftrightarrow \tilde{h} \trianglelefteq \tilde{a}_{\gamma_1} \vee \ldots \vee \tilde{a}_{\gamma_n} \quad \text{some } \gamma_1, \ldots, \gamma_n < \lambda.$$

This clearly gives the required sequence. □

<u>Proof of Theorem 1</u>. For the a_λ as in 1.6 set $B = [\{\bar{a}_\alpha \mid \alpha < \omega_1\}]_N$ and $L = \langle \{\tilde{a}_\lambda \mid \lambda < \omega_1\}, \triangleleft \rangle$. Then by lemma 0,

$$\langle \mathfrak{B}(B), \subset \rangle \cong \langle Id(L), \subset \rangle \cong \langle \omega_1, < \rangle. \quad \square$$

Notice there is no model B of T such that $\langle \mathfrak{B}(B), \subset \rangle \cong \langle \omega_1 + 1, < \rangle$ for suppose such a B did exist and $\theta : \omega_1 + 1 \longrightarrow \mathfrak{B}(B)$ were such an isomorphism. Then for $\alpha < \omega_1$, $|\theta(\alpha)| \subset |\theta(\alpha+1)|$ so B must be uncountable. Pick $a \in |B| - |\theta(\omega_1)|$. Then $[\{a\}]_B = B$ and since T is countable, $[\{a\}]_B$ must be countable which is a contradiction.

<u>Proof of Theorem 2</u>. The first step is to construct a model B of T such that $\langle \mathfrak{B}(B), \subset \rangle \cong \langle \mathcal{P}(\mathbb{N}), \subset \rangle$. ($\mathbb{N} = \omega_0 - \{0\}$.)

Let G_n, $n \in \omega_0$ enumerate $\bigcup_{i \in \mathbb{N}} \mathcal{S}_i$. We define a sequence B_n, $n \in \omega_0$, of M-definable sets as follows. Set $B_0 = |M|$ and suppose B_n

has been defined and is M-inf. Let $G_n = F^i_j$. By 1.3 pick B_{n+1} to be an M-inf subset of B_n such that for some $i_1, \dots, i_k \leq i$ and all $\langle a_1,\dots,a_i\rangle, \langle b_1,\dots,b_i\rangle \in [B_{n+1}]^i$

$$F^i_j(a_1,\dots,a_i) = F^i_j(b_1,\dots,b_i) \leftrightarrow \bigwedge_{t=1}^{k} a_{i_t} = b_{i_t}.$$

Now for $j \in \mathbb{N}$ define $g_j \in |M|^{\omega_0}$ by

$$g_j(m) = \text{the } j\text{'th element of } B_m.$$

Claim:

1) $\{\bar{g}_j \mid j \in \mathbb{N}\}$ are (distinct) indiscernibles in N (and so also in $[\{\bar{g}_j \mid j \in \mathbb{N}\}]_N$).

2) If $h \in |M|^{\omega_0}$ and $\tilde{h} \trianglelefteq \tilde{g}_1 \vee \dots \vee \tilde{g}_n$ then $\exists\, i_1, \dots, i_t \leq n$ such that $\tilde{h} = \tilde{g}_{i_1} \vee \dots \vee \tilde{g}_{i_t}$. (If $\{i_1,\dots,i_t\} = \emptyset$ we take $\tilde{g}_{i_1} \vee \dots \vee \tilde{g}_{i_t}$ to be $\tilde{\underline{0}}$.)

Proof of 1): Clearly the $\bar{g}_j$ are distinct. Let $\theta(x_1,\dots,x_n)$ be a formula of $\mathbb{L}$ and let m be such that for all $a_1, \dots, a_n \in |M|$,

$$G_m(a_1,\dots,a_n) = \begin{cases} 0 \text{ if } \models_M \theta(a_1,\dots,a_n) \\ 1 \text{ otherwise.} \end{cases}$$

Clearly G_m must be constant on $[B_{m+1}]^n$. Let $j_1 < j_2 < \dots < j_n$, $k_1 < k_2 < \dots < k_n$. Then since $\models_M g_i(t) < g_j(t)$ and $g_i(t) \in B_{m+1}$ for all $i < j$, $t \geq m+1$, $G_m(g_{j_1}(t),\dots,g_{j_n}(t)) = G_m(g_{k_1}(t),\dots,g_{k_n}(t))$ all $t \geq m+1$. $\therefore\ \theta(g_{j_1}(t),\dots,g_{j_n}(t)) \leftrightarrow \theta(g_{k_1}(t),\dots,g_{k_n}(t))$ all $t \geq m+1$. $\therefore \models_N \theta(\bar{g}_{j_1},\dots,\bar{g}_{j_n}) \leftrightarrow \theta(\bar{g}_{k_1},\dots,\bar{g}_{k_n})$.

Proof of 2): Let $h \in |M|^{\omega_0}$ and suppose $\tilde{h} \trianglelefteq \tilde{g}_1 \vee \dots \vee \tilde{g}_n$. Pick $m \in \omega_0$ such that

$$\bar{h} = \overline{G_m(g_1,\dots,g_n)}.$$

By choice of B_{m+1} there are $i_1, \dots, i_t \leq n$ such that for all $\langle a_1,\dots,a_n\rangle, \langle b_1,\dots,b_n\rangle \in [B_{m+1}]^n$,

$$G_m(a_1,\dots,a_n) = G_m(b_1,\dots,b_n) \leftrightarrow \bigwedge_{k=1}^{t} a_{i_k} = b_{i_k}.$$

$\therefore$ since $g_i(s) \in B_{m+1}$ all $i \in \omega_0$, $s \geq m+1$, $\{s \in \omega_0 \mid h(s) = \mu z: (\exists \langle z_1,\dots,z_n\rangle \in [B_{m+1}]^n)((\bigwedge_{k=1}^{t} z_{i_k} = g_{i_k}(s)) \wedge G_m(z_1,\dots,z_n) = z)\} \in \mathcal{U}$, and for $1 \leq k \leq t$, $\{s \in \omega_0 \mid g_{i_k}(s) = \mu z: (\exists \langle z_1,\dots,z_n\rangle \in [B_{m+1}]^n)(z_{i_k} = z \wedge h(s) = G_m(z_1,\dots,z_n)\} \in \mathcal{U}$. $\therefore$ $\tilde{h} \trianglelefteq \tilde{g}_{i_1} \vee \dots \vee \tilde{g}_{i_t}$ and $\tilde{g}_{i_k} \trianglelefteq \tilde{h}$ all $1 \leq k \leq t$. $\therefore$ $\tilde{h} = \tilde{g}_{i_1} \vee \dots \vee \tilde{g}_{i_t}$.

From 1), 2) it now follows that the map

$$\psi : \mathcal{P}(\mathbb{N}) \longrightarrow \mathfrak{B}([\{\bar{g}_i \mid i \in \mathbb{N}\}]_N)$$

defined by $\psi(e) = [\{\bar{g}_i \mid i \in e\}]_N$ is an isomorphism from $\langle \mathcal{P}(\mathbb{N}), \subset\rangle$ onto $\mathfrak{B}([\{\bar{g}_i \mid i \in \mathbb{N}\}]_N), \subset\rangle$. Now for A an arbitrary set let $\mathcal{J} = \langle A, <_{\mathcal{J}}\rangle$ be a linear ordering of A and set

$$\Phi(\mathcal{J}) = \{\theta(x_{a_1},\dots,x_{a_n}) \mid \models_N \theta(\bar{g}_1,\dots,\bar{g}_n) \;\&\; a_1,\dots,a_n \in A \;\&\; a_1 <_{\mathcal{J}} a_2 <_{\mathcal{J}} \dots <_{\mathcal{J}} a_n\}.$$

Clearly $\Phi(\mathcal{J})$ is consistent. Let $M(\mathcal{J})$ be the minimal model in which $\Phi(\mathcal{J})$ is satisfied. Then $M(\mathcal{J})$ is a model of T and by 1), 2) it follows that

$$\langle \mathcal{P}(A), \subset\rangle \cong \langle \mathfrak{B}(M(\mathcal{J})), \subset\rangle.$$

□

An Aside (Gaifman, Hodges)

If $\mathcal{J}_1$, $\mathcal{J}_2$ are linear orderings and h an embedding of $\mathcal{J}_1$ into $\mathcal{J}_2$ then h induces an elementary embedding $\bar{h}$ of $M(\mathcal{J}_1)$ into $M(\mathcal{J}_2)$ in the obvious way. Conversely if f is an elementary embedding of $M(\mathcal{J}_1)$ into $M(\mathcal{J}_2)$ then there is an embedding h of $\mathcal{J}_1$ into $\mathcal{J}_2$ such that $f = \bar{h}$. (Hint: notice that if $a \in |N|$ satisfies the same type as $\bar{g}_1$ in $[\{\bar{g}_j \mid j \in \mathbb{N}\}]_N$ then $a = \bar{g}_j$ some $j \in \mathbb{N}$.)

This gives the following interesting result.

LEMMA 1.7. Let λ be a cardinal and $\mathcal{J}_1$, $\mathcal{J}_2$ linear orderings such that for any linear ordering $\mathcal{J}_3$, $\mathcal{J}_3$ can be embedded into $\mathcal{J}_1$ and into $\mathcal{J}_2$ iff $\mathrm{card}(|\mathcal{J}_3|) < \lambda$. Then for P an elementary substructure of $M(\mathcal{J}_1)$, P can be embedded elementarily into $M(\mathcal{J}_2)$ iff P is generated in $M(\mathcal{J}_1)$ by less than λ elements. □

Remark. As shown by Gaifman [2] we can, by improving 1.4, show that for any $f_i \in |M|^{\omega_0}$, $i \in \mathbb{N}$ there are $g_j \in |M|^{\omega_0}$, $j \in \mathbb{N}$ such that

1) $\{\bar{g}_j \mid j \in \mathbb{N}\}$ are (distinct) indiscernibles in $(N, \bar{f}_i)_{i\in\mathbb{N}}$ (and so also in $[\{\bar{g}_j \mid j \in \mathbb{N}\}]_N$),

2) if $h \in |M|^{\omega_0}$, $m \in \mathbb{N}$ then $\tilde{h} \trianglelefteq \tilde{g}_1 \vee \dots \vee \tilde{g}_m \leftrightarrow \exists i_1, \dots, i_t \leq m$, $\tilde{h} = \tilde{g}_{i_1} \vee \dots \vee \tilde{g}_{i_t}$ or $\exists n \in \mathbb{N}$, $\tilde{h} \trianglelefteq \tilde{f}_1 \vee \dots \vee \tilde{f}_n$.

This gives the following generalization of theorem 2. "Let B be a countable model of T (so since N is ω_1-saturated we may assume $B \in \mathfrak{B}(N)$) and let A be any set. Then there is a model D of T such that

$$\langle \mathfrak{B}(B), \subset \rangle + \langle \mathcal{P}(A), \subset \rangle \cong \langle \mathfrak{B}(D), \subset \rangle."$$

Section 2

In this section we give a proof of theorem 3. Before giving the proof we recall some results and notation of lattice theory. (See [1].)

Finite distributive lattices

Let π be a finite distributive lattice with least element 0. For $a \in |\pi|$ define $\ell^{\pi}(a)$ (the length of a in π) to be the length of the longest chain from 0 to a in π. Then $a < b$ in π implies $\ell^{\pi}(a) < \ell^{\pi}(b)$.

$a \in |\pi|$ is join-irreducible (j-i) if $a \neq 0$ and there do not exist $b, c \in |\pi|$ distinct from a such that $a = b \vee c$.

Let $\alpha_1, \ldots, \alpha_n$ be the j-i elements of π in non decreasing order of length and define

$$\theta : \pi \to \mathcal{P}(\{1,\ldots,n\})$$

by $\theta(x) = \{j \mid \alpha_j \leq x \text{ in } \pi\}$. Then $\theta : \pi \cong \langle \theta`\pi, \cup, \cap \rangle$ and for $x \in |\pi|$, $x = \bigvee\{\alpha_j \mid j \in \theta(x)\}$. Notice $\theta(\alpha_j) \subseteq \{1,\ldots,j\}$ since $\alpha_i \leq \alpha_j$ in π $\to \ell^{\pi}(\alpha_i) \leq \ell^{\pi}(\alpha_j) \to i \leq j$.

Compactly generated lattices

Let L be a lattice.

DEFINITION.

i) $D \subseteq |L|$ is directed if $a \in \mathcal{P}_{<\omega_0}(D) \to \bigvee a \leq t$ for some $t \in D$.

ii) $c \in |L|$ is compact if for all $S \subseteq |L|$ if $c \leq \bigvee S$ then $c \leq \bigvee G$ some finite $G \subseteq S$.

iii) L is compactly generated if every element of L is a sup of compact elements of L.

iv) L is ω_α-compactly generated if L is compactly generated and there are $\leq \omega_\alpha$ compact elements in L.

PROPOSITION 2.1. ([1] p. 187) If L is a complete, compactly generated lattice, $a \in |L|$, $D \subseteq |L|$ and D directed then

$$a \wedge \bigvee_{s \in D} s = \bigvee_{s \in D} s \wedge a. \qquad \square$$

PROPOSITION 2.2. (Birkhoff-Frink, [1] p. 187) A lattice L is complete and ω_α-compactly generated iff L is isomorphic to the subalgebra lattice of some universal algebra [A,F] with $\|A\| \leq \omega_\alpha$. (Here $\|A\|$ is the cardinality of A.) $\square$

Since for B a model of T we have, $A \in \mathfrak{B}(B)$ iff A is a subalgebra of B,

COROLLARY 2.3. If B is a model of T then $\langle \mathfrak{B}(B), \subset \rangle$ is complete and $\|B\|$-compactly generated. $\square$

This corollary shows that theorem 3 is a best result, given that $\langle \mathfrak{B}(B), \subset \rangle$ is distributive.

Upper semi-lattices

Let K be an upper semi-lattice with least element O.

PROPOSITION 2.4. $\langle Id(K), \subset \rangle$ is a complete, $\|K\|$-compactly generated lattice (the compact elements being the principal ideals). $\square$

LEMMA 2.5. For L a lattice, L is complete ω_α-compactly generated iff L is isomorphic to $\langle Id(K), \subset \rangle$ for some bottomed upper semi-lattice K such that $\|K\| \leq \omega_\alpha$.

Proof. $\rightarrow$ Let K be the compact elements of L with $<_L$ restricted to $|K|$. Then $0_L \in |K|$, K forms an upper semi-lattice and the map

$$I \rightarrow \bigvee I \quad \text{for I an ideal in K}$$

is an isomorphism of $\langle Id(K), \subset \rangle$ to L. □

DEFINITION. An upper semi-lattice K is <u>distributive</u> if $\langle Id(K), \subset \rangle$ is.

PROPOSITION 2.6. If every countable distributive bottomed upper semi-lattice can be embedded as an initial segment of $\mathcal{L}$ then every complete distributive ω_0-compactly generated lattice can be embedded as an initial segment of $\mathfrak{P}(N)$.

<u>Proof</u>. By lemma 0. □

Thus the theorem will be proved if we can show that every countable distributive bottomed upper semi-lattice can be embedded as an initial segment of $\mathcal{L}$. Towards this end we show the following results.

LEMMA 2.7. Let L be a complete, compactly generated lattice, in which the compact elements can be well ordered. Let $e, f \in |L|$, $e < f$. Then $\exists\, e', f'$ such that $e \leq e' < f' \leq f$ and f' is minimal above e'. Furthermore $\langle e', f' \rangle$ may be picked uniformly.

<u>Proof</u>. Let $e, f \in |L|$, $e < f$. Pick a compact $c \in |L|$ such that $c \leq f$ and $\neg(c \leq e)$. Then $e < e \vee c$. We now pick a sequence $\{d_\nu\}$ of compact elements of L. Set $d_0 = 0_L$ and suppose d_ν picked for $\nu < \alpha$ such that for $\beta < \alpha$, $e \vee \bigvee_{\nu \leq \beta} d_\nu < e \vee c$. Now pick d_α, if possible, distinct from the d_ν for $\nu < \alpha$ such that

$$e \vee \bigvee_{\nu \leq \alpha} d_\nu < e \vee c.$$

If this is not possible stop. Suppose the sequence d_ν stops at stage λ. If $e \vee \bigvee_{\nu<\lambda} d_\nu = e \vee c$ then $c \leq e \vee (d_{\nu_1} \vee \dots \vee d_{\nu_n})$ some $\nu_1, \dots, \nu_n < \lambda$ which is clearly impossible. Thus $e \vee \bigvee_{\nu<\lambda} d_\nu < e \vee c$. But then there can be no $h \in L$ such that

$$e \vee \bigvee_{\nu<\lambda} d_\nu < h < e \vee c$$

so $e' = e \vee \bigvee_{\nu<\lambda} d_\nu$, $f' = e \vee c$ will do. Clearly e', f' can be picked uniformly. □

LEMMA 2.8. Let K be a distributive upper semi-lattice with least element 0, and let $\mathcal{S}$ be a finite subset of $|K|$. Then $\exists\, d_1, \ldots, d_n \in |K|$ such that

a) For $a \in \mathcal{S}$ $\exists\, i_1, \ldots, i_m \leq n$, $a = d_{i_1} \vee \ldots \vee d_{i_m}$.

b) If $t \subseteq \{1, \ldots, n\}$, $i \leq n$ then $d_i \leq \bigvee_{\nu \in t} d_\nu \rightarrow d_i \leq d_\nu$ some $\nu \in t$.

c) For $i, j \leq n$, $d_i \leq d_j \rightarrow i \leq j$.

<u>Proof</u>. Let $L = \langle Id(K), \subset \rangle$. We treat $|K|$ as a subset of $|L|$. Let G be the finite sublattice of L generated by $\mathcal{S}$, and let $H = \{\langle e,f \rangle \mid e, f \in G,\ e < f\}$. For each pair $\langle e,f \rangle \in H$ pick e', f' such that $e \leq e' < f' \leq f$ and f' is minimal above e' (by lemma 2.7).

Now define a relation $\sim_{ef}$ on L by

$$a \sim_{ef} b \leftrightarrow (a \vee e') \wedge f' = (b \vee e') \wedge f'.$$

Then $\sim_{ef}$ is a congruence relation and there are exactly two equivalence classes, one containing e, the other f.

Define $\Phi : L \longrightarrow \mathcal{P}(H)$ by

$$\Phi(a) = \{\langle e,f \rangle \mid a \sim_{ef} f\}.$$

Φ is a homomorphism onto $\langle \Phi L, \cup, \cap \rangle$ and is 1-1 on G.

We will now pick d's to satisfy a) - c) by a sequence of approximations.

For $s \in G$ pick D_s to be the set of elements of K less than or equal to s in L. Then D_s is directed and $s = \bigvee D_s$ in L. We now have the following two results:

i) Let $s \in G$. Then $\exists\, d_0 \in D_s$ such that for all $d \in D_s$, $d \geq d_0$, $\Phi(d) = \Phi(s)$.

Proof: Let $\langle e,f\rangle \in H$ and pick $d_{ef} \in D_s$ as follows: if $s \sim_{ef} e$ set d_{ef} to be any element of D_s. Since $d_{ef} \leq s$, $d_{ef} \sim_{ef} e$. Otherwise $s \sim_{ef} f$ so

$$\begin{aligned} f' = (e' \vee s) \wedge f' &= (\textstyle\bigvee_{d \in D_s} e' \vee d) \wedge f' \\ &= \textstyle\bigvee_{d \in D_s} (e' \vee d) \wedge f'. \end{aligned}$$

Thus we can pick $d_{ef} \in D_s$ such that $(e' \vee d_{ef}) \wedge f' = f'$ so $d_{ef} \sim_{ef} f$. Finally set $d_0 = \bigvee_{\langle e,f\rangle \in H} d_{ef}$ and let $d \in D_s$, $d \geq d_0$. Then for $\langle e,f\rangle \in H$, $s \sim_{ef} e \rightarrow d \sim_{ef} e$ since $d \leq s$, and $s \sim_{ef} f \rightarrow d_{ef} \sim_{ef} f$ $\rightarrow d \sim_{ef} f$ since $d \geq d_{ef}$. Thus $\Phi(d) = \Phi(s)$.

ii) Let $a \in \mathcal{S}$, $b_1, \ldots, b_j \in G$ and $a = \bigvee_{i=1}^{j} b_i$. Then $\exists\, \langle d_1^0, \ldots, d_j^0\rangle \in D_{b_1} \times \ldots \times D_{b_j}$ such that for all $\langle d_1, \ldots, d_j\rangle \in D_{b_1} \times \ldots \times D_{b_j}$ with $d_i \geq d_i^0$, $i = 1, \ldots, j$,

$$a = \bigvee_{i=1}^{j} d_i.$$

Proof: $$\begin{aligned} a = \textstyle\bigvee_{i=1}^{j} b_i &= \textstyle\bigvee_{i=1}^{j} \bigvee D_{b_i} \\ &= \textstyle\bigvee \{\bigvee_{i=1}^{j} d_i \mid \langle d_1, \ldots, d_j\rangle \in D_{b_1} \times \ldots \times D_{b_j}\} \end{aligned}$$

so since the set on the r.h.s. is directed the result follows by the compactness of a.

Now let $e_1, \ldots, e_n$ be the j-i elements of G in ascending order of length and for $i = 1, \ldots, n$ pick, by using i), ii), $d_i^0 \in D_{e_i}$ such that $\Phi(d_i^o) = \Phi(e_i)$ and if $t \subseteq \{1,\dots,n\}$, $a \in \mathcal{A}$ and $a = \bigvee_{i\in t} e_i$ then $a = \bigvee_{i\in t} d_i^o$. For $i \leqslant n$ set $d_i = \bigvee\{d_j^o \mid e_j \leqslant e_i\}$.

It remains to check a) - c) for these d_i. To show a) notice $d_i^o \leqslant d_i \leqslant e_i$ so since $\exists\ i_1, \dots, i_m \leqslant n$ such that $a = e_{i_1} \vee \dots \vee e_{i_m}$,

$$a = d_{i_1}^o \vee \dots \vee d_{i_m}^o \leqslant d_{i_1} \vee \dots \vee d_{i_m} \leqslant e_{i_1} \vee \dots \vee e_{i_m} = a$$

so a) is proved.

For b) suppose $t \subseteq \{1,\dots,n\}$ and

$$d_i \leqslant \bigvee_{\nu\in t} d_\nu .$$

Then since $\Phi(d_j) = \Phi(e_j)$ for $j \leqslant n$,

$$\Phi(e_i) = \bigcup_{\nu\in t} \Phi(e_\nu).$$

But Φ is 1-1 on G so

$$\begin{aligned} e_i \leqslant \bigvee_{\nu\in t} e_\nu &\rightarrow e_i \leqslant e_\nu \quad \text{some } \nu \in t \\ &\rightarrow d_i \leqslant d_\nu \quad \text{some } \nu \in t. \end{aligned}$$

Finally for $i, j \leqslant n$,

$$\begin{aligned} d_i \leqslant d_j &\rightarrow \Phi(d_i) \subseteq \Phi(d_j) \\ &\rightarrow \Phi(e_i) \subseteq \Phi(e_j) \\ &\rightarrow e_i \leqslant e_j \\ &\rightarrow i \leqslant j \end{aligned}$$

so c) is proved. □

Notice if we are given a well ordering of K then $d_1, \ldots, d_n$ can be picked uniformly.

Combinatorial lemmas

Throughout this section let π be a finite lattice with least element 0. Let $\alpha_1, \ldots, \alpha_n$ be non-zero elements of π such that

i) if $a \in |\pi|$ then $\exists\, i_1, \ldots, i_m \leq n$ such that $a = \bigvee_{k=1}^{m} \alpha_{i_k}$,

ii) if $t \subseteq \{1,\ldots,n\}$, $i \leq n$ and $\alpha_i \leq \bigvee_{\nu\in t} \alpha_\nu$ then $\alpha_i \leq \alpha_\nu$ some $\nu \in t$,

iii) $\alpha_i \leq \alpha_j \rightarrow i \leq j$

(i.e. the α_i are the j-i elements of π).

DEFINITIONS

i) For $B \subseteq [|M|]^n$, $\langle a_1,\ldots,a_m\rangle \in |M|^m$,

$$B[a_1,\ldots,a_m] = \{c \mid (\exists \langle c_1,\ldots,c_n\rangle \in B)(c = c_{m+1} \wedge \bigwedge_{j=1}^{m} a_j = c_j)\}.$$

ii) Let $s = \{i_1,\ldots,i_m\}$ (in ascending order) be a subset of $\{1,\ldots,n\}$. A map H with $\mathrm{dom}(H) \subseteq [|M|]^m$ is (s,n)-<u>defined</u> if H is defined on all $\langle a_1,\ldots,a_m\rangle$ such that $\exists \langle b_1,\ldots,b_n\rangle \in [|M|]^n$ such that $\bigwedge_{j=1}^{m} a_j = b_{i_j}$.

For H, s as above, $\langle b_1,\ldots,b_n\rangle \in [|M|]^n$ write $H(b_j : j \in s)$ for $H(b_{i_1},\ldots,b_{i_m})$.

DEFINITION 2.9. $B \subseteq [|M|]^n$ is π-<u>fat</u> if the following hold:

a) B is definable (in M) and $B \neq \emptyset$.

b) If $\langle a_1,\ldots,a_n\rangle \in B$, $i < n$ then $B[a_1,\ldots,a_i]$ is M-inf.

c) If $\langle a_1,\ldots,a_n\rangle$, $\langle b_1,\ldots,b_n\rangle \in B$, $i < n$, $a_i \leq b_i$ and $\bigwedge_{\alpha_j<\alpha_{i+1}} a_j = b_j$ then

$$c \in B[b_1,\dots,b_i] \leftrightarrow c > b_i \wedge c \in B[a_1,\dots,a_i].$$

d) If $\langle a_1,\dots,a_n\rangle$, $\langle b_1,\dots,b_n\rangle \in [|M|]^n$, $i < n$ then

$$\bigvee_{\alpha_j<\alpha_{i+1}} a_j \neq b_j \rightarrow B[a_1,\dots,a_i] \cap B[b_1,\dots,b_i] = \emptyset.$$

LEMMA 2.10. Let P be M-inf. Then there is a π-fat set $B \subseteq [P]^n$.

<u>Proof</u>. Define $B_0 = \{\langle a_1,\dots,a_n\rangle \mid \bigwedge_{j=1}^{n} (a_j$ is the $\langle a_k : k \in \{\nu \mid \alpha_\nu < \alpha_j\}\rangle^\frown\langle b\rangle$'th element of P for some $b)\}$. Then B_0 satisfies a), b), d) of 2.9 and for $\langle a_1,\dots,a_n\rangle$, $\langle b_1,\dots,b_n\rangle \in B_0$, $i < n$,

$$\bigwedge_{\alpha_j<\alpha_{i+1}} a_j = b_j \rightarrow B_0[a_1,\dots,a_i] = B_0[b_1,\dots,b_i].$$

Setting $B = B_0 \cap [P]^n$ now gives the required π-fat set. □

LEMMA 2.11. Let $F \in \mathscr{F}_n$ $(n > 0)$. Then $\exists$ a π-fat set B and $x \in |\pi|$ such that for $\langle a_1,\dots,a_n\rangle$, $\langle b_1,\dots,b_n\rangle \in B$,

$$\bigwedge_{\alpha_j \leq x} a_j = b_j \leftrightarrow F(a_1,\dots,a_n) = F(b_1,\dots,b_n).$$

<u>Proof</u>. By lemma 1.13 pick P M-inf such that for some $i_1 < \dots < i_m \leq n$, if $\langle a_1,\dots,a_n\rangle$, $\langle b_1,\dots,b_n\rangle \in [P]^n$ then

$$F(a_1,\dots,a_n) = F(b_1,\dots,b_n) \leftrightarrow \bigwedge_{k=1}^{m} a_{i_k} = b_{i_k}.$$

Let $x = \bigvee_{k=1}^{m} \alpha_{i_k}$ and, by lemma 2.10, pick $B \subseteq [P]^n$, B π-fat. Then for $\langle a_1,\dots,a_n\rangle$, $\langle b_1,\dots,b_n\rangle \in B$,

i) $\bigwedge_{\alpha_j \leq x} b_j = a_j \rightarrow \bigwedge_{k=1}^{m} a_{i_k} = b_{i_k} \rightarrow F(a_1,\dots,a_n) = F(b_1,\dots,b_n).$

ii) If $\bigvee_{\alpha_j \leq x} b_j \neq a_j$ pick j such that $\alpha_j \leq x$, $b_j \neq a_j$. Then for some $k = 1,\dots,m$, $\alpha_j \leq \alpha_{i_k}$. ∴ By 2.9 d), $a_{i_k} \neq b_{i_k}$ so $F(a_1,\dots,a_n) \neq F(b_1,\dots,b_n)$.

This proves the result. □

NOTATION. Let π be an upper semi-sublattice of η, η a distributive, finite lattice and 0_η, $1_\eta \in |\pi|$. Let $\beta_1, \dots, \beta_m$ be the join irreducible elements of η arranged so that there are $e_1 < e_2 < \dots < e_m = n$ such that for $i \leq n$, $\bigvee_{j=1}^{i} \alpha_j = \bigvee_{j=1}^{e_i} \beta_j$, and $j(i) = \{j \mid \beta_j \leq \alpha_i\} \subseteq \{1,\dots,e_i\}$. Notice $e_{i-1} < e_i$.

With this notation,

LEMMA 2.12. Let B be π-fat. Then for $i \leq n$ ∃ $(j(i),m)$-defined, 1-1 functions H_i such that for $\langle a_1,\dots,a_m\rangle \in [|M|]^m$,

$$\langle H_1(a_s : s \in j(1)),\dots,H_n(a_s : s \in j(n))\rangle \in B.$$

<u>Proof</u>. Suppose H_k defined for $k < i \leq n$ such that it is 1-1, $(j(k),m)$-defined and for $\langle a_1,\dots,a_{e_{i-1}}\rangle \in [|M|]^{e_{i-1}}$, ∃ $\langle d_i,\dots,d_n\rangle$ such that

$$\langle H_1(a_s : s \in j(1)),\dots,H_{i-1}(a_s : s \in j(i-1)),d_i,\dots,d_n\rangle \in B.$$

Now define H_i by: $H_i(b_s : s \in j(i)) = z$ iff z is the $\langle b_{e_{i-1}+1},\dots,b_{e_i}\rangle$'th element of $G(b_s : s \in j(i))$, where $G(b_s : s \in j(i)) = \{c \mid (\exists \langle c_1,\dots,c_n\rangle \in B)(\bigwedge_{\alpha_k<\alpha_i} c_k = H_k(b_s : s \in j(k)) \wedge c_i = c > H_k{}^{\text{`}}[\{a \mid a < b_{e_i}\}]^{\|j(k)\|}$ for all $k < i\}$.

H_i is $(j(i),m)$-defined by 2.9 b).

If $\langle a_1,\dots,a_{e_i}\rangle \in [|M|]^{e_i}$ then the following hold:

i) There is a $<c_1,\dots,c_n> \in B$ such that $\bigwedge_{\alpha_k \leq \alpha_i} c_k =$ $H_k(a_s : s \in j(k))$ and $c_i = c > H_{i-1}(a_s : s \in j(i-1))$.

ii) There is a $<d_1,\dots,d_n> \in B$ such that $\bigwedge_{k=1}^{i-1} d_k = H_k(a_s : s \in j(k))$.

$c_i > c_{i-1}, d_{i-1}$ so by 2.9 c),

$$
\begin{aligned}
c_i \in B[c_1,\dots,c_{i-1}] &\rightarrow c_i \in B[d_1,\dots,d_{i-1}] \\
&\rightarrow \exists\, h_{i+1},\dots,h_n \text{ such that } \\
&\quad <d_1,\dots,d_{i-1},c_i,h_{i+1},\dots,h_n> \in B \\
&\rightarrow \exists\, h'_{i+1},\dots,h_n \text{ such that } \\
&\quad <H_1(a_s : s \in j(1)),\dots,H_i(a_s : s \in j(i)), \\
&\quad h_{i+1},\dots,h_n> \in B.
\end{aligned}
$$

It only remains to show H_i is 1-1. Let $<b_s : s \in j(i)>$, $<a_s : s \in j(i)>$ be distinct elements of $\mathrm{dom}(H_i)$ and let $<c_1,\dots,c_n>$, $<d_1,\dots,d_n> \in B$ be such that

$$\bigwedge_{\alpha_k \leq \alpha_i} c_k = H_k(b_s : s \in j(k))$$

$$\bigwedge_{\alpha_k \leq \alpha_i} d_k = H_k(a_s : s \in j(k)).$$

If $b_s \neq a_s$ some $s \in j(k)$ and $\alpha_k < \alpha_i$ then since H_k is 1-1 $d_k \neq c_k$ so by 2.9 d), $c_i \neq d_i$. Otherwise $b_s = a_s$ for all $s \in j(i) \cap \{1,\dots,e_{i-1}\}$. Recall c_i is the $<b_{e_{i-1}+1},\dots,b_{e_i}>$'th element of $G(b_s : s \in j(i))$ and d_i is the $<a_{e_{i-1}+1},\dots,a_{e_i}>$'th element of $G(a_s : s \in j(i))$. Thus if $b_{e_i} = a_{e_i}$ then $G(b_s : s \in j(i)) = G(a_s : s \in j(i))$ so since $<b_{e_{i-1}+1},\dots,b_{e_i}> \neq <a_{e_{i-1}+1},\dots,a_{e_i}>$, $c_i \neq d_i$. Finally if $b_{e_i} \neq a_{e_i}$ assume without loss of generality that $b_{e_i} < a_{e_i}$. Then

$G(a_s : s \in j(i)) \subseteq G(b_s : s \in j(i))$ and $<b_{e_{i-1}+1},\dots,b_{e_i}> <_{e_i-e_{i-1}}$ $<a_{e_{i-1}+1},\dots,a_{e_i}>$ so again $c_i \neq d_i$.

This concludes the proof. □

From now on let K be a countable bottomed distributive upper semi-lattice. We shall show K can be embedded as an initial segment of $\mathcal{L}$.

Let $\{z_i \mid i < \omega_0\}$ enumerate the elements of $|K|$ and let 0 be the least element in K. Without loss of generality we may assume K has a greatest element 1, and $z_0 = 1$. We now define a sequence of upper semi-sublattices π_n of K and elements $\alpha_1^n, \dots ,\alpha_{q_n}^n$ of π_n as follows:

Set $\pi_0 = <\{0,1\},<>$ and $\alpha_1^0 = 1$. Suppose now π_n has been found and we have picked elements $\alpha_1^n, \dots ,\alpha_{q_n}^n$ in $|\pi_n|$ such that the following hold:

i) For $a \in |\pi_n|$ $\exists$ $i_1,\dots,i_t \leq q_n$ such that $a = \bigvee_{j=1}^t \alpha_{i_j}^n$ (in K).

ii) For $i, j \leq q_n$, $\alpha_i^n \leq \alpha_j^n \rightarrow i \leq j$.

iii) If $i \leq q_n$, $t \subseteq \{1,\dots,q_n\}$ and $\alpha_i^n \leq \bigvee_{\nu \in t} \alpha_\nu^n$ then $\alpha_i^n \leq \alpha_\nu^n$ some $\nu \in t$.

Set $\mathcal{S} = \{\alpha_1^n,\dots,\alpha_{q_n}^n, z_{n+1}\}$ and by lemma 2.8 pick $d_1, \dots ,d_{q_{n+1}}$ satisfying a), b), c) of lemma 2.8 for this $\mathcal{S}$. Let π_{n+1} be the upper semi-sublattice generated by $\{d_1,\dots,d_{q_{n+1}}\}$. Notice π_n is an upper semi-sublattice of π_{n+1}, and, viewed in isolation, π_{n+1} is a lattice. Hence the previous remarks will hold with π_n, π_{n+1} in place of π, η etc.

We define $\alpha_1^{n+1},\dots,\alpha_{q_{n+1}}^{n+1}$ as follows: suppose $i \leq q_n$ and we have picked $\alpha_1^{n+1}, \dots ,\alpha_{e_{i-1}}^{n+1}$ from $d_1, \dots,d_{q_{n+1}}$ such that

$$\bigvee_{j=1}^{i-1} \alpha_j^n = \bigvee_{j=1}^{e_{i-1}} \alpha_j^{n+1}.$$

Let $\alpha^{n+1}_{e_{i-1}+1}, \ldots, \alpha^{n+1}_{e_i}$ be the new elements c of $d_1, \ldots, d_{q_{n+1}}$ such that $c \leq \alpha^n_i$ arranged in ascending order of length in π_{n+1}. Then clearly

$$\bigvee_{j=1}^{i} \alpha_j^n = \bigvee_{j=1}^{e_i} \alpha_j^{n+1}.$$

Since $\bigvee_{j=1}^{q_n} \alpha_j^n = 1$, $e_{q_n} = q_{n+1}$.

Clearly i) - iii) above hold with n + 1 in place of n.

DEFINITION. For $m \leq n$, $j \leq q_m$ set $j_n^m(j) = \{i \mid \alpha_i^n \leq \alpha_j^m\}$.

Construction

Let V^i, $i < \omega_0$, enumerate all "words" of the form $F(\alpha^n_{i_1}, \ldots, \alpha^n_{i_j})$ where $F \in \mathcal{F}_j$, $i_1 < \ldots < i_j \leq q_n$, in such a way that if V^i is $F(\alpha^n_{i_1}, \ldots, \alpha^n_{i_j})$ then $n \leq i$.

We shall now construct π_n-fat sets B_n and 1-1 definable $(j_n^m(i), q_n)$-defined maps ${}^mH_i^n$ for $m < n$, $i \leq q_m$ such that

$$\langle a_1, \ldots, a_{q_n} \rangle \in [|M|]^n \rightarrow$$
$$\langle {}^mH_1^n(a_s : s \in j_n^m(1)), \ldots, {}^mH_{q_m}^n(a_s : s \in j_n^m(q_m)) \rangle \in B_m.$$

First pick B_0 to be any π_0-fat set, (one exists by lemma 2.10). Now suppose B_n, ${}^mH_i^n$ have been successfully defined for $m < n$, $i \leq q_m$. By lemma 2.12, pick ${}^nH_i^{n+1}$ for $i \leq q_n$ which are 1-1, definable, $(j_{n+1}^n(i), q_{n+1})$-defined such that for $\langle a_1, \ldots, a_{q_{n+1}} \rangle \in [|M|]^{q_{n+1}}$,

$$\langle {}^nH_1^{n+1}(a_s : s \in j_{n+1}^n(1)), \ldots, {}^nH_{q_n}^{n+1}(a_s : s \in j_{n+1}^n(q_n) \rangle \in B_n.$$

For $m < n$, $i \leq q_m$ define

$${}^{m}H_i^{n+1}(a_s : s \in j_{n+1}^m(i)) = {}^{m}H_i^n({}^{n}H_s^{n+1}(a_t : t \in j_{n+1}^n(s)) : s \in j_n^m(i)).$$

It is easy to check ${}^{m}H_i^{n+1}$ is 1-1, definable, $(j_{n+1}^m(i), q_{n+1})$-defined and satisfies the inductive hypothesis.

Now let V^n be $F(\alpha_{i_1}^m, \ldots, \alpha_{i_j}^m)$, $i_1 < \ldots < i_j \leq q_m$. (So $n \geq m$.) Define $E_n : [|M|]^{q_{n+1}} \longrightarrow |M|$ by $E_n(a_1, \ldots, a_{q_{n+1}}) =$ $F({}^{m}H_{i_1}^{n+1}(a_s : s \in j_{n+1}^m(i_1)), \ldots, {}^{m}H_{i_j}^{n+1}(a_s : s \in j_{n+1}^m(i_j)))$ and by lemma 2.11 pick B_{n+1} to be π_{n+1}-fat such that for some $x \in \pi_{n+1}$, if $\langle a_1, \ldots, a_{q_{n+1}} \rangle$, $\langle b_1, \ldots, b_{q_{n+1}} \rangle \in B_{n+1}$ then

$$E_n(a_1, \ldots, a_{q_{n+1}}) = E_n(b_1, \ldots, b_{q_{n+1}}) \leftrightarrow \bigwedge_{\alpha_j^{n+1} \leq x} a_j = b_j.$$

<u>Construction of embedding</u>

For $m \in \omega_0$ pick $\langle z_1^m, \ldots, z_{q_m}^m \rangle \in B_m$ and for $n \in \omega_0$, $i \leq q_n$ pick $x_i^n \in |M|^{\omega_0}$ such that for all $m \geq n$,

$$x_i^n(m) = {}^{n}H_i^m(z_s^m : s \in j_m^n(i)).$$

We now define $\tau : |K| \longrightarrow |M|^{\omega_0}$ as follows. For $y \in |K|$ let n be minimal such that $y \in \pi_n$ and pick $\tau(y) \in |M|^{\omega_0}$ such that

$$\widetilde{\tau(y)} = \bigvee_{\alpha_i^n \leq y} \widetilde{x_i^n}.$$

We now claim $G = \{\widetilde{\tau(y)} \mid y \in K\}$ is an initial segment of $\mathcal{L}$ and the map $y \to \widetilde{\tau(y)}$ is an isomorphism. It remains to show:

1) If $x_1, \ldots, x_n \in |K|$, $F \in \mathcal{S}_n$ then $\exists\, x \in |K|$ such that $\widetilde{\tau(x)} = \widetilde{F(\tau(x_1), \ldots, \tau(x_n))}$. (This shows G is an initial segment of $\mathcal{L}$.)

2) $x, y \in |K|,\ x \not\leq y \rightarrow \tau(x) \not\trianglelefteq \tau(y).$

3) $x, y \in |K|,\ x < y \rightarrow \tau(x) \triangleleft \tau(y).$

Before proving 1) - 3) it will be useful to make some observations.

Let $m < n$, $i \leq q_m$. Then for $n \leq s$, ${}^{m}H_i^n(x_t^n(s) : t \in j_n^m(i)) = x_i^m(s)$. $\therefore$ Since ${}^{m}H_i^n$ is 1-1,

$$\widetilde{x_i^m} = \bigvee_{t \in j_n^m(i)} \widetilde{x_t^n}.$$

This result also holds for $n = m$ since if $\langle a_1, \ldots, a_{q_n} \rangle \in B_n$ then, since B_n is π_n-fat, a_i will uniquely determine the a_t such that $t \in j_n^m(i)$. Now let $y \in |K|$, m minimal such that $y \in \pi_m$. Then,

$$\begin{aligned} \widetilde{\tau(y)} &= \bigvee_{\alpha_i^m \leq y} \widetilde{x_i^m} \\ &= \bigvee_{\alpha_i^m \leq y} \bigvee_{\alpha_j^n \leq \alpha_i^m} \widetilde{x_j^n} \quad \text{for } n \geq m \\ &= \bigvee_{\alpha_j^n \leq y} \widetilde{x_j^n}. \end{aligned}$$

(Thus the choice of m is really immaterial.)

Proof of 1): Pick i such that $x_1, \ldots, x_n \in \pi_i$. Since

$$\widetilde{\tau(x_k)} = \bigvee_{\alpha_j^i \leq x_k} \widetilde{x_j^i} \quad \text{for } 1 \leq k \leq n$$

we can pick a definable function H such that

$$H(x_1^i, \ldots, x_{q_i}^i) \equiv F(\tau(x_1), \ldots, \tau(x_n)).$$

Let V^m be the condition $H(\alpha_1^i, \ldots, \alpha_{q_i}^i)$. By construction of B_{m+1} there is an $x \in \pi_{m+1}$ such that for $<a_1,\dots,a_{q_{m+1}}>$, $<b_1,\dots,b_{q_{m+1}}> \in B_{m+1}$,

i) $$\bigwedge_{\alpha_i^{m+1} \leq x} a_i = b_i \leftrightarrow E_m(a_1,\dots,a_{q_{m+1}}) = E_m(b_1,\dots,b_{q_{m+1}})$$

ii) $$E_m(a_1,\dots,a_{q_{m+1}}) = H({}^iH_1^{m+1}(a_s : s \in j_{m+1}^i(1)),\dots, {}^iH_{q_i}^{m+1}(a_s : s \in j_{m+1}^i(q_i))).$$

From i) we have for $s > m$,

$$<x_i^{m+1}(s) : \alpha_i^{m+1} \leq x> = \text{least } <c_i : \alpha_i^{m+1} \leq x> \text{ such that } (\exists <d_1,\dots,d_{q_{m+1}}> \in B_{m+1}) (\bigwedge_{\alpha_i^{m+1} \leq x} c_i = d_i \wedge E_m(d_1,\dots,d_{q_{m+1}}) = E_m(x_1^{m+1}(s),\dots,x_{q_{m+1}}^{m+1}(s))).$$

So $\lambda s: <x_i^{m+1}(s) : \alpha_i^{m+1} \leq x> \trianglelefteq E_m(x_1^{m+1},\dots,x_{q_{m+1}}^{m+1})$. Also from i), for $s > m$,

$$E_m(x_1^{m+1}(s),\dots,x_{q_{m+1}}^{m+1}(s)) = \text{least } c \text{ such that } (\exists <d_1,\dots,d_{q_{m+1}}> \in B_{m+1}) ((\bigwedge_{\alpha_i^{m+1} \leq x} d_i = x_i^{m+1}(s)) \wedge F(d_1,\dots,d_{q_{m+1}}) = c)$$

so $\lambda s: <x_i^{m+1}(s) : \alpha_i^{m+1} \leq x> \trianglerighteq E_m(x_1^{m+1},\dots,x_{q_{m+1}}^{m+1})$. Thus

$$\begin{aligned}\widetilde{\tau(x)} &= \bigvee_{\alpha_i^{m+1} \leq x} \widetilde{x_i^{m+1}} \\ &= \widetilde{E_m(x_1^{m+1},\dots,x_{q_{m+1}}^{m+1})} \\ &= \widetilde{H(x_1^i,\dots,x_{q_i}^i)} \quad \text{(by ii))}\end{aligned}$$

$$= \widetilde{F(\tau(x_1),\ldots,\tau(x_n))}.$$

Proof of 2): Let x, y $\in |K|$, $x \not\leq y$ and pick m such that x, y $\in |\pi_n|$. Since $x \not\leq y$ pick $f \leq q_m$ such that $\alpha_f^m \leq x$, $\alpha_f^m \not\leq y$. We shall show $x_f^m \not\trianglelefteq \tau(y)$ and this will clearly give the result since $x_f^m \trianglelefteq \tau(x)$. Let $p = \{k \mid \alpha_k^m \leq y\}$.

Now suppose $x_f^m \trianglelefteq \tau(y)$. (We shall derive a contradiction.) Pick a function F' such that $\{s \mid x_f^m(s) = F'(x_i^m(s) : i \in p)\} \in \mathcal{U}$, and let F be the function on $[|M|]^{q_n}$ defined by

$$F(a_1,\ldots,a_{q_m}) = \begin{cases} 0 & \text{if } a_f = F'(a_i : i \in p) \\ 1 & \text{otherwise.} \end{cases}$$

Let V_n be the condition $F(\alpha_1^m,\ldots,\alpha_{q_m}^m)$. Then for $\langle a_1,\ldots,a_{q_{n+1}}\rangle$,

$$E_n(a_1,\ldots,a_{q_{n+1}}) = \begin{cases} 0 & \text{if } {}^mH_f^{n+1}(a_s : s \in j_{n+1}^m(f)) = \\ & F'({}^mH_k^{n+1}(a_s : s \in j_{n+1}^m(k)) : k \in p) \\ 1 & \text{otherwise.} \end{cases}$$

Since $\alpha_f^m \not\leq \bigvee_{k\in p} \alpha_k^m$ we can pick $t \in j_{n+1}^m(f) - \bigcup_{k\in p} j_{n+1}^m(k)$. $\therefore$ Since E_n must be constant on B_{n+1} and ${}^mH_f^{n+1}$ is 1-1, E_n must have constant value 1 on B_{n+1}. $\therefore$ For $\langle a_1,\ldots,a_{q_{n+1}}\rangle \in B_{n+1}$,

$${}^mH_f^{n+1}(a_s : s \in j_{n+1}^m(f)) \neq F'({}^mH_k^{n+1}(a_s : s \in j_{n+1}^m(k)) : k \in p)$$

so $x_f^m(s) \neq F'(x_k^m(s) : k \in p)$ all $s \geq n+1$ - contradiction.

Our assumption is false and so we must have $\tau(x) \not\trianglelefteq \tau(y)$.

Proof of 3): Let x, y $\in |K|$, $x < y$. Pick m such that x, y $\in$

$|\pi_m|$. Then since $\{j \mid \alpha_j^m \leq x\} \subseteq \{j \mid \alpha_j^m \leq y\}$, $\tau(x) \trianglelefteq \tau(y)$. By 2), $\tau(x) \ntrianglelefteq \tau(y)$ so we must have $\tau(x) \triangleleft \tau(y)$ and this completes the proof.

□

COROLLARY 2.13. Any bottomed distributive countable upper semi-lattice can be embedded as an initial segment of $\mathcal{L}$. □

Theorem 3 now follows by lemma 0. □

Section 3

In this section we state some small results related to the problem "for what non-distributive lattices L are there models B of T such that $\langle \mathfrak{B}(B), \subset \rangle \cong L$?"

We can show that the pentagon and 1-n-1 lattices ($1 \leq n < \omega_0$) are sublattices of $\langle \mathfrak{B}(N), \subset \rangle$ but no isomorphism results are known.

In the opposite direction Gaifman and myself have shown that if the standard model of arithmetic, with added functions etc., is a model of T then there is no model B of T such that $\langle \mathfrak{B}(B), \subset \rangle \cong$ 1-n-1 lattice for $3 \leq n < \omega_0$.

REFERENCES

[1] G. Birkhoff, *Lattice Theory*, Amer. Math. Soc. Colloq. Publications Vol. xxv, 3rd edition (1967).

[2] H. Gaifman, Uniform Extension Operators for Models and their Applications, in *Sets, Models and Recursion Theory*, ed. Crossley, North-Holland (1967).

$\tilde{\Delta}_1$-DEFINABILITY IN SET THEORY[1]

Moto-o Takahashi

Rikkyo University, Tokyo

§0. $\tilde{\Delta}_1$-definability

As a generalization of Lévy's concept of restricted formulas ([5]), we introduced the concept of quasi-bounded formulas in set theory in [10].

A quasi-bounded formula (abbreviated by q.b.f. or by $\tilde{\Delta}_0$-formula) is a formula of set theory which is constructed by a finite number of applications of the following formation rules:

(i) If x and y are variables, then $x \in y$ is a q.b.f.;

(ii) If ϕ and ψ are q.b.f.'s, so are $\neg\phi$, $\phi \wedge \psi$, $\phi \vee \psi$, $\phi \supset \psi$, $\phi \equiv \psi$;

(iii) If ϕ is a q.b.f., and if x and y are variables, then

$$\forall x \in y[\phi], \quad \exists x \in y[\phi],$$

$$\forall x \subseteq y[\phi], \quad \exists x \subseteq y[\phi]$$

are q.b.f.'s.

Note. In a formula $\forall x \in y[\phi]$, for example, the variable x is bound but the y is free, even if x and y are syntactically identical.

---

[1] This work was supported by the Sakkokai Foundation.

Let Q be an axiomatic system of set theory formulated in the first order predicate calculus with the binary predicate symbol ϵ as its only non-logical symbol.

A formula ϕ of the theory Q is called $\tilde{\Sigma}_1^Q$ ($\tilde{\Pi}_1^Q$) iff

$$\vdash_Q \phi \equiv \exists x[\psi] \quad (\vdash_Q \phi \equiv \forall x[\psi])$$

for some q.b.f. ψ. Also ϕ is called $\tilde{\Delta}_1^Q$ iff it is both $\tilde{\Sigma}_1^Q$ and $\tilde{\Pi}_1^Q$. The notions of $\tilde{\Sigma}_n^Q$, $\tilde{\Pi}_n^Q$, $\tilde{\Delta}_n^Q$ are similarly defined.

An n-ary operation is a mapping which maps each n-tuple of sets to a set. Constants are construed as 0-ary operations.

A notion $\mathfrak{R}(a_1, \ldots, a_n)$ or an operation $\mathcal{F}(a_1, \ldots, a_n)$ is called $\tilde{\Sigma}_1$ ($\tilde{\Pi}_1$, $\tilde{\Delta}_1$, etc.) -definable in Q iff its defining formula ϕ is $\tilde{\Sigma}_1^Q$ ($\tilde{\Pi}_1^Q$, $\tilde{\Delta}_1^Q$, etc.), that is to say, $\mathfrak{R}$ or $\mathcal{F}$ is defined by

$$\mathfrak{R}(a_1, \ldots, a_n) \equiv \phi(a_1, \ldots, a_n)$$

or

$$b = \mathcal{F}(a_1, \ldots, a_n) \equiv \phi(b, a_1, \ldots, a_n)$$

for some $\tilde{\Sigma}_1^Q$ ($\tilde{\Pi}_1^Q$, $\tilde{\Delta}_1^Q$, etc.) -formula ϕ.

The aim of this paper is: (i) to show that almost all notions and operations that appear in set theory and other branches of mathematics (as formalized in Zermelo-Fraenkel set theory in a certain natural way) are $\tilde{\Delta}_1$-definable, (ii) to characterize this notion of $\tilde{\Delta}_1$-definability in somewhat semantical ways, and (iii) to investigate the relationship between Lévy's notion and ours.

The axiom of choice is not essential except in the proof of

Theorem 6. Through these investigations it would seem to be interesting to consider another new axiom system which includes the axiom schema of replacement for only $\tilde{\Delta}_1$-formulas besides certain other axioms. This will not be discussed here but presented elsewhere.

For the remainder of this paper we shall deal mainly with the system ZFC (Zermelo-Fraenkel set theory with the axiom of choice). So $\tilde{\Delta}_1^{ZFC}$, $\tilde{\Sigma}_1^{ZFC}$ etc. will be written simply as $\tilde{\Delta}_1$, $\tilde{\Sigma}_1$ etc. Similarly, notions and operations which are $\tilde{\Delta}_1$-definable in ZFC will be referred to simply as $\tilde{\Delta}_1$-definable (or $\tilde{\Delta}_1$-notions or $\tilde{\Delta}_1$-operations).

§1. Fundamental theorems on $\tilde{\Delta}_1$-definability

We shall first investigate some general closure properties of $\tilde{\Delta}_1$-definability. These are non-effective analogues of ones in recursion theory.

THEOREM 1.

(I) If $\mathfrak{A}(a_1, \ldots, a_n)$ and $\mathfrak{B}(a_1, \ldots, a_n)$ are $\tilde{\Delta}_1$-notions, so are $\neg\mathfrak{A}(a_1, \ldots, a_n)$, $\mathfrak{A}(a_1, \ldots, a_n) \supset \mathfrak{B}(a_1, \ldots, a_n)$, etc.

(II) If $\mathfrak{A}(b, a_1, \ldots, a_n)$ is a $\tilde{\Delta}_1$-notion, so are

$$\forall x \in a_i\ \mathfrak{A}(x, a_1, \ldots, a_n), \quad \exists x \in a_i\ \mathfrak{A}(x, a_1, \ldots, a_n),$$
$$\forall x \subseteq a_i\ \mathfrak{A}(x, a_1, \ldots, a_n), \quad \exists x \subseteq a_i\ \mathfrak{A}(x, a_1, \ldots, a_n),$$

and, more generally, if in addition $\mathcal{F}(a_1, \ldots, a_n)$ is a $\tilde{\Delta}_1$-operation, then

$$\forall x \in \mathcal{F}(a_1, \ldots, a_n)\ [\mathfrak{A}(x, a_1, \ldots, a_n)],$$
$$\exists x \in \mathcal{F}(a_1, \ldots, a_n)\ [\mathfrak{A}(x, a_1, \ldots, a_n)],$$
$$\forall x \subseteq \mathcal{F}(a_1, \ldots, a_n)\ [\mathfrak{A}(x, a_1, \ldots, a_n)],$$
$$\exists x \subseteq \mathcal{F}(a_1, \ldots, a_n)\ [\mathfrak{A}(x, a_1, \ldots, a_n)]$$

are $\tilde{\Delta}_1$.

(III) If $\mathcal{A}(b, a_1, \ldots, a_n)$ and $\mathcal{B}(b, a_1, \ldots, a_n)$ are $\tilde{\Delta}_1$-notions and if

$$\vdash_{ZFC} \exists x\ \mathcal{A}(x,a_1,\ldots,a_n) \equiv \forall x\ \mathcal{B}(x,a_1,\ldots,a_n),$$

then the notion $\mathcal{C}$ defined by

$$\mathcal{C}(a_1,\ldots,a_n) \equiv \exists x\ \mathcal{A}(x,a_1,\ldots,a_n)$$
$$(\equiv \forall x\ \mathcal{B}(x,a_1,\ldots,a_n))$$

is $\tilde{\Delta}_1$.

(IV) If an operation $\mathcal{F}(a_1,\ldots,a_n)$ is $\tilde{\Sigma}_1$-definable, then it is $\tilde{\Delta}_1$.

(V) If $\mathcal{F}_1(a_1, \ldots, a_n), \ldots, \mathcal{F}_m(a_1, \ldots, a_n)$ and $\mathcal{G}(b_1, \ldots, b_m)$ are $\tilde{\Delta}_1$-operations, so is the composed operation $\mathcal{H}$ defined by

$$\mathcal{H}(a_1,\ldots,a_n) = \mathcal{G}(\mathcal{F}_1(a_1,\ldots,a_n),\ldots,\mathcal{F}_m(a_1,\ldots,a_n)).$$

(VI) If $\mathcal{F}_1(a_1, \ldots, a_n), \ldots, \mathcal{F}_m(a_1, \ldots, a_n)$ are $\tilde{\Delta}_1$-operations and $\mathcal{A}(b_1, \ldots, b_m)$ is a $\tilde{\Delta}_1$-notion, then the notion $\mathcal{B}$ defined by

$$\mathcal{B}(a_1,\ldots,a_n) \equiv \mathcal{A}(\mathcal{F}_1(a_1,\ldots,a_n),\ldots,\mathcal{F}_m(a_1,\ldots,a_n))$$

is $\tilde{\Delta}_1$.

(VII) If $\mathcal{F}(b, a_1, \ldots, a_n)$ is a $\tilde{\Delta}_1$-operation, so is the operation $\mathcal{F}^*$ defined by

$$\mathcal{F}^*(b,a_1,\ldots,a_n) = \{\mathcal{F}(x,a_1,\ldots,a_n) \mid x \in b\}.$$

Proof. (I), (III) and the first half of (II) of the theorem follow immediately from the following equivalences with their duals:

(1.1) $$\neg\exists x\ \phi(x) \equiv \forall x\ \neg\phi(x);$$

(1.2) $$\exists x\ \phi(x) \wedge \exists y\ \psi(y) \equiv \exists x\ \exists y\ [\phi(x) \wedge \psi(y)];$$

(1.3) $$\exists x\ \exists y\ \phi(x,y) \equiv \exists z\ \exists x{\in}z\ \exists y{\in}z\ \phi(x,y);$$

(1.4) $$\exists x\ \phi(x) \vee \exists y\ \psi(y) \equiv \exists z\ [\phi(z) \vee \psi(z)];$$

(1.5) $$\forall t{\in}x\ \exists y\ \phi(t,x,y) \equiv \exists z\ \forall t{\in}x\ \exists y{\in}z\ \phi(t,x,y);$$

(1.6) $$\forall t{\subseteq}x\ \exists y\ \phi(t,x,y) \equiv \exists z\ \forall t{\subseteq}x\ \exists y{\subseteq}z\ \phi(t,x,y).$$

The second half of (II) follows from the equivalences

(1.7) $$\begin{aligned}\forall x{\in}\mathcal{F}(a_1,\ldots,a_n)\ [\mathfrak{A}(x,a_1,\ldots,a_n)]\\ \equiv \exists y\ [y = \mathcal{F}(a_1,\ldots,a_n) \wedge \forall x{\in}y\ \mathfrak{A}(x,a_1,\ldots,a_n)]\\ \equiv \forall y\ [y = \mathcal{F}(a_1,\ldots,a_n) \supset \forall x{\in}y\ \mathfrak{A}(x,a_1,\ldots,a_n)]\end{aligned}$$

etc., with the use of results already proved.

In order to prove (IV), assume that $\mathcal{F}(a_1, \ldots, a_n)$ is $\tilde{\Sigma}_1$. Then the notion "$b = \mathcal{F}(a_1, \ldots, a_n)$" is $\tilde{\Sigma}_1$. We have to prove that it is also $\tilde{\Pi}_1$. This follows from

(1.8) $$b = \mathcal{F}(a_1,\ldots,a_n) \equiv \forall y\ [y = \mathcal{F}(a_1,\ldots,a_n) \supset y = b].$$

Note that the equality is Δ_0- (and <u>a fortiori</u> $\tilde{\Delta}_1$-) definable:

$$a = b \equiv \forall x{\in}a\ [x \in b] \wedge \forall x{\in}b\ [x \in a],$$

and that $\mathcal{F}(a_1, \ldots, a_n)$ is defined for all sets $a_1, \ldots, a_n$. Similarly we can prove (V), (VI) and (VII) of the theorem using the following equivalences.

(1.9) $b = \mathcal{H}(a_1,\dots,a_n)$
$$\equiv \exists x_1 \dots \exists x_m \, [\bigwedge_{i=1}^{m}[x_i = \mathcal{F}_i(a_1,\dots,a_n)] \wedge b = \mathcal{G}(x_1,\dots,x_m)];$$

(1.10) $\mathfrak{b}(a_1,\dots,a_n)$
$$\equiv \exists x_1 \dots \exists x_m \, [\bigwedge_{i=1}^{m}[x_i = \mathcal{F}_i(a_1,\dots,a_n)] \wedge \mathfrak{A}(x_1,\dots,x_m)]$$
$$\equiv \forall x_1 \dots \forall x_m \, [\bigwedge_{i=1}^{m}[x_i = \mathcal{F}_i(a_1,\dots,a_n)] \supset \mathfrak{A}(x_1,\dots,x_m)];$$

(1.11) $c = \mathcal{F}^*(b,a_1,\dots,a_n)$
$$\equiv \forall y \epsilon c \; \exists x \epsilon b \; [y = \mathcal{F}(x,a_1,\dots,a_n)] \wedge \forall x \epsilon b \; \exists y \epsilon c \; [y = \mathcal{F}(x,a_1,\dots,a_n)].$$

q.e.d.

§2. Examples of $\tilde{\Delta}_1$-definable notions and operations

(I) Every Δ_1-definable (i.e. recursive in the sense of Kripke [4] and Platek [7]) notion and operation is $\tilde{\Delta}_1$-definable. The following are such examples:

"$a \subseteq b$", "$a = b$", "a is transitive",

"a is an ordinal", "a is a limit ordinal",

"$\alpha < \beta$" - the ordering on ordinals,

"$\{a_1,\dots,a_n\}$", "$\langle a_1,\dots,a_n\rangle$",

"$\cup a$" - the union of (elements of) a,

"$\cap a$" - the intersection of (elements of) a (with $\cap 0 = 0$),

"$a - b$",

"$\mathcal{D}(a)$" - the domain of a,

"$\mathcal{W}(a)$" - the range of a,

"$f`a$" - the application operation,

"$f\restriction a$", "$a + 1$", "$a \times b$",

"Tc(a)" - the transitive closure of a,

"a is of rank α",

"Ra(a)" - the rank of a,

"0", "1", ... , "ω",

Gödel's fundamental operations $\mathcal{F}_i$ $(1 \leq i \leq 8)$,

Gödel's F operation,

"$[f : a \longrightarrow b]$" - f is a function from a into b,

"$[f : a \xrightarrow{inj} b]$" - " injection "

"$[f : a \xrightarrow{sur} b]$" - " surjection "

"$[f : a \xrightarrow{bij} b]$" - " bijection "

(II) The simplest example of $\tilde{\Delta}_1$-definable operation which is not Δ_1 is the power set operation $\mathcal{P}$ defined by

$$b = \mathcal{P}(a) \equiv \forall x \epsilon b \, [x \subseteq a] \wedge \forall x \subseteq a \, [x \epsilon b].$$

Then Theorem 1 enables us to construct many other $\tilde{\Delta}_1$-notions and operations by combining $\mathcal{P}$ with ones already shown to be $\tilde{\Delta}_1$.

Examples are as follows.

1°. "$\mathcal{P}^n(a)$", "$\mathcal{P}^n(a \times b)$" etc. where $\mathcal{P}^n$ is defined as $\mathcal{P}^0(a) = a$, $\mathcal{P}^{n+1}(a) = \mathcal{P}(\mathcal{P}^n(a))$. (Use (V) of Theorem 1)

2°. "${}^a b$" - the set of all functions from a into b.

In fact,

$$c = {}^a b \equiv \forall x \subseteq (b \times a) \, [[x : a \longrightarrow b] \supset x \epsilon c] \wedge \forall x \epsilon c \, [x : a \longrightarrow b].$$

3°. "a ~ b" - a is equi-potent with b.

For

$$a \sim b \equiv \exists x \subseteq (b \times a)\ [x : a \xrightarrow{bij} b].$$

Similarly,

4°. "α is an initial ordinal (cardinal)" (≡ "α is an ordinal" $\wedge\ \forall \beta \in \alpha\ [\neg \beta \sim \alpha]$).

5°. The cardinality operation "$\bar{\bar{\ }}$".

$$\alpha = \bar{\bar{a}} \equiv \text{"}\alpha \text{ is a cardinal"} \wedge \alpha \sim a.$$

6°. "α is a regular ordinal"
≡ "α is a limit ordinal" $\wedge\ \forall \beta \in \alpha\ \forall f \in {}^{\beta}\alpha\ \exists \gamma \in \alpha\ [f``\beta \subseteq \gamma]$.

7°. "α is a limit cardinal"
≡ "α is a cardinal" $\wedge\ \forall \beta \in \alpha\ \exists \gamma \in \alpha\ [\beta < \gamma \wedge$ "γ is a cardinal"].

8°. Hartogs' $\aleph$ operation.

$$\alpha = \aleph(a) \equiv$$
$$\text{"}\alpha \text{ is an ordinal"} \wedge \forall \beta \in \alpha\ \exists f \in {}^{\beta}a\ [f : \beta \xrightarrow{inj} a] \wedge \neg \exists f \in {}^{\alpha}a\ [f : \alpha \xrightarrow{inj} a].$$

We also write α^+ instead of $\aleph(\alpha)$. If α is a cardinal, then α^+ is the next cardinal.

9°. "2^{α}" (the cardinal power operation) = $\overline{\overline{\mathcal{P}(\alpha)}}$.

10°. "$\rho^*(a)$" = $\overline{\overline{Tc(a)}}$.

11°. "α is a strong limit cardinal"
≡ "α is a cardinal" $\wedge\ \forall \beta \in \alpha\ [2^{\beta} < \alpha]$.

12°. "α is a weakly (strongly) inaccessible cardinal"
$\equiv$ "α is a (strong) limit cardinal" $\wedge$ "α is a regular ordinal".

13°. "α is a measurable cardinal".

To prove the $\tilde{\Delta}_1$-definability of it, let $U(u,a) \equiv$ "u is a non-principal ultrafilter over a". U is $\tilde{\Delta}_1$, since

$$U(u,a) \equiv u \subseteq \mathcal{P}(a) \wedge 0 \notin a \wedge \forall x \epsilon u\, \forall y \epsilon u\, [x \cap y \in u]$$
$$\wedge \forall x \subseteq a\, [x \in u \vee a - x \in u] \wedge \forall y \epsilon a\, [\{y\} \notin u].$$

Then

"α is a measurable cardinal"
$$\equiv \text{"}\alpha \text{ is a cardinal"} \wedge \exists u \subseteq \mathcal{P}(\alpha)\, [U(u,\alpha) \wedge \forall x \subseteq u\, [0 < \bar{\bar{x}} < \alpha \supset \cap x \in u]].$$

14°. Sc(a) ("a is super-complete")
$\equiv \forall x \epsilon a\, [\forall y \epsilon x\, [y \in a] \wedge \forall y \subseteq x\, [y \in a]]$.

15°. "a is constructible" $\equiv \exists \alpha \epsilon \rho^*(a)^+\, [a = F`\alpha]$.

Now we prove another closure property of $\tilde{\Delta}_1$-definability.

THEOREM 2. If $\mathcal{F}(b, a_1, \ldots, a_n)$ is a $\tilde{\Delta}_1$-operation, so is the operation $\mathcal{H}$ defined by

$$\mathcal{H}(b,a_1,\ldots,a_n) = \mathcal{F}(\mathcal{H}\restriction(b,a_1,\ldots,a_n),a_1,\ldots,a_n),$$

where $\mathcal{H}\restriction(b, a_1, \ldots, a_n) = \{\langle \mathcal{H}(x,a_1,\ldots,a_n),x\rangle \mid x \in b\}$.

__Proof__. $c = \mathcal{H}(b,a_1,\ldots,a_n) \equiv \exists f\, [c = f`b \wedge \forall x \epsilon Tc(b)\, [f`x = \mathcal{F}(f\restriction x,a_1,\ldots,a_n)]$. So $\mathcal{H}$ is a $\tilde{\Sigma}_1$-operation. Now use (IV) of Theorem 1.

q.e.d.

16°. "R(α)"

As an application of Theorem 2, we shall show that the R(α) operation is $\tilde{\Delta}_1$-definable. Our convention requires that an operation is defined for <u>all</u> sets. So R is regarded as an operation whose restriction to ordinal numbers is the usual R(α) operation. We adopt the following inductive definition for R:

$$R(x) = \cup\{\mathcal{P}(R(y)) \mid y \in x\},$$

for all sets x. Of course this equality well-defines an operation R and is consistent with the usual definition of R on ordinals. Now R is $\tilde{\Delta}_1$. For

$$\begin{aligned} R(x) &= \cup\{\mathcal{P}(y) \mid y \in R``x\} \\ &= \cup\,\mathcal{P}^*(R``x) \\ &= \cup\,\mathcal{P}^*\,\mathcal{W}(R\restriction x). \end{aligned}$$

(For the definition of *, see (VII) of Theorem 1.) Let $\mathcal{F}(y) = \cup\,\mathcal{P}^*\,\mathcal{W}(y)$. $\mathcal{F}$ is $\tilde{\Delta}_1$ and we have $R(x) = \mathcal{F}(R\restriction x)$. So by Theorem 2, R is $\tilde{\Delta}_1$.

17°. Next we define a $\tilde{\Delta}_1$-notion "Acc(b, a_1, ... , a_n)" (read "b is accessible from a_1, ... , a_n") by

$$\forall\alpha\in Ra(b)\ [\{a_1,\dots,a_n\} \subseteq R(\alpha) \supset \text{"}\alpha \text{ is not strongly inaccessible"}],$$

where Ra(b) is the rank of b.

18°. "ω_α" $= \omega \cup \{\omega_\beta^+ \mid \beta < \alpha\}$ is $\tilde{\Delta}_1$-definable.

From this, "ω", "ω_1", "ω_2", ... , "ω_ω", "ω_{ω_1}" etc. are $\tilde{\Delta}_1$-constants (i.e. 0-ary $\tilde{\Delta}_1$-operations).

(III) Examples from other branches of classical mathematics.

Algebraic notions such as groups, rings, fields, homomorphisms between these, etc., are $\tilde{\Delta}_1$-definable. For instance,

$\mathrm{Gr}(g,a,f)$ ("g is a group with the universe a and the multiplication f") $\equiv g = \langle a,f\rangle \wedge a \neq 0 \wedge [f : a^2 \longrightarrow a] \wedge \forall x\epsilon a\ \forall y\epsilon a\ \forall z\epsilon a$ $[f`\langle f`\langle x,y\rangle,z\rangle = f`\langle x,f`\langle y,z\rangle\rangle \wedge \exists u\epsilon a\ [y = f`\langle x,u\rangle] \wedge$ $\exists u\epsilon a\ [y = f`\langle u,x\rangle]]$.

"g is a group" $\equiv \exists a\epsilon \mathrm{Tc}(g)\ \exists f\epsilon \mathrm{Tc}(g)\ \mathrm{Gr}(g,a,f)$.

"h is a homomorphism from a group g into a group g' " $\equiv \exists a\epsilon \mathrm{Tc}(g) \wedge \exists f\epsilon \mathrm{Tc}(g)\ \exists a'\epsilon \mathrm{Tc}(g')\ \exists f'\epsilon \mathrm{Tc}(g')\ [\mathrm{Gr}(g,a,f) \wedge$ $\mathrm{Gr}(g',a',f') \wedge [[h : a \longrightarrow a'] \wedge \forall x\epsilon a\ \forall y\epsilon a\ [h`f`\langle x,y\rangle =$ $f`\langle h`x,h`y\rangle]]$.

More generally, the notion of "a structure $\mathcal{M} = \langle A,\ldots\rangle$ of a given theory $\mathcal{T}$ of the first order language" is $\tilde{\Delta}_1$. If the theory is finitely axiomatized, it is indeed Δ_0-definable, since it can be defined, as in the above example, by describing first the similarity type and then the finitely many non-logical axioms of the theory in question restricted to the universe of the structure. In the general case it can be Δ_1-defined via formal descriptions of "formulas", "sentences", "satisfaction" etc., each of which is obviously Δ_1-definable. Also using these formal descriptions, we easily see that the notions of "1st order definability" and

"$\mathrm{Def}_1(\mathcal{M})$" - the set of all the 1st order definable subsets of the universe of $\mathcal{M}$

are $\tilde{\Delta}_1$ (in fact Δ_1).

Further we shall observe that not only the first-order but also the higher-order structures are $\tilde{\Delta}_1$-definable. For example, we can $\tilde{\Delta}_1$-define topological spaces as follows:

"Top($\mathcal{T}$,X,S)" ("$\mathcal{T}$ is a topological space with the universe X and the family of open sets S")
$\equiv \mathcal{T} = \langle X,S\rangle \wedge X \in S \wedge S \subseteq \mathcal{P}(X) \wedge \forall U{\in}S\ \forall V{\in}S\ [U\cap V \in S] \wedge$
$\forall W{\subseteq}S\ [\bigcup W \in S]$,

"$\mathcal{T}$ is a topological space" $\equiv \exists X{\in}Tc(\mathcal{T})\ \exists S{\in}Tc(\mathcal{T})\ Top(\mathcal{T},X,S)$.

"$[f : \mathcal{T} \xrightarrow{cont} \mathcal{T}']$" ("f is a continuous function from $\mathcal{T}$ into $\mathcal{T}'$")
$\equiv \exists X{\in}Tc(\mathcal{T})\ \exists S{\in}Tc(\mathcal{T})\ \exists X'{\in}Tc(\mathcal{T}')\ \exists S'{\in}Tc(\mathcal{T}')\ [Top(\mathcal{T},X,S) \wedge$
$Top(\mathcal{T}',X',S') \wedge [f : X \longrightarrow X'] \wedge \forall U{\in}S'\ [f^{-1}``U \in S]]$.

Similarly one can $\tilde{\Delta}_1$-define, in certain fashions, various other fundamental notions and operations of mathematics such as the real and the complex number system with related operations, differentiation, holomorphic functions, analytic sets, Lebesgue measure, manifolds and so on. On the basis of these, one would easily see how he can $\tilde{\Delta}_1$-define notions and operations appearing in deeper development of mathematics.

§3. Preservation theorems for $\tilde{\Delta}_1$-definability

Let M be a unary predicate symbol. By $\phi_{(M)}$, $\mathcal{O}\!\iota_{(M)}$, $\mathcal{F}_{(M)}$ etc., we shall denote the relativizations to M of a formula ϕ, a notion $\mathcal{O}\!\iota$, an operation $\mathcal{F}$ etc. of ZFC, when they are well-defined. Similarly $ZFC_{(M)}$ denotes the set of all $\phi_{(M)}$ with ϕ an axiom of ZFC. Moreover by ZFC^{+M} we shall denote the same axiom system as ZFC except that in the axiom schema of replacement

$$\forall u\ \exists! v\ \phi(u,v) \supset \forall x\ \exists y\ \forall v\ [v \in y \equiv \exists u{\in}x\ \phi(u,v)]$$

ϕ may contain the predicate symbol M besides ϵ. (Here we assume that ZFC consists of the axiom schema of replacement plus a finite number of axioms.)

Now consider the following axiom system γ with two predicate symbols ϵ and M:

$$ZFC^{+M} \cup ZFC_{(M)} \cup \{\forall x \forall y\ [[x \in y \vee x \subseteq y] \wedge M(y) \supset M(x)]\}.$$

Intuitively this axiom system expresses the following situation. M is a super-complete submodel of the universe and M is admissible in the axiom schema of replacement for the universe. The following lemma can easily be proved by induction on the definition of q.b.f.s.

LEMMA 3. Let ψ be a q.b.f. Then

(3.1)
$$\vdash_{ZFC} \forall x \forall y\ [[x \in y \vee x \subseteq y] \wedge M(y) \supset M(x)] \wedge M(a_1) \wedge \ldots \wedge M(a_n) \supset [\psi_{(M)}(a_1,\ldots,a_n) \equiv \psi(a_1,\ldots,a_n)],$$

and a fortiori

$$\vdash_{\gamma} M(a_1) \wedge \ldots \wedge M(a_n) \supset [\psi_{(M)}(a_1,\ldots,a_n) \equiv \psi(a_1,\ldots,a_n)].$$

Now we shall prove

THEOREM 4.

(I) In order for a notion $\mathfrak{A}$ (defined in ZFC) to be $\tilde{\Sigma}_1$ it is necessary and sufficient that

$$\vdash_{\gamma} M(a_1) \wedge \ldots \wedge M(a_n) \wedge \mathfrak{A}_{(M)}(a_1,\ldots,a_n) \supset \mathfrak{A}(a_1,\ldots,a_n).$$

(II) In order for a notion $\mathfrak{A}$ (defined in ZFC) to be $\tilde{\Delta}_1$ it is necessary and sufficient that

$$\vdash_{\mathcal{T}} M(a_1) \wedge \ldots \wedge M(a_n) \supset [\mathfrak{A}_{(M)}(a_1,\ldots,a_n) \equiv \mathfrak{A}(a_1,\ldots,a_n)].$$

(III) In order for an operation $\mathcal{F}$ (defined in ZFC) to be $\tilde{\Delta}_1$ it is necessary and sufficient that

$$\vdash_{\mathcal{T}} M(a_1) \wedge \ldots \wedge M(a_n) \supset \mathcal{F}_{(M)}(a_1,\ldots,a_n) = \mathcal{F}(a_1,\ldots,a_n).$$

Note. Here we identify the notion $\mathfrak{A}$ with its defining formula. An obvious abbreviation is used also for the case of operation $\mathcal{F}$.

Note. A similar theorem with Σ_1 or Δ_1 instead of $\tilde{\Sigma}_1$ or $\tilde{\Delta}_1$ is proved in [12].

Proof. (II) and (III) follow immediately from (I). To prove (I), assume first that the notion $\mathfrak{A}$ is $\tilde{\Sigma}_1$, that is, $\mathfrak{A}$ is defined by

$$\mathfrak{A}(a_1,\ldots,a_n) \equiv \exists x\ \phi(a_1,\ldots,a_n,x),$$

where ϕ is a q.b.f. We must give a formal proof in $\mathcal{T}$ of

$$M(a_1) \wedge \ldots \wedge M(a_n) \wedge \mathfrak{A}_{(M)}(a_1,\ldots,a_n) \supset \mathfrak{A}(a_1,\ldots,a_n),$$

that is,

$$(3.2) \qquad M(a_1) \wedge \ldots \wedge M(a_n) \wedge \exists x\ [M(x) \wedge \phi_{(M)}(a_1,\ldots,a_n,x)] \supset \exists x\ \phi(a_1,\ldots,a_n,x).$$

But by lemma 3 we have

$$M(a_1) \wedge \ldots \wedge M(a_n) \wedge M(x) \wedge \phi_{(M)}(a_1,\ldots,a_n,x) \supset \phi(a_1,\ldots,a_n,x).$$

From this (3.2) easily follows.

Conversely assume that

$$(3.3) \quad M(a_1) \wedge \ldots \wedge M(a_n) \wedge \mathfrak{A}_{(M)}(a_1,\ldots,a_n) \supset \mathfrak{A}(a_1,\ldots,a_n)$$

is provable in $\mathcal{T}$.

Then there is a finite set of axioms U of ZFC such that (3.3) is deducible from the system $\mathcal{T}'$:

$$ZFC^{+M} \cup \{\chi_{(M)}\} \cup \{\forall x \forall y\ [[x \in y \vee x \subseteq y] \wedge M(y) \supset M(x)]\},$$

where χ is the conjunction of formulas of U. Now we shall show in ZFC that

$$(3.4) \quad \mathfrak{A}(a_1,\ldots,a_n) \equiv$$
$$\exists m\ [\chi_{(m)} \wedge Sc(m) \wedge a_1 \in m \wedge \ldots \wedge a_n \in m \wedge \mathfrak{A}_{(m)}(a_1,\ldots,a_n)],$$

where $Sc(m) \equiv$ "m is super-complete" (cf. 14° of §2) and where m does not appear in χ and $\mathfrak{A}$, and $\chi_{(m)}$ and $\mathfrak{A}_{(m)}$ denote the formulas obtained from χ and $\mathfrak{A}$ respectively by restricting each quantifier in them to $\in m$.

Since the right-hand side of (3.4) is $\tilde{\Sigma}_1$, it remains only to show the equivalence (3.4).

($\Leftarrow$) Replacing each occurrence of $M(*)$ by $* \in m$ in the proof of (3.3) from $\mathcal{T}'$ we obtain a proof of

$$a_1 \in m \wedge \ldots \wedge a_n \in m \wedge \mathfrak{A}_{(m)}(a_1,\ldots,a_n) \supset \mathfrak{A}(a_1,\ldots,a_n)$$

from $\mathcal{T}''$:

$$ZFC^{+m} \cup \{\chi_{(m)}\} \cup \{\forall x \forall y\ [[x \in y \vee x \subseteq y] \wedge y \in m \supset x \in m]\}.$$

In this case ZFC^{+m} becomes the axiom system consisting of a finite number of axioms of ZFC plus the axiom schema of replacement in which m may occur. But since m is merely a set variable, this schema is included in the original axiom schema of replacement in ZFC. Hence each axiom in ZFC^{+m} is in ZFC. Moreover

$$\begin{aligned}\forall x \forall y\ [[x \in y \vee x \subseteq y] \wedge y \in m \supset x \in m]\\ &\equiv \forall y{\in}m\ [\forall x{\in}y\ [x \in m] \wedge \forall x{\subseteq}y\ [x \in m]]\\ &\equiv Sc(m) \quad \text{(m is super-complete).}\end{aligned}$$

From these, with the use of the deduction theorem, we have in ZFC that

$$\begin{aligned}\chi_{(m)} \wedge Sc(m) \wedge a_1 \in m \wedge \ldots \wedge a_n \in m \wedge \mathfrak{A}_{(m)}(a_1,\ldots,a_n)\\ \supset \mathfrak{A}(a_1,\ldots,a_n),\end{aligned}$$

and so

$$\begin{aligned}(3.5)\quad \exists m\ [\chi_{(m)} \wedge Sc(m) \wedge a_1 \in m \wedge \ldots \wedge a_n \in m \wedge \mathfrak{A}_{(m)}(a_1,\ldots,a_n)]\\ \supset \mathfrak{A}(a_1,\ldots,a_n).\end{aligned}$$

($\Rightarrow$) On the other hand, by the partial reflection theorem for ZFC we can prove in ZFC that

$$\begin{aligned}\exists m\ [Sc(m) \wedge \chi \equiv \chi_{(m)} \wedge a_1 \in m \wedge \ldots \wedge a_n \in m\\ \wedge [\mathfrak{A}(a_1,\ldots,a_n) \equiv \mathfrak{A}_{(m)}(a_1,\ldots,a_n)]].\end{aligned}$$

Since $\vdash_{ZFC} \chi$, we have in ZFC

$$\mathfrak{A}(a_1,\ldots,a_n) \supset \exists m\, [Sc(m) \wedge \chi_{(m)} \wedge a_1 \in m \wedge \ldots \wedge a_n \in m \wedge [\mathfrak{A}(a_1,\ldots,a_n) \equiv \mathfrak{A}_{(m)}(a_1,\ldots,a_n)]],$$

or

$$(3.6) \qquad \mathfrak{A}(a_1,\ldots,a_n) \supset \exists m\, [\chi_{(m)} \wedge Sc(m) \wedge a_1 \in m \wedge \ldots \wedge a_n \in m \wedge \mathfrak{A}_{(m)}(a_1,\ldots,a_n)].$$

By (3.5) and (3.6), we have (3.4) in ZFC.

q.e.d.

COROLLARY 5. For any $\tilde{\Delta}_1$-operation $\mathcal{F}$, we have

$$\vdash_{\mathcal{T}} M(a_1) \wedge \ldots \wedge M(a_n) \supset M(\mathcal{F}(a_1,\ldots,a_n))$$

and

$$\vdash_{ZFC} \text{"}\mathcal{F}(a_1,\ldots,a_n) \text{ is accessible from } a_1,\ldots,a_n\text{"}.$$

Proof. The first half of the corollary is immediate from the theorem. The second half follows from the fact

$$\vdash_{ZFC} \text{"}\alpha \text{ is strongly inaccessible"} \supset \phi_{(R(\alpha))},$$

for each axiom ϕ of ZFC.

The theorem characterizes $\tilde{\Delta}_1$-notions and operations as those definable in set theory which are absolute for all super-complete sub-models. This together with the corollary suggests a local property of $\tilde{\Delta}_1$-notions and operations. In order to make this point clearer we shall make the following observation.

Let $\mathfrak{A}(a)$ be a $\tilde{\Delta}_1$-notion. Then

$$\mathfrak{A}(a) \equiv \exists x\ \phi(a,x) \equiv \forall x\ \psi(a,x)$$

for some q.b.f.'s ϕ and ψ. So,

$$\vdash_{ZFC} \exists x\ [\phi(a,x) \vee \neg\psi(a,x)].$$

Hence, given a set a we can find an $x \in R(\alpha)$ for some α. Once we can find such an x, we can examine (not necessarily effectively but locally without viewing the whole universe) whether or not $\phi(a,x)$ holds in $R(\alpha)$. If and only if $\phi(a,x)$ is the case we have $\mathfrak{A}(a)$.

§4. Connection with Lévy's hierarchy

We shall prove the following theorem which gives the exact relationship between Lévy's hierarchy and the one introduced in §0 on the basis of q.b.f's. The proof requires the axiom of choice. We do not know whether the theorem holds without the axiom of choice.

THEOREM 6. For $n \geq 1$, $\tilde{\Pi}_n = \Pi_{n+1}$ and hence $\tilde{\Sigma}_n = \Sigma_{n+1}$ and $\tilde{\Delta}_n = \Delta_{n+1}$.

Proof. We have only to treat the case n = 1. Other cases are obtained from this by prefixing quantifiers. In order to prove $\tilde{\Pi}_1 = \Pi_2$, it suffices to prove that

$$(4.1) \qquad \Sigma_1 \subseteq \tilde{\Delta}_1,$$

and

$$(4.2) \qquad \tilde{\Delta}_0 \subseteq \Delta_2.$$

For suppose we have both (4.1) and (4.2). Then, prefixing a universal quantifier to them we obtain both

$$\Pi_2 = \forall\Sigma_1 \subseteq \forall\tilde{\Delta}_1 = \tilde{\Pi}_1,$$

and

$$\tilde{\Pi}_1 = \forall\tilde{\Delta}_0 \subseteq \forall\Delta_2 = \Pi_2.$$

So $\tilde{\Pi}_1 = \Pi_2$.

To prove (4.1) we invoke Theorem 36 of Lévy's monograph [5] p. 52 which says:

For any Σ_1-formula (and <u>a fortiori</u> for any Δ_0-formula) ϕ having $x, a_1, \dots, a_n$ as its only free variables, it holds in ZFC that

$$\exists x\ \phi(x,a_1,\dots,a_n) \equiv$$
$$\exists x'[\rho^*(x) \leq \max(\aleph_0,\rho^*(a_1),\dots,\rho^*(a_n)) \wedge \phi(x,a_1,\dots,a_n)],$$

where $\rho^*(x) = \overline{\overline{Tc(x)}}$.

From this theorem we easily have the equivalence

(4.3)
$$\exists x\ \phi(x,a_1,\dots,a_n) \equiv$$
$$\exists x \in R(\max(\aleph_0,\rho^*(a_1),\dots,\rho^*(a_n))^+)\ [\phi(x,a_1,\dots,a_n)],$$

for each Δ_0-formula ϕ. By virtue of (II) of Theorem 1, we have only to check that the operation $\mathcal{F}$ defined by

$$\mathcal{F}(a_1,\dots,a_n) = R(\max(\aleph_0,\rho^*(a_1),\dots,\rho^*(a_n))^+)$$

is $\tilde{\Delta}_1$. (The right hand side of (4.3) is then shown to be $\tilde{\Delta}_1$.) But,

$$R(\max(\aleph_0, \rho^*(a_1), \ldots, \rho^*(a_n))^+) = R(\bigcup\{\aleph_0, \rho^*(a_1), \ldots, \rho^*(a_n)\}^+)$$

and each of the operations R, $\bigcup$, $\{\ldots\}$, ρ^*, $^+$ has already been shown to be $\tilde{\Delta}_1$. Hence $\mathcal{F}(a_1,\ldots,a_n)$ is $\tilde{\Delta}_1$ by (V) of Theorem 1. The proof of (4.1) is now complete.

To get (4.2): every q.b.f. is Δ_2, we shall use induction on the definition of q.b.f. Since other cases are similar or trivial, we only treat the case where ϕ is of the form $\forall x \subseteq a\ [\psi]$. By the induction hypothesis, ψ is equivalent both to a Π_2-formula ($\forall u\ \exists v\ \psi'$, say) and to a Σ_2-formula ($\exists u\ \forall v\ \psi''$, say). Then ϕ is Π_2-definable thus:

$$\begin{aligned}\phi &\equiv \forall x\ [x \subseteq a \supset \forall u\ \exists v\ \psi'] \\ &\equiv \forall x\ \forall u\ \exists v\ [x \subseteq a \supset \psi'] \\ &\equiv \forall s\ \exists t\ \forall x \in s\ \forall u \in s\ \exists v \in t\ [x \subseteq a \supset \psi'].\end{aligned}$$

Also ϕ is Σ_2-definable thus:

$$\begin{aligned}\phi &\equiv \exists z\ [z = \mathcal{P}(a) \wedge \forall x \in z\ \exists u\ \forall v\ \psi''] \\ &\equiv \exists z\ [\forall y \in z\ [y \subseteq a] \wedge \forall y\ [y \subseteq a \supset y \in z] \wedge \forall x \in z\ \exists u\ \forall v\ \psi''] \\ &\equiv \exists s\ \forall t\ \exists z \in s\ [\forall y \in z\ [y \subseteq a] \wedge \forall y \in t\ [y \subseteq a \supset y \in z] \wedge \\ &\qquad \forall x \in z\ \exists u \in s\ \forall v \in t\ \psi''].\end{aligned}$$

This completes the proof of (4.2) and hence of the theorem.

q.e.d.

By the proof of the theorem we easily have

COROLLARY 7. For $n \geq 1$, $\tilde{\Delta}_n = \Delta_{n+1} = [\Delta_n \text{ in } \mathcal{P}]$.

In particular $\tilde{\Delta}_1 = \Delta_2 = [\Delta_1 \text{ in } \mathcal{P}]$. So $\tilde{\Delta}_1$-notions and operations are characterized as those notions and operations which are recursive in the power set operation $\mathcal{P}$.

Note. If we modify our situation and restrict our attention to the notions and operations on ordinals and if we assume V = L, then by the result of [9], $\tilde{\Delta}_1$-notions and operations are those notions and operations which are recursive (in the sense of Takeuti [11]) in 2^{α} (the cardinal power operation) (which coincides with α^+, the cardinal successor operation, by virtue of V = L).

§5. Examples of notions which are not $\tilde{\Delta}_1$-definable

(I) The usual diagonal method would be the most natural method to obtain a notion which is definable but not $\tilde{\Delta}_1$-definable in ZFC. First we present the following parametrization theorem.

THEOREM 8. For $n \geq 0$, there is a $\tilde{\Sigma}_1$-formula $\mathcal{T}_n(b, a_1, \ldots, a_n)$, with b, a_1, ... , a_n as its only free variables, such that, for every $\tilde{\Sigma}_1$-formula $\phi(a_1, \ldots, a_n)$ with a_1, ... , a_n as its only free variables, we can find a natural number e such that

$$\vdash_{ZFC} \phi(a_1, \ldots, a_n) \equiv \mathcal{T}_n(\underline{e}, a_1, \ldots, a_n),$$

where $\underline{e}$ is the constant denoting e.

The proof of the theorem is omitted here. By the diagonal method we can easily show that the notion $\mathcal{O}\!\mathcal{l}$ defined by

$$\mathcal{O}\!\mathcal{l}(a) \equiv \mathcal{T}_1(a, a)$$

is not $\tilde{\Pi}_1$-definable and a fortiori not $\tilde{\Delta}_1$-definable.

(II) There can be found a few notions appearing in the recent development of set theory which are shown not to be $\tilde{\Delta}_1$-definable. Among these are the notion of ordinal definability ([6]) and the notion of <u>second</u> (or <u>finite</u>) <u>order cardinal characterizability</u> ([3]). Here we only treat the notion of ordinal definability.

The notion OD(a) defined by

$$\mathrm{OD}(a) \text{ (a is ordinal definable)} \equiv \exists\alpha\ [a \in \mathrm{Def}_1(R(\alpha),\epsilon)]$$

is $\tilde{\Sigma}_1$. ($\mathrm{Def}_1(R(\alpha),\epsilon)$ is the set of all 1st order definable subsets of $R(\alpha)$.) We shall show that OD is not $\tilde{\Pi}_1$. Suppose for <u>reductio ad absurdum</u> that OD were $\tilde{\Pi}_1$-defined:

(5.1) $$\mathrm{OD}(a) \equiv \forall u\ \chi(a,u),$$

where χ is a q.b.f. Here we use the Cohen method.

Cohen (cf. [1]) constructed a model $M = L[a]$ by adjoining to L a <u>generic</u> subset a of ω. It is known that in the model M the hereditarily ordinal definable sets are exactly the constructible sets. Since a is not constructible but $a \subseteq \omega$, a is not ordinal definable in M. So by (5.1), there must exist a set b in M such that

$$M \models \neg\chi(a,b).$$

Next take an ordinal α such that $a, b \in R(\alpha)$. Since $\neg\chi$ is a q.b.f., by lemma 3 at the beginning of §3,

(5.2) $$R(\alpha)_{(M)} \models \neg\chi(a,b).$$

On the other hand, by the method of Solovay and Easton we can extend the model M to N such that N is a Cohen extension of M,

$$R(\alpha)_{(N)} = R(\alpha)_{(M)} \quad \text{and} \quad N \models \mathrm{OD}(a).$$

For instance we can take such an N that

$$N \models a = \{n \in \omega \mid 2^{\aleph_{\beta+n}} = \aleph_{\beta+n+1}\},$$

where β is a regular cardinal (in M) > $\overline{\overline{R(\alpha)}}$. But then by (5.1), $N \models \forall u\ \chi(a,u)$ and hence $N \models \chi(a,b)$. Since a, b $\in R(\alpha)_{(N)}$ by lemma 3 again we have

(5.3) $$R(\alpha)_{(N)} \models \chi(a,b).$$

But since $R(\alpha)_{(N)} = R(\alpha)_{(M)}$, (5.2) and (5.3) are inconsistent. This contradiction shows that OD is not $\tilde{\Pi}_1$-definable.

q.e.d.

REFERENCES

[1] Cohen, P. J., The independence of the continuum hypothesis, PNAS 50 (1963), 1143-1148; 51 (1964), 105-110.

[2] Easton, W. B., Powers of regular cardinals, Annals of Math. Logic 1 (1970), 139-178.

[3] Garland, S. J., Second-order cardinal characterizability, preprint of a paper presented at the Summer Institute for Set Theory, Los Angeles 1967.

[4] Kripke, S., Transfinite recursions, constructible sets and analogues of cardinals, preprint of a paper presented at the Summer Institute for Set Theory, Los Angeles 1967.

[5] Lévy, A., A hierarchy of formulas in set theory, Memoirs of American Mathematical Society No. 57 (1965).

[6] Myhill, J., and Scott, D., Ordinal definability, in Axiomatic Set Theory, Part I, Dana S. Scott (ed.), Amer. Math. Soc. (1971).

[7] Platek, R., Foundations of recursion theory, Doctoral dissertation and Supplement, Stanford University (1966).

[8] Solovay, R., $2^{\aleph_0}$ can be anything it ought to be, in The theory of models, North-Holland, Amsterdam (1965), 435 (abstract).

[9] Takahashi, M., Recursive functions of ordinal numbers and Lévy's hierarchy, Comment. Math. Univ. St. Paul. 17 (1968), 21-29.

[10] Takahashi, M., An induction principle in set theory I, Yokohama Mathematical Journal 17 (1969), 53-59.

[11] Takeuti, G., On the recursive functions of ordinal numbers, J. Math. Soc. Japan 12 (1960), 119-128.

[12] Feferman, S., and Kreisel, G., Persistent and invariant formulas relative to theories of higher order, BAMS 72 (1966), 480-485.

[13] Karp, C. R., A proof of the relative consistency of the continuum hypothesis, in Sets, Models and Recursion Theory, John N. Crossley (ed.), North-Holland (1967), 1-32.

[14] Kruse, A. H., Localization and iteration of axiomatic set theory, Wayne State University Press, Detroit.

INITIAL SEGMENTS AND IMPLICATIONS FOR THE STRUCTURE OF DEGREES

C. E. M. Yates
Manchester, England

It has been proved by Lachlan [3] and Lerman [4], respectively, that all countable distributive lattices (with least element) and all finite lattices are embeddable as initial segments of the upper semilattice $\mathcal{D}$ of degrees of recursive unsolvability; these are the two most significant steps so far taken towards an understanding of the general theory of initial segments of $\mathcal{D}$. Both of these theorems extend the special case of finite distributive lattices, which is particularly interesting because, as Thomason [9] has recently observed, it implies the undecidability of the elementary theory of $\mathcal{D}$ (Lachlan had previously observed that this followed from his theorem). Lachlan's theorem also generalised a particularly useful special case previously obtained by Hugill [2]: every countable linear ordering (with least element) is embeddable as an initial segment of $\mathcal{D}$. The most striking consequence of this is the refutation of the strong homogeneity-conjecture; this was first noticed by Feiner [1] and we shall indicate how to considerably strengthen his observation in §5 of this paper. The homogeneity-conjecture, which remains open, asserts the existence, for any degree $\underset{\sim}{c}$, of an isomorphism between $(\mathcal{D}, \leqslant)$ and $(\mathcal{D}(\geqslant \underset{\sim}{c}), \leqslant)$; the stronger conjecture asserted the existence of an isomorphism between $(\mathcal{D}, \leqslant, J)$ and $(\mathcal{D}(\geqslant \underset{\sim}{c}), \leqslant, J)$, where J is the jump operator.

The present paper is devoted to presenting new proofs of the two special cases mentioned above along the lines initiated in [8], [10],

[11]; hence, it is essentially expository in nature. This material was originally scheduled to appear in [12] but the developments mentioned above indicate that it should be of more general interest and so receive special treatment. A brief discussion of the homogeneity-conjectures is appended at the end of the paper. The methods of this paper can be used with only slight modification to embed countable Boolean algebras as initial segments, but rather messy alterations at present seem unavoidable in dealing with arbitrary countable distributive lattices. (I am grateful to Mrs. Dina Cohen-Kulka and Gordon MacNair for pointing out an error in the first hastily-written draft of this paper, written under the assumption that such alterations are avoidable.) In [12] we shall continue our programme for formalising all these results in a framework which is as uniform as possible, then constructivising them by means of priority arguments in order to push them below an arbitrary nonzero recursively enumerable (Σ_1^0) degree, in particular $\underset{\sim}{0}^{(1)}$.

After the necessary preliminaries in the Introduction, we sketch in §2 the basic framework of results concerning $\underset{\sim}{S}$-treemaps (called trees in [7], [9] and [10]) which are needed for the two principal results and the later work in [12]. Finite distributive lattices are dealt with in §3, countable linear orderings in §4 and the homogeneity-conjectures in §5.

§1. Introduction

Although the development of relative recursiveness in terms of strings and treemaps can be found in [10] and [11], it seems best to make the present paper as self-contained as possible, especially as we are changing some notation and terminology (these changes will be firmly indicated). A string is just a finite sequence of zeros and ones; the set of all strings is denoted by $\underset{\sim}{S}$. We use 'ϕ' to denote the null string and '0', '1' to denote the two single-element strings. The number of elements of a string σ will be called its <u>length</u> and denoted by $|\sigma|$ (the first change in notation). The $j+1$-st element of σ, for $j < |\sigma|$, will be denoted by $\sigma(j)$ and the initial segment of σ which has length $j+1$ will be denoted by $\sigma[j]$. We let $\sigma * \tau$ be the string obtained by adding τ to the right-hand side of σ; in particular, $\sigma * \phi = \sigma$. We write $\sigma \subset \tau$ to mean that τ is a proper extension of σ, i.e. $\tau = \sigma * \rho$ for some $\rho \neq \phi$. If $\sigma(n) \neq \tau(n)$ for some $n < \min(|\sigma|, |\tau|)$ then we say that σ, τ are <u>incompatible</u>, written $\sigma|\tau$; otherwise, they are <u>compatible</u>. Since we identify a set $X \subseteq \underset{\sim}{N}$ (the set of natural numbers) with its characteristic function, and since this in turn can be regarded as an infinite sequence of zeros and ones, we also write $\sigma \subset X$ to mean that $\sigma = X[n]$ for some n, where $X[n]$ is the initial segment of (the characteristic function of) X which has length $n+1$.

We now come to the most basic definitions of the theory.

DEFINITION 1.1. F is a <u>partial $\underset{\sim}{S}$-map</u> if its domain and image are subsets of $\underset{\sim}{S}$; if its domain is $\underset{\sim}{S}$ then it is simply called an <u>$\underset{\sim}{S}$-map</u>. F is a <u>partial $\underset{\sim}{S}$-treemap</u> if in addition it is order-preserving, i.e. if

$$\sigma \subseteq \tau \rightarrow F(\sigma) \subseteq F(\tau)$$

for all $\sigma, \tau \in \mathrm{dom}(F)$; again it is simply called an <u>$\underset{\sim}{S}$-treemap</u> (or even just <u>treemap</u>) if its domain is $\underset{\sim}{S}$. Lastly, a partial $\underset{\sim}{S}$-treemap is <u>invertible</u> if

$$\sigma|\tau \rightarrow F(\sigma)|F(\tau)$$

for all $\sigma, \tau \in \mathrm{dom}(F)$; this terminology will be explained immediately.

If F is a partial $\underset{\sim}{S}$-treemap then we may define a partial functional $F^* : \mathcal{P}(\underset{\sim}{N}) \longrightarrow \mathcal{P}(\underset{\sim}{N})$ by setting

$$F^*(X) = \lim\{F(\sigma) : \sigma \subset X \;\&\; \sigma \in \mathrm{dom}(F)\}$$

for all $X \subseteq \underset{\sim}{N}$ such that the R.H.S. is infinite. In particular, if F is invertible then F^* is one-one and so $(F^*)^{-1}$ exists; this explains our terminology. (Note that we are not requiring F to be one-one.) Moreover, if F is a partial recursive invertible treemap then there is a partial recursive treemap G such that $(F^*)^{-1} = G^*$: set $G(\sigma) =$ some τ $(F(\tau) = \sigma)$ for all $\sigma \in \mathrm{im}(F)$. The significance of all this for relative recursiveness lies in the following proposition:

PROPOSITION 1.2. Let X, Y both belong to $\mathcal{P}(\underset{\sim}{N})$. Then X is recursive in Y iff $X = F^*(Y)$ for some partial recursive $\underset{\sim}{S}$-treemap F.

It is easily shown (cf. the introduction to [11]) that there is a recursive enumeration $\underset{\sim}{F}_0, \underset{\sim}{F}_1, \ldots,$ of recursive $\underset{\sim}{S}$-treemaps such that for any partial recursive $\underset{\sim}{S}$-treemap F there exists an e such that $F^* = \underset{\sim}{F}_e^*$; this is in fact best proved directly in the course of proving the proposition above. This enumeration is very convenient for diagonalisation-arguments when it is required to prove some set is not recursive in another set. The following corollary plays a very useful role in Lemma 2.9.

COROLLARY 1.3. Let F be a partial recursive invertible $\underset{\sim}{S}$-treemap. Then $X \equiv_T F^*(X)$ for all $X \in \mathrm{dom}(F^*)$.

(Remark: in [8], [10] and [11] the term "tree" was used instead of "treemap", and it was subjected to extra conditions that are superfluous in the present context. In any case, it seems advisable to reserve the term "tree" for its more familiar usage to denote a special type of partial ordering; for example, any subset of $\underset{\sim}{S}$ forms a tree under the usual lexicographic ordering of $\underset{\sim}{S}$.)

A subset $\mathcal{I}$ of $\mathcal{D}$ is an initial segment of $\mathcal{D}$ if $\underset{\sim}{x} \in \mathcal{I}$ whenever $\underset{\sim}{x} \leq \underset{\sim}{y}$ and $\underset{\sim}{y} \in \mathcal{I}$. A partial ordering $\mathcal{P}$ is <u>embeddable as an initial segment</u> of $\mathcal{D}$, written $\mathcal{P} \overset{*}{\Rightarrow} \mathcal{D}$, if $\mathcal{P}$ is isomorphic to $(\mathcal{I}, \leq)$ for some initial segment $\mathcal{I}$ of $\mathcal{D}$. If a partial ordering has a largest (smallest) element then we shall call it <u>topped</u> (<u>bottomed</u>). Clearly, only bottomed partial orderings are $\overset{*}{\Rightarrow} \mathcal{D}$, and because any partial ordering has a simple topped extension we shall only need to consider topped partial orderings (for, if the simple topped extension is $\overset{*}{\Rightarrow} \mathcal{D}$ then the original ordering is $\overset{*}{\Rightarrow} \mathcal{D}$). Moreover, any topped initial segment of $\mathcal{D}$ is a countable upper semilattice, which immediately restricts our attention to these structures. It is probable that every countable bottomed upper semilattice is $\overset{*}{\Rightarrow} \mathcal{D}$, but the answer is not yet known even for lattices. Lachlan's theorem asserts that every countable bottomed distributive lattice is $\overset{*}{\Rightarrow} \mathcal{D}$. Distributive lattices are easier to deal with because of the availability of various nice representations for these lattices. The particular representations we shall use will be of a very simple and natural type.

DEFINITION 1.4. A <u>distributive representation</u> is any sublattice of the power-set algebra $(\mathcal{P}(\underset{\sim}{N}), \subseteq)$ composed of recursive sets and containing $\underset{\sim}{N}$, ϕ but no finite sets other than ϕ.

The convenience of this approach lies in that we append ourselves to the task of constructing a single function $B : \underset{\sim}{N} \to \{0,1\}$ such that the functions representing the lower elements of the required initial segment can be "read off" in a natural way from B and the elements X of the representation. The technique used for this is the following.

DEFINITION 1.5. Let $B : \underset{\sim}{N} \to \{0,1\}$ and let X be an infinite recursive set with elements x_0, x_1, ..., listed in their natural order. $B\langle X\rangle : \underset{\sim}{N} \to \{0,1\}$ is then defined by setting

$$B\langle X\rangle(n) = B(x_n)$$

for all $n \in \underset{\sim}{N}$.

An immediate and important observation is that for all infinite recursive sets X, Z:

$$X \subseteq Z \rightarrow B\langle X\rangle \leq_T B\langle Z\rangle.$$

For, there is a recursive function $F : \underset{\sim}{N} \rightarrow \underset{\sim}{N}$ such that

$$x_n = z_{F(n)}$$

for all n, where $z_0, z_1, \ldots,$ is an enumeration of the elements of Z in their natural order; it follows that

$$B\langle X\rangle(n) = B(z_{F(n)}) = B\langle Z\rangle(F(n))$$

and so B<X> is in fact many-one reducible to B<Z>. If $\mathcal{L}$ is a countable topped and bottomed distributive lattice with representation $\mathbb{L}$, then our procedure in the present paper (for special cases) and in [12] will be to arrange that

$$X \subseteq Z \leftrightarrow B\langle X\rangle \leq_T B\langle Z\rangle$$

for all X, Z $\in \mathbb{L}$ (of course, only $\leftarrow$ requires attention because of our observations above), and

$$C \leq_T B \rightarrow (\exists Y)(Y \in \mathbb{L} \;\&\; C \equiv_T B\langle Y\rangle).$$

This ensures that if $\underset{\sim}{b}\langle X\rangle$ is defined to be the degree of B<X> then $\mathcal{I} = \{\underset{\sim}{b}\langle X\rangle : X \in \mathbb{L}\}$ is an initial segment of $\mathcal{D}$ isomorphic to $\mathcal{L}$. It is not, however, quite as simple as this might indicate: the special feature of the countable (as distinct from finite) case is that the representation has to be constructed alongside the function B. In the present paper we only have to meet this problem in the relatively simple case when $\mathcal{L}$ is a linear ordering, but messy combinatorial problems arise in the general case.

Finally, if σ is a string and X is an infinite recursive set, then $\sigma\langle X\rangle$ may be defined in essentially the same way as for the total functions. Also it is possible to extend the definition to $\underset{\sim}{S}$-treemaps as follows: if T is an $\underset{\sim}{S}$-treemap then we set $T\langle X\rangle(\sigma) = T(\sigma)\langle X\rangle$ for all σ; $T\langle X\rangle$ is again an $\underset{\sim}{S}$-treemap, and it can be seen that $T\langle X\rangle^*(Y) = T^*(Y)\langle X\rangle$ for all $Y \in \mathrm{dom}(T^*)$.

§2. Various special recursive $\underset{\sim}{S}$-treemaps

Much variation is possible in presenting constructions of initial segments, but in all of these the central concept (in some form or another) is that of a partial recursive $\underset{\sim}{S}$-treemap. Certain basic lemmas concerning splitting and splittingmaps were proved in [8] and [11] in order to deal with minimal degrees. For more complicated initial segments, some refinement and generalisation is required and the purpose of this section is to reorganise this groundwork. All of the essential difficulties involved in embedding finite distributive lattices or bottomed linear orderings in $\mathcal{D}$ (or in $\mathcal{D}(\leq \underset{\sim}{0}^{(2)})$) are surmounted in this section.

Since we are not here concerned with the more delicate priority arguments necessary for embedding lattices in $\mathcal{D}(\leq \underset{\sim}{c})$ where $\underset{\sim}{c}$ is a non-zero Σ^0_1 degree, our attention may be confined to recursive (as distinct from merely partial recursive) $\underset{\sim}{S}$-treemaps. $\underset{\sim}{S}$-treemaps are adequate for constructing minimal degrees but the following more restricted concept considerably simplifies the construction of more complicated initial segments.

DEFINITION 2.1. An $\underset{\sim}{S}$-treemap T is uniform if for each $n > 0$ there exist strings $(T)_n^0$, $(T)_n^1$ such that $|(T)_n^0| = |(T)_n^1|$ and for all τ with $|\tau| = n - 1$ we have:

$$T(\tau*0) = T(\tau)*(T)_n^0$$
$$T(\tau*1) = T(\tau)*(T)_n^1.$$

We write $(T)_n$ for the pair $((T)_n^0, (T)_n^1)$ and $|T|_n$ for the common length of $T(\tau)$ when $|\tau| = n$. Finally, we call $\{(T)_n\}_{n>0}$ the treequence associated with T.

We do in fact require a much more refined object: the $\mathbb{L}$-independent $\underset{\sim}{S}$-treemap, where $\mathbb{L}$ is a distributive representation. First, we need some notions concerning pairs of strings.

DEFINITION 2.2. Let $X \subseteq \underset{\sim}{N}$ be infinite. Two strings σ, τ are X-compatible if $\sigma\langle X\rangle$, $\tau\langle X\rangle$ are compatible; otherwise they are X-incompatible.

Before the next definition concerning strings we need some notation and terminology concerning representations.

DEFINITION 2.3. Let $\mathbb{L}$ be a finite distributive representation. $\mathcal{B}(\mathbb{L})$ is the Boolean algebra generated by $\mathbb{L}$. For any atom A of $\mathcal{B}(\mathbb{L})$ we denote the largest element of $\mathbb{L}$ disjoint from A by $A^{\mathbb{L}}$. Finally, we say that two atoms A, A' of $\mathcal{B}(\mathbb{L})$ are separated by X, Z $\in \mathbb{L}$ if $X \supseteq A$, $X \cap A' = \phi$, $Z \supseteq A'$, $Z \cap A = \phi$ (so that $X \cap Z$ is infinite).

Now we can introduce a restriction on pairs of strings which plays an important role in the embedding of distributive lattices.

DEFINITION 2.4. Let $\mathbb{L}$ be a finite distributive representation. Two strings σ, τ are $\mathbb{L}$-acceptable if whenever σ, τ are A-incompatible, A'-incompatible and A, A' are distinct atoms of $\mathcal{B}(\mathbb{L})$ separated by

$X, Z \in \mathbb{L}$ then σ, τ are $X \cap Z$-incompatible.

Notice that if $\mathbb{L}$ is a linear representation then any two strings are $\mathbb{L}$-acceptable by default. Also when $\mathbb{L}$ is a Boolean algebra, i.e. $\mathbb{L} = \mathcal{B}(\mathbb{L})$, two strings are $\mathbb{L}$-acceptable if they are either incompatible on all the atoms or incompatible on at most one atom. We shall only deal here with the former case when $\mathbb{L}$ is countable, but in both cases it is the relative simplicity of this notion which in turn simplifies the proof of the corresponding embedding theorem.

We are at last in a position to introduce the important concept mentioned some way above.

DEFINITION 2.5. Let $\mathbb{L}$ be a finite distributive representation. An $\mathcal{S}$-treemap T is $\mathbb{L}$-<u>independent</u> if it is uniform and:

(a) for each atom A of $\mathcal{B}(\mathbb{L})$, there exist infinitely many n such that $T(\tau * 0)$, $T(\tau * 1)$ are A-incompatible, $A^{\mathbb{L}}$-compatible whenever $|\tau| = n$.

(b) $T(\tau * 0)$, $T(\tau * 1)$ are $\mathbb{L}$-acceptable for all τ.

The purpose of this definition is to arrange that B<A> is independent of $B\langle A^{\mathbb{L}}\rangle$ for all $B \in im(T^*)$ and all atoms A of $\mathcal{B}(\mathbb{L})$. This plays an essential role in the embedding of $\mathbb{L}$ as an initial segment of $\mathcal{D}$. Notice that when $\mathbb{L}$ is either a linear ordering or a Boolean algebra, clause (b) becomes redundant for reasons mentioned after Definition 2.4. It is this which simplifies the proof in these two cases.

The construction of initial segments hinges on two special operations for forming, from a given treemap T, another treemap T' such that $im(T') \subseteq im(T)$. The first and more trivial of these operations is used in the situation that forms the hypothesis of Lemma 2.8 below: this operation consists of forming a treemap T' called "T above σ", for some $\sigma \in im(T)$, and defined by:

$$T'(\tau) = T(\sigma' * \tau)$$

for all τ, where $T(\sigma') = \sigma$; so in particular, $T'(\phi) = \phi$. Notice of course that im(T') _is_ a subset of im(T).

The second operation is much less trivial and is in fact the central concept around which any construction of an initial segment is evolved. This is the formation of splittingmaps, which were used for dealing with minimal degrees in [8], [10] and [11] but now have to be appropriately generalised. A string σ is said to be $\underset{\sim}{F}_e$-_split_ by a pair of strings σ_0, σ_1 if they both extend σ and $\underset{\sim}{F}_e(\sigma_0) | \underset{\sim}{F}_e(\sigma_1)$. (In the earlier papers, σ_0 and σ_1 were said to split for e, but this becomes clumsy when generalised to more and more complex situations.)

DEFINITION 2.6. Let $Y \subseteq \underset{\sim}{N}$ be infinite. A string σ is said to be $(\underset{\sim}{F}_e, Y)$-_split_ by two strings σ_0, σ_1 if they both extend σ and $\sigma\langle Y \rangle$ is $\underset{\sim}{F}_e$-split by $\sigma_0\langle Y \rangle$, $\sigma_1\langle Y \rangle$.

DEFINITION 2.7. Let $Y \subseteq \underset{\sim}{N}$ be infinite. An $\underset{\sim}{S}$-treemap T is an $(\underset{\sim}{F}_e, Y)$-_splittingmap_ if it is uniform and

(i) $T(\tau * 0)$, $T(\tau * 1)$ are Y-incompatible for infinitely many τ,

(ii) $T(\tau)$ is $(\underset{\sim}{F}_e, Y)$-split by $T(\tau * 0)$ and $T(\tau * 1)$ whenever the latter are Y-incompatible.

The lemmas which now follow contain the essence of the proofs of the theorems in §3 and §4. The first of these lemmas corresponds to the situation in which we use the first and more trivial of the operations described above.

LEMMA 2.8. Let T be a recursive treemap and let Y, Z be recursive sets such that $Z \subseteq Y$. Suppose that $T(\phi)$ is not $(\underset{\sim}{F}_e, Y)$-split by any pair of Z-compatible strings in im(T). Then there exists a number h such that $\underset{\sim}{F}^*_e(B\langle Y \rangle) = \underset{\sim}{F}^*_h(B\langle Z \rangle)$ for all $B \in \mathrm{im}(T^*)$ such that

$B\langle Y\rangle \in \mathrm{dom}(\underset{\sim}{F}_e^*)$.

Proof. For each string σ, let δ_σ be a string in im(T) such that $\delta_\sigma\langle Z\rangle = \sigma$ and chosen so that $|\underset{\sim}{F}_e(\delta_\sigma\langle Y\rangle)|$ is as large as possible and such that $\delta_\sigma\langle Y\rangle \supseteq \delta_\rho\langle Y\rangle$ for all $\rho \subseteq \sigma$. Next define

$$D(\sigma) = \underset{\sim}{F}_e(\delta_\sigma\langle Y\rangle)$$

for all σ such that δ_σ is defined. D is a partial $\underset{\sim}{S}$-treemap because if $\sigma \subseteq \tau$ then δ_σ, δ_τ are Z-compatible and so $\underset{\sim}{F}_e(\delta_\sigma\langle Y\rangle)$, $\underset{\sim}{F}_e(\delta_\tau\langle Y\rangle)$ are compatible: hence, $\underset{\sim}{F}_e(\delta_\sigma\langle Y\rangle) \subseteq \underset{\sim}{F}_e(\delta_\tau\langle Y\rangle)$ because $\delta_\sigma\langle Y\rangle \subseteq \delta_\tau\langle Y\rangle$ and $\underset{\sim}{F}_e$ is an $\underset{\sim}{S}$-treemap.

We now claim that

$$\underset{\sim}{F}_e^*(B\langle Y\rangle) = D^*(B\langle Z\rangle),$$

which proves the lemma because there exists an h such that $D^* = \underset{\sim}{F}_h^*$ (see both the present introduction and that in [11]). We proceed to prove this claim. Let $\sigma_0 \subset \sigma_1 \subset \ldots$ be chosen so that $\sigma_n \in \mathrm{im}(T\langle Z\rangle)$, $\sigma_n \subset B\langle Z\rangle$ and $|\underset{\sim}{F}_e(\delta_{\sigma_{n+1}}\langle Y\rangle)| > |\underset{\sim}{F}_e(\delta_{\sigma_n}\langle Y\rangle)|$ for all n; this can be done because $B\langle Y\rangle \in \mathrm{dom}(\underset{\sim}{F}_e^*)$. Now it is easy to see that

$$D(\sigma_n) = \underset{\sim}{F}_e(\delta_{\sigma_n}\langle Y\rangle) \subset \underset{\sim}{F}_e^*(B\langle Y\rangle)$$

for all n. Suppose, in order to obtain a contradiction, that this is not so. Let $\delta \in \mathrm{im}(T)$ be such that $\delta \subset B$ and $|\underset{\sim}{F}_e(\delta\langle Y\rangle)| > |\underset{\sim}{F}_e(\delta_{\sigma_n}\langle Y\rangle)|$. Then δ, δ_{σ_n} are Z-compatible but $\underset{\sim}{F}_e(\delta\langle Y\rangle)$, $\underset{\sim}{F}_e(\delta_{\sigma_n}\langle Y\rangle)$ are incompatible which contradicts our basic assumption. Finally, since $D(\sigma_n) \subset \underset{\sim}{F}_e^*(B\langle Y\rangle)$ for all n and because of the other properties of the chain $\sigma_0 \subset \sigma_1 \subset \ldots$, we conclude that

$$D^*(B\langle Z\rangle) = \underset{\sim}{F}_e^*(B\langle Y\rangle). \qquad \square$$

(Note: it is easier to give a heuristic proof of this lemma using Church's Thesis, but it will be important in [12] to have a convincing proof that h can be computed effectively from e.)

The next lemma expresses the importance of splittingmaps.

LEMMA 2.9. Let T be a recursive $(\underset{\sim}{F}_e, Y)$-splittingmap. If $X \in \mathrm{im}(T^*\langle Y\rangle)$ then $X \in \mathrm{dom}(\underset{\sim}{F}_e^*)$ and $X \equiv_T \underset{\sim}{F}_e^*(X)$.

Proof. First, to prove that $X \in \mathrm{dom}(\underset{\sim}{F}_e^*)$, let τ be such that $T\langle Y\rangle(\tau) \subseteq X$; it will be sufficient to show that there is a $\sigma \supseteq T\langle Y\rangle(\tau)$ such that $\sigma \subseteq X$ and $\underset{\sim}{F}_e(\sigma) \supset \underset{\sim}{F}_e(T\langle Y\rangle(\tau))$. Now, because T is an $(\underset{\sim}{F}_e, Y)$-splittingmap there exists a $\tau' \supseteq \tau$ such that $T\langle Y\rangle(\tau') = T(\tau')$ is $(\underset{\sim}{F}_e, Y)$-split by $T(\tau' * 0)$ and $T(\tau' * 1)$. Let σ be whichever of $T(\tau' * 0)$ and $T(\tau' * 1)$ is a substring of X. Then $\underset{\sim}{F}_e(\sigma) \supset \underset{\sim}{F}_e(T\langle Y\rangle(\tau')) \supseteq \underset{\sim}{F}_e(T\langle Y\rangle(\tau))$ as required.

Now, $\underset{\sim}{F}_e$ restricted to $\mathrm{im}(T\langle Y\rangle)$ is an invertible partial recursive $\underset{\sim}{S}$-treemap, and we have just shown that $\mathrm{dom}(\underset{\sim}{F}_e^*)$ is exactly $\mathrm{im}(T^*\langle Y\rangle)$. It follows from Corollary 1.3 that $X \equiv_T \underset{\sim}{F}_e^*(X)$ for all $X \in \mathrm{im}(T^*\langle Y\rangle)$.

$\square$

We now sketch briefly how these lemmas will be used; although the detailed constructions appear in §3 and §4, this is the nub of both proofs and so is worthy of emphasis. There are two basic cases, each being associated, as we have already mentioned, with an operation for forming, from a given recursive treemap T, a treemap T' such that $\mathrm{im}(T') \subseteq \mathrm{im}(T)$. Also associated with each such situation there will be a finite representation $\mathcal{L}$ and a pair e, Y with $Y \in \mathcal{L}$.

Case A: there is a $\sigma \in \mathrm{im}(T)$ and atom $A \subseteq Y$ in $\mathcal{B}(\mathcal{L})$ such that σ is not $(\underset{\sim}{F}_e, Y)$-split by a pair of A-incompatible, $A^{\mathcal{L}} \cap Y$-compatible strings in im(T).

<u>Case B</u>: otherwise.

When Case A holds, the appropriate operation consists simply of forming T above σ, so that Lemma 2.8 can be used with $Z = A^{\mathcal{L}} \cap Y$. On the other hand, when Case B holds, the aim is to form an $(\underset{\sim}{F}_e, Y)$-splittingmap T' with $im(T') \subseteq im(T)$. If no other restrictions were needed, then this would be easy enough to arrange. We require, however, that T' is $\mathcal{L}$-independent and, in particular, uniform. The first lemma below makes uniformity possible, whereas the subsequent lemma enables Case B to be replaced by:

<u>Case B'</u>: for each atom $A \subseteq Y$ in $\mathcal{B}(\mathcal{L})$, every string in im(T) is $(\underset{\sim}{F}_e, Y)$-split by a pair of $\mathcal{L}$-acceptable, A-incompatible and $A^{\mathcal{L}}$-compatible strings in im(T).

As we shall see, this makes $\mathcal{L}$-independence possible. <u>The second lemma follows immediately from the first in the case of linear representations</u>

LEMMA 2.10. Let $\mathcal{L}$ be a finite distributive representation, Y an infinite element of $\mathcal{L}$ and A an atom of $\mathcal{B}(\mathcal{L})$ which is $\subseteq$ Y. Let T be a uniform treemap such that every $\sigma \in im(T)$ is $(\underset{\sim}{F}_e, Y)$-split by a pair of A-incompatible, $A^{\mathcal{L}} \cap Y$-compatible strings in im(T). Then for each n there exists a pair of strings ψ_0, ψ_1 of equal length such that for all τ with $|\tau| = n$:

(a) $T(\tau)*\psi_0$, $T(\tau)*\psi_1$ are A-incompatible, $A^{\mathcal{L}} \cap Y$-compatible and belong to im(T),

(b) $T(\tau)$ is $(\underset{\sim}{F}_e, Y)$-split by $T(\tau)*\psi_0$, $T(\tau)*\psi_1$.

<u>Proof</u>. Let $\tau_1, \ldots, \tau_{2^n}$ be the strings of length n. We prove by induction on k, for $1 \leq k \leq 2^n$, that there exist strings ψ_0, ψ_1 satis-

fying (a), (b) for $\tau \in \{\tau_1,\dots,\tau_k\}$. It is easy to see that it is true for $k = 1$. Supposing that it is true for $k < 2^n$, we now prove it true for $k + 1$. Let θ_0, θ_1 be a pair of strings satisfying (a), (b) for $\tau \in \{\tau_1,\dots,\tau_k\}$, and let $\pi_i = T(\tau_{k+1})*\theta_i$ for $i \in \{0,1\}$. Let ρ_0, ρ_1 be strings of equal length such that π_0 is $(\mathcal{E}_e,Y)$-split by $\pi_0*\rho_0$, $\pi_0*\rho_1$ and these latter strings are A-incompatible, $A^{\mathcal{L}} \cap Y$-compatible and belong to im(T). Now, we examine $\pi_1*\rho_0$, $\pi_1*\rho_1$; these strings $\in$ im(T) because T is uniform and $|\pi_0| = |\pi_1|$. It is clear that for some i_0, i_1 not necessarily unequal, $T(\tau_{k+1})$ must be $(\mathcal{E}_e,Y)$-split by $\pi_0*\rho_{i_0}$, $\pi_1*\rho_{i_1}$. Moreover, these strings are A-incompatible because π_0, π_1 are A-incompatible, and $A^{\mathcal{L}} \cap Y$-compatible because π_0, π_1 are $A^{\mathcal{L}} \cap Y$-compatible and $\pi_0*\rho_0$, $\pi_0*\rho_1$ are $A^{\mathcal{L}} \cap Y$-compatible. If we now set $\psi_0 = \theta_0*\rho_{i_0}$, $\psi_1 = \theta_1*\rho_{i_1}$ then it is clear that ψ_0, ψ_1 satisfy (a), (b) for $\tau \in \{\tau_1,\dots,\tau_{k+1}\}$.

LEMMA 2.11. Let $\mathcal{L}$ be a finite distributive representation, Y an infinite element of $\mathcal{L}$ and A an atom of $\mathcal{B}(\mathcal{L})$ which is $\subseteq$ Y. Let T be an $\mathcal{L}$-independent treemap such that every $\sigma \in$ im(T) is $(\mathcal{E}_e,Y)$-split by a pair of A-incompatible, $A^{\mathcal{L}} \cap Y$-compatible strings in im(T). Then for each n there exists a pair of strings ρ_0, ρ_1 of equal length such that for all τ with $|\tau| = n$:

(a)' $T(\tau)*\rho_0$, $T(\tau)*\rho_1$ are A-incompatible, $A^{\mathcal{L}}$-compatible and belong to im(T),

(b) $T(\tau)$ is $(\mathcal{E}_e,Y)$-split by $T(\tau)*\rho_0$, $T(\tau)*\rho_1$,

(c) $T(\tau)*\rho_0$, $T(\tau)*\rho_1$ are $\mathcal{L}$-acceptable.

Proof. Let π_0, π_1 be strings of the same length such that $T(\tau)*\pi_0$, $T(\tau)*\pi_1$ belong to im(T), are $\mathcal{L}$-acceptable, A-incompatible and $A^{\mathcal{L}}$-compatible whenever $|\tau| = n$; such strings exist because T is $\mathcal{L}$-independent. It follows easily from the preceding lemma that there exist θ_0, θ_1 with $|\theta_0| = |\theta_1|$ such that $T(\tau)$ is $(\mathcal{E}_e,Y)$-split by $T(\tau)*\pi_0*\theta_0$, $T(\tau)*\pi_1*\theta_1$ whenever $|\tau| = n$; moreover, θ_0, θ_1 can be chosen so that these splitting strings belong to im(T) and are $A^{\mathbb{L}} \cap Y$-compatible. We now describe a method for transforming θ_0, θ_1 to ψ_0, ψ_1 in such a way that $T(\tau)*\pi_0*\psi_0$, $T(\tau)*\pi_1*\psi_1$ are $\mathbb{L}$-acceptable, $A^{\mathbb{L}}$-compatible in addition to retaining the other properties possessed by $T(\tau)*\pi_0*\theta_0$, $T(\tau)*\pi_1*\theta_1$. Then $\pi_i*\theta_i$ is the required ρ_i, for $i \in \{0,1\}$.

Let p, q be such that $\theta_i = (T)_p^{c_i(p)} *\ldots* (T)_q^{c_i(q)}$ for $i \in \{0,1\}$; define $\theta_{im} = (T)_p^{c_i(p)} *\ldots* (T)_m^{c_i(m)}$ for $p \leq m \leq q$. We shall define a sequence of pairs ψ_{0m}, ψ_{1m} from θ_{0m}, θ_{1m}; the required pair ψ_0, ψ_1 will be ψ_{0q}, ψ_{1q}.

<u>Case 1</u>: $T(\mu)*(T)_m^{c_0(m)}$, $T(\mu)*(T)_m^{c_1(m)}$ are $(A^{\mathbb{L}}-Y)$-compatible whenever $|\mu| = m - 1$. In this case, let $\psi_{im} = \psi_{i,m-1}*(T)_m^{c_i(m)}$ for $i \in \{0,1\}$.

<u>Case 2</u>: $T(\mu)*(T)_m^{c_0(m)}$, $T(\mu)*(T)_m^{c_1(m)}$ are $(A^{\mathbb{L}}-Y)$-incompatible whenever $|\mu| = m - 1$. In this case, simply let $\psi_{im} = \psi_{i,m-1}*(T)_m^{c_0(m)}$ for $i \in \{0,1\}$.

We begin the whole procedure by setting $\psi_{i,p-1} = \phi$ for $i \in \{0,1\}$.

It is easy to see by induction on m that

$$T(\tau)*\pi_0*\psi_{0m}, \quad T(\tau)*\pi_1*\psi_{1m}$$

are $A^{\mathbb{L}}$-compatible. When Case 2 applies, this is immediate from the induction hypothesis (which we recall holds for m = p by assumption). If Case 1 applies then $(T)_m^{c_0(m)}$, $(T)_m^{c_1(m)}$ cannot give rise to $A^{\mathbb{L}}$-incompatibility, since $T(\tau)*\pi_0*\psi_0$, $T(\tau)*\pi_1*\psi_1$ are $A^{\mathbb{L}} \cap Y$-compatible.

Next, we prove by induction on m that

$$T(\tau)*\pi_0*\psi_{0m}, \quad T(\tau)*\pi_1*\psi_{1m}$$

are $\mathbb{L}$-acceptable. If Case 2 applies, this is again immediate from the induction hypothesis. To deal with Case 1, let A′, A″ be two atoms of $\mathcal{B}(\mathbb{L})$ such that the two strings above are A′-incompatible, A″-incompatible. Then both A′ and A″ must be $\subseteq \underset{\sim}{N} - A^{\mathbb{L}}$ because the two strings are $A^{\mathbb{L}}$-compatible. It follows that if A′, A″ are separated by X, Z $\in \mathbb{L}$ then $X \cap Z \supseteq A$ and so the two strings are $X \cap Z$-incompatible, as required.

Finally, we claim that $(T(\tau)*\pi_i*\psi_{im})\langle Y\rangle = (T(\tau)*\pi_i*\theta_{im})\langle Y\rangle$ for $i \in \{0,1\}$ and $p \leq m \leq q$. For, if Case 2 applies then $(T)_m^{c_0(m)}$, $(T)_m^{c_1(m)}$ cannot give rise to $(Y-A^{\mathbb{L}})$-incompatibility because T is $\mathbb{L}$-independent; hence, by replacing $(T)_m^{c_1(m)}$ by $(T)_m^{c_0(m)}$ we do not falsify the equation above. Under Case 1, we certainly do nothing to falsify this equation and so we conclude, inductively, that it holds true. But then it follows that $T(\tau)$ is $(\underset{\sim}{E}_e,Y)$-split by $T(\tau)*\pi_0*\psi_0$, $T(\tau)*\pi_1*\psi_1$ whenever $|\tau| = n$, as required. □

This lemma is immediate when $\mathbb{L}$ is a linear ordering because then $A^{\mathbb{L}} = \phi$ and so (a)′ is implied by (a) of Lemma 2.10; also, in this case, all strings are $\mathbb{L}$-acceptable so that (c) holds trivially. When $\mathbb{L}$ is a boolean algebra, the lemma can be simplified slightly because then (c) is implied by (a)′.

We are at last in a position to describe the operation

$$T \longrightarrow Sp^{\mathbb{L}}_{e,Y}(T)$$

which will be used for forming an $\mathbb{L}$-independent $(\underset{\sim}{E}_e,Y)$-splittingmap from a given ($\mathbb{L}$-independent) treemap T. In order to simplify the notation we shall temporarily abbreviate $Sp^{\mathbb{L}}_{e,Y}(T)$ by T′. Let A_0, A_1, ..., be a recursive list of the (finitely many) atoms of $\mathcal{B}(\mathbb{L})$ in which each atom appears infinitely often. We define $T'(\tau)$ by induction on

$|\tau|$. To define $T'(\tau*0)$, $T'(\tau*1)$ for all τ such that $|\tau| = n$ there are two cases.

Case 1: $A_n \subseteq Y$. Then we let $(T')_n^0$, $(T')_n^1$ be the least strings ψ_0, ψ_1 asserted to exist by Lemma 2.11.

Case 2: $A_n \not\subseteq Y$. In this case, we simply let $(T')_n^0$, $(T')_n^1$ be the least two strings satisfying conditions (a)', (c) of Lemma 2.11; the existence of such strings follows immediately from the $\mathbb{L}$-independence of T and does not need Lemma 2.11 itself.

It is clear that if T is a recursive $\mathbb{L}$-independent treemap then so is T'. T' is an $(\underset{\sim}{F}_e, Y)$-splittingmap because (i) $T'(\tau*0)$, $T'(\tau*1)$ are Y-incompatible for infinitely many τ, namely all those τ such that $|\tau| = n$ for some $A_n \subseteq Y$, and (ii) whenever $T'(\tau*0)$, $T'(\tau*1)$ do not $(\underset{\sim}{F}_e, Y)$-split $T'(\tau)$ then we must have $|\tau| = n$ where $A_n \not\subseteq Y$ so that $T'(\tau*0)$, $T'(\tau*1)$ are $A_n^{\mathbb{L}}$-compatible, hence Y-compatible.

§3. Finite distributive lattices

As was explained in [10], the embeddability of finite distributive lattices in $\mathcal{D}$ was uncovered over a period of years beginning with Spector's construction of a set of minimal degree. The intermediate steps were taken by Titgemeyer and Sacks (see [7], §11), Shoenfield (unpublished) and concluded with Lachlan's important paper [3].

THEOREM 3.1. (Lachlan) If $\mathcal{L}$ is a finite distributive lattice then $\mathcal{L} \overset{*}{\Longrightarrow} \mathcal{D}(\Delta_2^0)$.

Proof. Let $\mathbb{L}$ be a distributive representation for $\mathcal{L}$. We construct a function $B : \underset{\sim}{N} \longrightarrow \{0,1\}$ of degree $\leq \underset{\sim}{0}^{(2)}$ and satisfying

(i) $Z \subseteq X \leftrightarrow B\langle Z\rangle \leq_T B\langle X\rangle$ for all $X, Z \in \mathbb{L}$,

(ii) if $C \leq_T B$ then $C \equiv_T B\langle Y\rangle$ for some $Y \in \mathbb{L}$.

This ensures that if we define $\underset{\sim}{b}\langle X\rangle$, for each $X \in \mathcal{L}$, to be the degree of $B\langle X\rangle$, then $\mathcal{I} = \{\underset{\sim}{b}\langle X\rangle : X \in \mathcal{L}\}$ is an initial segment of $\mathcal{D}(\leq \underset{\sim}{0}^{(2)})$ isomorphic to $\mathcal{L}$.

The conditions (i) and (ii) above are met as usual by dealing with two infinite lists of more specialised requirements. Let $A_1, \ldots, A_M$ be a list of the atoms of $\mathcal{B}(\mathcal{L})$ and let $Y_1, \ldots, Y_N$ be a list of the nonempty (hence infinite) elements of $\mathcal{L}$. The more specialised requirements are:

$\Delta(e,m)$: if $\underset{\sim}{E}_e^*(B\langle A_m^{\mathcal{L}}\rangle)$ is defined then $B\langle A_m\rangle \neq \underset{\sim}{E}_e^*(B\langle A_m^{\mathcal{L}}\rangle)$,

for $1 \leq m \leq M$, and

$\Sigma(e,n)$: if $\underset{\sim}{E}_e^*(B\langle Y_n\rangle)$ is defined then either it has the same degree as $B\langle Y_n\rangle$ or it is recursive in $B\langle W\rangle$ for some $W \in \mathcal{L}$ such that $W \subset Y_n$,

for $1 \leq n \leq N$.

To see that the former conditions are sufficient, observe first that $\rightarrow$ in (i) holds trivially; secondly, let A_m be an atom of $\mathcal{B}(\mathcal{L})$ which is $\subseteq Z - X$ and note then that $X \subseteq A_m^{\mathcal{L}}$. The sufficiency of the second list is easily verified by induction over $\subset$. (<u>Note</u>: we could replace $\{\Sigma(e,n)\}$ by the tighter conditions $\{\Sigma(e)\}$ where $\Sigma(e)$ is:

if $\underset{\sim}{E}_e^*(B)$ is defined then it is $\equiv_T B\langle Y\rangle$ for some $Y \in \mathcal{L}$.

This does not rely on an inductive argument and is necessary for the general countable case (see §4 and [12]).)

Now, let $\Gamma_1, \Gamma_2, \ldots,$ be a single list (of type ω) containing _all_ the requirements which we have just mentioned. For each $k > 0$, we shall associate Γ_k with a string β_k and the formation of a treemap T_k. If Γ_k is of the form $\Delta(e,m)$ then β_k will be defined by a 'diagonalisation' intended to satisfy $\Delta(e,m)$, and if Γ_k is of the form $\Sigma(e,n)$ then the treemap T_k will be designed to satisfy $\Sigma(e,n)$. The treemaps will form a contracting chain in the sense that $im(T_k) \supseteq im(T_{k+1})$ for all k; the strings will form an ascending chain $\beta_0 \subset \beta_1 \subset \ldots$ with limit B which will also be the sole element of $\bigcap_{k\geq 0} im(T_k^*)$. Each treemap T_k will be recursive and $\mathbb{L}$-independent; the $\mathbb{L}$-independence of the treemaps is necessary in order to satisfy the conditions of the form $\Delta(e,m)$. We shall indicate at the end of the proof why B is of degree $\leq 0^{(2)}$.

We begin by letting T_0 be the identity treemap and $\beta_0 = \phi$. We then have, for each $k > 0$, the two tasks of defining first β_k and then T_k. There are various cases, the main subdivision being dictated by the nature of the condition Γ_k.

Case 1: Γ_k is of the form $\Delta(e,m)$. Let β_k be the least $\beta \supset \beta_{k-1}$ in $im(T_{k-1})$ for which $\underset{\sim}{F}_e(\beta\langle A_m^{\mathbb{L}}\rangle)$, $\beta\langle A_m\rangle$ are incompatible, if such a β exists; otherwise, let $\beta_k = T_{k-1}(0)$. Then define

$$T_k = T_{k-1} \text{ above } \beta_k.$$

Case 2: Γ_k is of the form $\Sigma(e,n)$.

Subcase 2a: for each atom A of $\mathcal{B}(\mathbb{L})$ such that $A \subseteq Y$, every $\sigma \in im(T_{k-1})$ is $(\underset{\sim}{F}_e, Y_n)$-split by a pair of A-incompatible, $A^{\mathbb{L}} \cap Y_n$-compatible strings in $im(T_{k-1})$. In this case we set $\beta_k = \beta_{k-1}$ and

$$T_k = Sp^{\mathbb{L}}_{e,Y_n}(T_{k-1}).$$

It follows from Lemma 2.11 that T_k is well-defined.

<u>Subcase 2b</u>: otherwise. Let β_k be the least string σ in $\mathrm{im}(T_{k-1})$ which acts as a counterexample to Subcase 2a, and set $T_k = T_{k-1}$ above β_k. This completes the construction.

It is immediately clear that $T_k(\phi) = \beta_k$ for all k, and $\beta_{k+1} \in \mathrm{im}(T_k)$, so that $\beta_k \subseteq \beta_{k+1}$ for all k. It is also obvious that T_k is total for all k. One consequence of this is that $\beta_k \subset \beta_{k-1}$ whenever Γ_k is of the form $\Delta(e,m)$. Another consequence, which can be easily derived by induction on k, is that T_k is recursive and $\mathbb{L}$-independent for all k.

Now, we have to prove that Γ_k is satisfied for all k. First, to prove that $\Delta(e,m)$ holds for all e and $1 \leq m \leq M$. Suppose that $\mathbb{E}^*_e(B\langle A^{\mathbb{L}}_m\rangle)$ is defined. Let $\Gamma_k = \Delta(e,m)$. We claim that there is a $\beta \supset \beta_{k-1}$ which belongs to $\mathrm{im}(T_{k-1})$ and is such that $\mathbb{E}_e(\beta\langle A^{\mathbb{L}}_m\rangle)$ and $\beta\langle A_m\rangle$ are incompatible. This implies that $\beta_k\langle A_m\rangle$ and $\mathbb{E}_e(\beta_k\langle A^{\mathbb{L}}_m\rangle)$ are incompatible, whence $\Delta(e,m)$ follows immediately. To prove this claim, let $(T_{k-1})_r$ be the first component in T_{k-1} such that $\beta_{k-1} \subseteq T_{k-1}(\tau)$ for some τ with $|\tau| = r - 1$, and $(T_{k-1})^0_r$, $(T_{k-1})^1_r$ are A_m-incompatible, $A^{\mathbb{L}}_m$-compatible; they exist because T_{k-1} is $\mathbb{L}$-independent. Let τ_0 have length $r - 1$ and be such that $T_{k-1}(\tau_0) \subseteq B$, and let γ_0, γ_1 be $T_{k-1}(\tau_0)*(T_{k-1})^0_r$, $T_{k-1}(\tau_0)*(T_{k-1})^1_r$ respectively; note that $\gamma_0\langle A^{\mathbb{L}}_m\rangle = \gamma_1\langle A^{\mathbb{L}}_m\rangle \subseteq B\langle A^{\mathbb{L}}_m\rangle$. Next, let δ_0, δ_1 be extensions of γ_0, γ_1 respectively such that $\delta_0\langle A^{\mathbb{L}}_m\rangle = \delta_1\langle A^{\mathbb{L}}_m\rangle$, and $\mathbb{E}_e(\delta_0\langle A^{\mathbb{L}}_m\rangle)$ - which is also $\mathbb{E}_e(\delta_1\langle A^{\mathbb{L}}_m\rangle)$ - has length $\geq |\gamma_0| = |\gamma_1|$. These exist because there is certainly some $\delta \subset B$ such that $|\mathbb{E}_e(\delta\langle A^{\mathbb{L}}_m\rangle)| \geq |\gamma_0| = |\gamma_1|$ and either $\delta \supset \gamma_0$ or $\delta \supset \gamma_1$. Now, either $\mathbb{E}_e(\delta_0\langle A^{\mathbb{L}}_m\rangle)$ is incompatible with $\delta_0\langle A_m\rangle$ or $\mathbb{E}_e(\delta_1\langle A^{\mathbb{L}}_m\rangle)$ is incompatible with $\delta_1\langle A_m\rangle$, because $\gamma_0\langle A_m\rangle$, $\gamma_1\langle A_m\rangle$ are incompatible. This proves our claim, and hence $\Delta(e,m)$ is satisfied.

In proving that $\Sigma(e,n)$ holds for all e and $1 \leq n \leq N$, there are two cases. Let $\Gamma_k = \Sigma(e,n)$. If T_k is defined through Subcase 2a then

$B\langle Y_n\rangle$ is recursive in $\underset{\sim}{F}_e^*(B\langle Y_n\rangle)$ by Lemma 2.9. On the other hand, if T_k is defined through Subcase 2b then there is a pair σ, A with $A \subseteq Y_n$ which form a counterexample to Subcase 2a and such that $\beta_k = \sigma$, T_k is T_{k-1} above β_k. Hence, it follows from Lemma 2.8 that $\underset{\sim}{F}_e^*(B\langle Y_n\rangle)$ is recursive in $B\langle A^{\ell} \cap Y_n\rangle$. Since $A \subseteq Y_n$ it follows that $A^{\ell} \cap Y_n \subset Y_n$ and $\Sigma(e,n)$ is satisfied.

Finally, we have to indicate why B is of degree $\leq \underset{\sim}{0}^{(2)}$. As in the construction of a function of minimal degree, the most nonconstructive step is the division between Subcases 2a and 2b. This is a two-quantifier predicate and so the resulting function is Δ_3^0, i.e. of degree $\leq \underset{\sim}{0}^{(2)}$. □

The most important consequence of the above theorem is the following. It was derived by Lachlan [3] from the embeddability of countable distributive lattices; the present derivation from the embeddability of finite distributive lattices is due to Thomason [9].

COROLLARY 3.2. (Lachlan-Thomason) $\underset{\sim}{Th}(\mathcal{D})$ is undecidable, in fact non-axiomatizable.

<u>Proof</u>. Let $\underset{\sim}{T}$ be the set of sentences, in the first-order language $\underset{\sim}{L}_{\leq}$ with one additional binary relation $\leq$, which are true in all distributive lattices. Let $\underset{\sim}{T}'$ be the superset of $\underset{\sim}{T}$ consisting of all those sentences which are true in all finite distributive lattices. It can be shown that $\underset{\sim}{T}$ and $Sn(\underset{\sim}{L}_{\leq}) - \underset{\sim}{T}'$ are recursively inseparable. Now, if we define

$$\underset{\sim}{T}'' = \{\Phi : \Phi \in Sn(\underset{\sim}{L}_{\leq}) \;\&\; (\forall \underset{\sim}{b})(\text{if } \mathcal{D}(\leq \underset{\sim}{b}) \text{ is a distributive lattice then } \mathcal{D}(\leq \underset{\sim}{b}) \models \Phi)\},$$

then it follows from Theorem 3.1 that $\underset{\sim}{T} \subseteq \underset{\sim}{T}'' \subseteq \underset{\sim}{T}'$; this is because every finite distributive lattice features as $\mathcal{D}(\leq \underset{\sim}{b})$ for some $\underset{\sim}{b}$. Hence,

$\underset{\sim}{T}''$ cannot be recursive, and in fact, since $\underset{\sim}{T}''$ is complete, $\underset{\sim}{T}''$ cannot even be Σ_1^0. But $\underset{\sim}{T}''$ is clearly one-one reducible to $\underset{\sim}{Th}(\mathcal{D})$: this is because the defining predicate for $\underset{\sim}{T}''$ can be recursively translated into $\underset{\sim}{L}_{\leqslant}$ and then interpreted in $\mathcal{D}$. Hence, $\underset{\sim}{T}''$ is not Σ_1^0. □

COROLLARY 3.3. $\underset{\sim}{Th}(\mathcal{D}(\Delta_2^0))$ is non-axiomatizable.

<u>Proof</u>. As above except that the full strength of the theorem is used. □

In [12] we shall compress all the work in this section below $\underset{\sim}{0}^{(1)}$, hence proving that $\underset{\sim}{Th}(\mathcal{D}(\Delta_2^0))$ is non-axiomatizable.

§4. <u>Countable Linear Orderings</u>

The main theorem of this section was first proved by Hugill [2] using a quite different system of representation. The present method seems considerably easier; we recall that much of the work in §2 is not needed in this particular case.

THEOREM 4.1. (Hugill) If $\mathcal{L}$ is a bottomed linear ordering then $\mathcal{L} \overset{*}{\Rightarrow} \mathcal{D}$.

<u>Proof</u>. It is sufficient to prove that every infinite bottomed and topped linear ordering is $\overset{*}{\Rightarrow} \mathcal{D}$, and so let $\mathcal{L}$ be such an ordering. As we have already mentioned in the Introduction, the special feature of the <u>infinite</u> countable case is that, instead of using a fixed representation of $\mathcal{L}$, it is necessary to construct this representation alongside the various treemaps and functions which we are setting out to define. Our intention then is to simultaneously produce a linear representation $\mathbb{L}$ (consisting of various infinite recursive sets plus ϕ) such that $\mathcal{L} \cong \mathbb{L}$, together with a function $B : \underset{\sim}{N} \longrightarrow \{0,1\}$ such that:

(i) $X \subseteq Z \leftrightarrow B\langle X\rangle \leq_T B\langle Z\rangle$ for all $X, Z \in \mathbb{L}$,

(ii) if $C \leq_T B$ then $C \equiv_T B\langle Y\rangle$ for some $Y \in \mathbb{L}$.

This ensures that if we define $\underset{\sim}{b}\langle X\rangle$, for each $X \in \mathbb{L}$, to be the degree of $B\langle X\rangle$ then $\mathcal{I} = \{\underset{\sim}{b}\langle X\rangle : X \in \mathbb{L}\}$ is an initial segment of $\mathcal{D}$ isomorphic to $\mathcal{L}$.

Let $\mathcal{L}_0, \mathcal{L}_1, \ldots,$ be an ascending chain of finite suborderings of $\mathcal{L}$ with limit $\mathcal{L}$; it is convenient to assume that $\mathcal{L}_0$ consists of the top and bottom elements of $\mathcal{L}$, and also that $\mathcal{L}_{k+1}$ is obtained from $\mathcal{L}_k$ by the addition of exactly one element. It is our intention at stage k in the construction to define an isomorphism H_k of $\mathcal{L}_k$ onto a finite linear representation $\mathbb{L}_k$ (in the sense of Definition 1.4). The atoms of $\mathcal{B}(\mathbb{L}_k)$ may be assumed to be exactly the sets of the form $Z - X$ where $X \subset Z$ and X, Z are consecutive elements in the ordering of $\mathbb{L}_k$. Clearly, in moving from $\mathcal{L}_k$ to $\mathcal{L}_{k+1}$, a unique atom of $\mathcal{B}(\mathbb{L}_k)$ will have to be split to form atoms of $\mathcal{B}(\mathbb{L}_{k+1})$; we shall denote this atom by A_k. The division of A_k cannot be done in an arbitrary fashion, and this step is the only non-trivial one in moving from finite to infinite linear orderings.

We shall succeed in satisfying (i) and (ii) above if we attend to the infinite list of conditions $\{\Gamma_k\}_{k>0}$, where

Γ_{2e}: if $Z - X$ is an atom of $\mathcal{B}(\mathbb{L}_{2e-1})$ and $\underset{\sim}{F}_e^*(B\langle X\rangle)$ is defined then $B\langle Z\rangle \neq \underset{\sim}{F}_e^*(B\langle X\rangle)$.

Γ_{2e+1}: if $\underset{\sim}{F}_e^*(B)$ is defined then it is $\equiv_T B\langle Y\rangle$ for some $Y \in \mathbb{L}_{2e}$.

For each $k > 0$, we associate Γ_k with a string β_k and treemap T_k as in §3. The only difference now is that we want T_k to be $\mathbb{L}_k$-independent: this has to be arranged at stage k <u>in the definition of</u> $\mathbb{L}_k$. Once again, the $\mathbb{L}_k$-independence of T_k is necessary in order to satisfy the 'diagonalisation' conditions.

We begin by letting T_0 be the identity tree, $\mathbb{L}_0$ the trivial Boolean algebra $\{\phi, \underset{\sim}{N}\}$ and $\beta_0 = \phi$. For $k > 0$, we then have two separate tasks: the construction of β_k and T_k, and the construction of H_k and $\mathbb{L}_k$.

<u>Construction of</u> β_k <u>and</u> T_k. There are various cases, the main subdivision being related to the nature of the condition Γ_k. Let Y_0, ... ,Y_k be a list of the (distinct) elements of $\mathbb{L}_{k-1}$ with $Y_j \subset Y_{j+1}$ for $j < k$.

<u>Case 1</u>: $k = 2e$. β_k will be the last string in an ascending chain $\beta_{k0} \subset \dots \subset \beta_{kk}$ with $\beta_{k0} = \beta_{k-1}$. For $1 \leqslant j \leqslant k$, $\beta_{k,j}$ is constructed from $\beta_{k,j-1}$ as follows: it is the least string $\beta \supset \beta_{k,j-1}$ such that $\beta \in \mathrm{im}(T_{k-1})$ and $\underset{\sim}{F}_e(\beta\langle Y_{j-1}\rangle)$, $\beta\langle Y_j\rangle$ are incompatible, if such a β exists; otherwise, it is $\beta_{k,j-1}$. Finally, we set $\beta_k = \beta_{kk}$ and

$$T_k = T_{k-1} \text{ above } \beta_k.$$

<u>Case 2</u>: $k = 2e + 1$. T_k will be defined as the last member in a contracting chain $\{T_{ki}\}$ such that

$$\mathrm{im}(T_{k0}) \supseteq \dots \supseteq \mathrm{im}(T_{\ ,g(k)}),$$

where T_{k0} is T_{k-1}. For $0 \leqslant i < g(k)$, there will be associated with T_{ki} a set $Y_{m_i} \in \mathbb{L}_{k-1}$ and a number $h(k,i)$ such that $\underset{\sim}{F}^*_e(C) = \underset{\sim}{F}^*_{G(k,i)}(C\langle Y_{m_i}\rangle)$ for all $C \in \mathrm{im}(T^*_{ki})$; this procedure is initiated by setting $Y_{m_0} = \underset{\sim}{N}$ and $h(k,0) = e$. In the definition of $T_{k,i+1}$, $Y_{m_{i+1}}$ and $h(k,i+1)$ from T_{ki}, Y_{m_i} and $h(k,i)$, there are two subcases.

<u>Subcase (a)</u>: for each atom $Y_m - Y_{m-1}$ of $\mathcal{B}(\mathbb{L}_{k-1})$ with $m \leq m_i$, every $\sigma \in \text{im}(T_{ki})$ is $(\underset{\sim}{E}_{h(k,i)}, Y_{m_i})$-split by a pair of $(Y_m - Y_{m-1})$-incompatible, Y_{m-1}-compatible strings in $\text{im}(T_{ki})$. In this case, we define $T_{k,i+1}$ to be $Sp^{\mathbb{L}_{k-1}}_{h(k,i),Y_{m_i}}(T_{ki})$. Also, we set $g(k) = i + 1$ and so the procedure terminates.

<u>Subcase (b)</u>: otherwise. Let $Y_m - Y_{m-1}$ and σ constitute a counterexample to subcase (a). Define $T_{k,i+1}$ to be T_{ki} above σ. We know through Lemma 2.8 that $\underset{\sim}{E}^*_{h(k,i)}(C\langle Y_{m_i}\rangle)$ is recursive in $C\langle Y_{m-1}\rangle$ for all $C \in \text{im}(T^*_{k,i+1})$. If $Y_{m-1} = \phi$ then we simply set $g(k) = i + 1$ so that the procedure terminates. Otherwise, we set $Y_{m_{i+1}} = Y_{m-1}$ and let $h(k,i+1)$ be such that $\underset{\sim}{E}^*_{h(k,i)}(C\langle Y_{m_i}\rangle) = \underset{\sim}{E}^*_{h(k,i+1)}(C\langle Y_{m_{i+1}}\rangle)$ for all $C \in \text{im}(T^*_{k,i+1})$; the existence of $h(k,i+1)$ follows from Lemma 2.8. Finally, we set $T_k = T_{k,g(k)}$ and $\beta_k = T_k(0)$, which completes this part of the construction. Clearly $\beta_k \supset \beta_{k-1}$ and if T_{k-1} is an $\mathbb{L}_{k-1}$-independent recursive treemap then so is T_k.

<u>Construction of</u> $\mathbb{L}_k$. This is the crucial problem pertaining to the countable case, for we wish to define $\mathbb{L}_k$ in such a way that if the previously-defined treemap T_k is $\mathbb{L}_{k-1}$-independent then it is in fact $\mathbb{L}_k$-independent. Hence, the definition of $\mathbb{L}_k$ depends very heavily on that of T_k, which is why we could not start with a fixed representation $\mathbb{L}$.

We assume that at stage $k - 1$ an isomorphism H_{k-1} was established between $\mathcal{L}_{k-1}$ and a linear representation $\mathbb{L}_{k-1}$. As already mentioned, in order to extend H_{k-1} to an isomorphism H_k between $\mathcal{L}_k$ and a representation $\mathbb{L}_k$, it is necessary to partition some atom A_k of $\mathcal{B}(\mathbb{L}_{k-1})$ into two infinite recursive sets A^0_k, A^1_k. To make the situation clearer, the new element of $\mathcal{L}_k$ lies between y_{m-1} and y_m, say, where these are elements of $\mathcal{L}_{k-1}$. This new element has to be mapped to a set between

Y_{m-1} and Y_m, where $Y_{m-1} = H_{k-1}(y_{m-1})$, $Y_m = H_{k-1}(y_m)$. So our intention is that $\mathbb{L}_k$ be formed by adding $Y_{m-1} \cup A_k^0$, say, to $\mathbb{L}_{k-1}$. Note that $Y_{m-1} = A_k^{\mathbb{L}_{k-1}}$ in the sense of §2.

We form $\{A_k^0, A_k^1\}$ as follows. First, define:

$$N_k = \{n : T_k(\tau*0),\ T_k(\tau*1) \text{ are } A_k\text{-incompatible, } A_k^{\mathbb{L}_{k-1}}\text{-compatible whenever } |\tau| = n\}.$$

Then let N_k^0, N_k^1 be a partition of N_k into two disjoint infinite recursive sets; note that if T_k, A_k and $A_k^{\mathbb{L}}$ are recursive then so is N_k. Lastly, we have the desired partition:

$$A_k^0 = \{x : x \in A_k \ \& \ (\exists n)(n \in N_k^0 \ \& \ |T_k|_n < x \leq |T_k|_{n+1})\},$$
$$A_k^1 = A_k - A_k^0.$$

This completes the entire construction and now we have to prove that it works.

The purpose of the special construction of $\mathbb{L}_k$, which we have just described, is to make the following lemma possible.

LEMMA 4.2. For each $k > 0$, if T_k is $\mathbb{L}_{k-1}$-independent then T_k is $\mathbb{L}_k$-independent.

<u>Proof</u>. As explained in §2, clause (b) in the definition of $\mathbb{L}_k$-independence is immediate when $\mathbb{L}_k$ is linear. So we only have to verify clause (a). But $T_k(\tau*0)$, $T_k(\tau*1)$ are A_k^0-incompatible, $(A_k^0)^{\mathbb{L}_k}$-compatible whenever $|\tau| = n$ with $n \in N_k^0$, because $(A_k^0)^{\mathbb{L}_k} = A_k^{\mathbb{L}_{k-1}}$; also $T_k(\tau*0)$, $T_k(\tau*1)$ are A_k^1-incompatible, $(A_k^1)^{\mathbb{L}_k}$-compatible whenever $|\tau| = n$ with $n \in N_k^1$, because $(A_k^1)^{\mathbb{L}_k} = A^{\mathbb{L}_{k-1}} \cup A_k^0$. □

Since T_0 is trivially $\mathbb{L}_0$-independent, it is now easy to prove by induction on k that T_k is $\mathbb{L}_k$-independent for all k; this of course uses the observation made during the construction that if T_{k-1} is $\mathbb{L}_{k-1}$-independent then T_k is $\mathbb{L}_{k-1}$-independent. It is in fact possible to see that T_k is $\mathbb{L}_j$-independent for all j, k: if $j > k$ then this is because the existence of an $\mathbb{L}_j$-independent treemap T', such that $\mathrm{im}(T') \subseteq \mathrm{im}(T_k)$, implies $\mathbb{L}_j$-independence for T_k. Hence, T_k is in fact $\mathbb{L}$-independent for all k. We do not, however, need to make explicit use of this stronger assertion.

It remains to prove that B possesses the properties required of it; in other words, that Γ_k is satisfied for all $k > 0$. Let $Y_0, \ldots, Y_k$ be the elements of $\mathbb{L}_{k-1}$. First we deal with the case $k = 2e$, our task being to show that if $\underset{\sim}{F}^*_e(B<Y_{m-1}>)$ is defined then $\underset{\sim}{F}^*_e(B<Y_{m-1}>) \not\subseteq B<Y_m>$. Let $(T_{k-1})_r$ be the first component in T_k such that $\beta_{k,m-1} \subseteq T_{k-1}(\tau)$ for all τ with $|\tau| = r - 1$, and $(T_{k-1})^0_r$, $(T_{k-1})^1_r$ are Y_m-incompatible, Y_{m-1}-compatible; they exist because T_{k-1} is $\mathbb{L}_{k-1}$-independent. Let τ_0 have length $r - 1$ and be such that $T_{k-1}(\tau_0) \subseteq B$, and let γ_0, γ_1 be $T_{k-1}(\tau_0)*(T_{k-1})^0_r$, $T_{k-1}(\tau_0)*(T_{k-1})^1_r$ respectively; note that $\gamma_0<Y_{m-1}> = \gamma_1<Y_{m-1}> \subseteq B<Y_{m-1}>$. Next, let δ_0, δ_1 be extensions of γ_0, γ_1 respectively such that $\delta_0<Y_{m-1}> = \delta_1<Y_{m-1}>$, and $\underset{\sim}{F}_e(\delta_0<Y_{m-1}>)$ - which is also $\underset{\sim}{F}_e(\delta_1<Y_{m-1}>)$ - has length $\geq |\gamma_0| = |\gamma_1|$. These exist because there is certainly some $\delta \subseteq B$ such that $|\underset{\sim}{F}_e(\delta<Y_{m-1}>)| \geq |\gamma_0| = |\gamma_1|$ and either $\delta \supset \gamma_0$ or $\delta \supset \gamma_1$. Now, either $\underset{\sim}{F}_e(\delta_0<Y_{m-1}>)$ is incompatible with $\delta_0<Y_m>$ or $\underset{\sim}{F}_e(\delta_1<Y_{m-1}>)$ is incompatible with $\delta_1<Y_m>$, because $\gamma_0<Y_m>$, $\gamma_1<Y_m>$, are incompatible. This shows that Γ_k is satisfied.

Next, suppose that $k = 2e + 1$; there are two cases. If g(k) is defined through subcase (a) then, by the discussion under that part of the construction, $B<Y_{m_{g(k)-1}}>$ is recursive in $\underset{\sim}{F}^*_{h(k,g(k)-1)}(B<Y_{m_{g(k)-1}}>) = \underset{\sim}{F}^*_e(B)$, and so $\underset{\sim}{F}^*_e(B) \equiv_T B<Y_{m_{g(k)-1}}>$ (by Lemma 2.9). If g(k) is defined through subcase (b) then $\underset{\sim}{F}_e^*(B) = \underset{\sim}{F}^*_{h(k,g(k)-1)}(B<Y_{m_{g(k)-1}}>)$ is recursive in $B<Y_{m_{g(k)}}>$ and $Y_{m_{g(k)}} = \phi$ so that $\underset{\sim}{F}_e^*(B)$ is recursive (by Lemma 2.8). This concludes the proof of the theorem. □

This useful theorem has of course many corollaries in terms of particular linear orderings. Also, since it implies that the ordering $(1,\omega^*)$ is $\overset{*}{\Rightarrow} \mathcal{D}$, it implies the existence of degrees with no minimal predecessors; as mentioned in [10] there are various proofs of the latter result. Subsequently, a particularly powerful application has been noticed by Feiner [1] and a refinement of this is the subject of the discussion in the next and final section.

To conclude the present section, it is worth noting the exact problem that arises with arbitrary countable distributive lattices. This lies in pushing through Lemma 4.2; for, there is no a priori reason why strings which are $\mathcal{L}_k$-acceptable should be $\mathcal{L}_{k+1}$-acceptable.

§5. Jump-preserving isomorphisms

Let a distributive lattice be called $\underset{\sim}{a}$-presentable, where $\underset{\sim}{a} \in \mathcal{D}$, if it is isomorphic to a distributive lattice $(\underset{\sim}{N},\leq,\cap,\cup)$ in which the relation $\leq$ and functions $\cap$, $\cup$ are of degree $\leq \underset{\sim}{a}$. Careful examination of the construction in the preceding section shows, along the lines briefly outlined at the end of Theorem 3.2, that every $\underset{\sim}{a}^{(2)}$-presentable distributive lattice is embeddable as an initial segment of $\mathcal{D}(\underset{\sim}{a},\underset{\sim}{a}^{(2)})$. We showed on p. 77 of [10], using the familiar chain of lines and diamonds, that there is a distributive lattice which is $\underset{\sim}{a}^{(2)}$-presentable but not $\underset{\sim}{a}$-presentable. This was essentially followed by the observation that the partial ordering $\mathcal{D}(\leq\underset{\sim}{a})$ is $\underset{\sim}{a}^{(3)}$-presentable for any $\underset{\sim}{a}$; hence, if $\mathcal{D}(\leq\underset{\sim}{a})$ is in fact a lattice then it is $\underset{\sim}{a}^{(4)}$-presentable as a lattice. It follows from these observations that, on the one hand every lattice which is an initial segment of $\mathcal{D}(\leq\underset{\sim}{0}^{(2)})$ is $\underset{\sim}{0}^{(6)}$- presentable, yet there is an $\underset{\sim}{0}^{(8)}$-presentable distributive lattice which is not $\underset{\sim}{0}^{(6)}$-presentable. Since this lattice is $\overset{*}{\Rightarrow} \mathcal{D}(\underset{\sim}{0}^{(6)},\underset{\sim}{0}^{(8)})$ we conclude that there is no jump-preserving isomorphism of $\mathcal{D}(\geq \underset{\sim}{0}^{(6)})$ onto $\mathcal{D}$; it immediately follows that there is no jump-preserving isomorphism of $\mathcal{D}(\geq \underset{\sim}{0}^{(1)})$ onto $\mathcal{D}$, since it could be iterated. In fact, with little trouble one can prove that there is no jump-preserving isomorphism of $\mathcal{D}(\geq \underset{\sim}{0}^{(n)})$ onto $\mathcal{D}$. The impossibility of a jump-preserving isomorphism from $\mathcal{D}(\geq \underset{\sim}{0}^{(6)})$ to $\mathcal{D}$ was first noticed by Feiner [1];[1] he made use of a linear ordering which is $\underset{\sim}{0}^{(6)}$-presentable but not $\underset{\sim}{0}^{(5)}$-presentable. The method which we have just outlined lends itself, however, to a striking generalisation.

THEOREM 5.1. Let $\underset{\sim}{a}$ be any degree such that $\underset{\sim}{a}^{(1)} \not\leq \underset{\sim}{0}^{(6)}$. Then there is <u>no</u> jump-preserving isomorphism from $\mathcal{D}(\geq \underset{\sim}{a})$ to $\mathcal{D}$.

<u>Proof</u>. It is easy to prove, using the "chain of lines and diamonds" method, that there is an $\underset{\sim}{a}^{(1)}$-presentable lattice which is not $\underset{\sim}{0}^{(5)}$-presentable. Such a lattice is $\overset{*}{\Rightarrow} \mathcal{D}(\underset{\sim}{a},\underset{\sim}{a}^{(1)})$ by the main result to appear in [12], but is not $\overset{*}{\Rightarrow} \mathcal{D}(\leq \underset{\sim}{0}^{(1)})$ because of one of the observations made at the beginning of this section. □

Hence, if there exist jump-preserving isomorphisms of $\mathcal{D}$ onto $\mathcal{D}(\geq \underset{\sim}{a})$ then $\underset{\sim}{a}$ is certainly $\leq \underset{\sim}{0}^{(6)}$ and so the possibilities for $\underset{\sim}{a}$ are severely limited. It should be possible to replace $\underset{\sim}{0}^{(6)}$ by $\underset{\sim}{0}^{(5)}$; this would require a proof that if $\underset{\sim}{a}^{(1)} \not\leq \underset{\sim}{0}^{(5)}$ then there is an $\underset{\sim}{a}^{(1)}$-presentable distributive lattice which is not $\underset{\sim}{0}^{(5)}$-presentable. We suspect, however, that it will be difficult to substantially strengthen Theorem 5.1, because though the degree of $\underset{\sim}{a}$ may be raised by finite iterations there is no way of iterating this operation over transfinite ordinals

---

[1]The strong homogeneity conjecture was also refuted independently by Jockusch (written communication) who noticed, using work of Sacks and Putnam on the hyperarithmetical hierarchy, that there is a jump-preserving isomorphism from $\mathcal{D}$ to $\mathcal{D}(\geq \underset{\sim}{a})$ only when $\rho_1^{\underset{\sim}{a}} = \rho_1$ (the least non-recursive ordinal). This result follows from our Theorem 5.1.

and hence no obvious way of breaking out of $\mathcal{D}(\leq \underline{0}^{(6)})$ in order to obtain the required contradiction.

More challenging than the various possible refinements in the refutation of the strong homogeneity-conjecture, is the homogeneity-conjecture (omitting the jump operator). A refutation now seems more likely than a proof but virtually nothing is known about this problem. Before trying to obtain an affirmative answer, it would be more natural to try and prove first that $\mathcal{D}$ and $\mathcal{D}(\geqslant \underline{a})$ are indiscernible for all $\underline{a}$ (with or without the jump operator): some discussion of this 'easier' problem occurs at the end of [10], where we suggested (on rather flimsy evidence) that it might be independent of the usual axioms for set theory.

REFERENCES

[1] L. Feiner, _The strong homogeneity conjecture_, Jour. Symb. Logic, 35 (1970), 375-377.

[2] D. F. Hugill, _Initial segments of Turing degrees_, Proc. Lond. Math. Soc., 19 (1969), 1-15.

[3] A. H. Lachlan, _Distributive initial segments of the degrees of unsolvability_, Zeits. für math. Logik und Grund. der Math., 14 (1968), 457-472.

[4] M. Lerman, _Initial segments of the degrees of unsolvability_, Annals of Math. 93 (1971), 365-389.

[5] D. A. Martin, _Category, measure and the degrees of unsolvability_, (unpublished manuscript).

[6] H. Rogers, _Theory of recursive functions and effective computability_, McGraw Hill (1967).

[7] G. E. Sacks, _Degrees of Unsolvability_, Annals of Mathematics Study No. 55, Princeton (1963).

[8] J. R. Shoenfield, _A theorem on minimal degrees_, Jour. Symb. Logic 31 (1966), 539-544.

[9] S. K. Thomason, _On initial segments of hyperdegrees_, Jour. Symb. Logic 35 (1970), 189-197.

[10] C. E. M. Yates, Initial segments of the degrees of unsolvability, Part I: A Survey, *Mathematical Logic and the Foundations of Set Theory*, North-Holland (1970), 63-83.

[11] C. E. M. Yates, Initial segments of the degrees of unsolvability, Part II: Minimal Degrees, Jour. Symb. Logic 35 (1970), 243-266.

[12] C. E. M. Yates, Initial segments of the degrees, Parts III and IV (in preparation).

Erratum.

The restriction to atoms on page 327, line 6 from bottom, should be replaced by the looser restriction to nonzero elements (unnecessary in the finite case). This necessitates a number of completely trivial modifications in the proof that follows.

ABSTRACTS OF CONTRIBUTED PAPERS

1. PETER ACZEL: The ordinals of the superjump and related functionals

If T is a functional of finite type, ω_1^T is the sup of the order types of well-orderings of ω recursive in T. The superjump S is a total type three functional, introduced by Gandy, that diagonalises recursion in total type two objects. It may be formulated as an operator that maps $F : {}^{\omega}\omega \longrightarrow \omega$ to $S(F) : {}^{\omega}\omega \longrightarrow \omega$ given by $S(F)(\alpha) = \{\alpha(0)\}(F, \lambda n\alpha(n+1)) + 1$ if this is defined, and 0 otherwise. Using techniques and results of Platek concerning S and Richter concerning the first recursively Mahlo ordinal we obtain:

THEOREM 1. ω_1^S is the first recursively Mahlo ordinal.

Computations of $\{e\}(F,\alpha)$ may be carried out even if F is not defined on the whole of ${}^{\omega}\omega$, except that $\{e\}(F,\alpha)$ may be undefined because a value of $F(\gamma)$ is required when $F(\gamma)$ is not defined. If $\{e\}(F,\alpha)$ is undefined but not for this reason write $\{e\}(F,\alpha)\uparrow$. S has an extension to a consistent functional $S^{\#}$ such that for possibly partial F, $S^{\#}(F)(\alpha) = \{\alpha(0)\}(F,\lambda n\alpha(n+1)) + 1$ if this is defined, $S^{\#}(F)(\alpha) = 0$ if $\{\alpha(0)\}(F,\lambda n\alpha(n+1))\uparrow$ and $S^{\#}(F)(\alpha)$ is undefined in the remaining case. Let $E_1^{\#}(f) = 0$ if $\forall\alpha\exists n f(\bar{\alpha}n) = 0$, $E_1^{\#}(f) = 1$ if $\exists\alpha\forall n f(\bar{\alpha}n) > 0$ and $E_1^{\#}(f)$ is undefined otherwise. $|\Sigma_1^1\text{-mon}|$ and $|\Delta_1^1|$ are defined as in [1].

THEOREM 2. $\omega_1^{S^{\#}} = \omega_1^{E_1^{\#}} = |\Sigma_1^1\text{-mon}|$.

Note that $|\Sigma_1^1\text{-mon}| \geqslant |\Delta_1^1| > |\Pi_2^0| > \omega_1^S$ so that $\omega_1^{S^{\#}} > \omega_1^S$.

THEOREM 3. $|\Sigma_1^1\text{-mon}|$ is the first admissible ordinal λ such that if $R \subseteq \lambda \times \lambda$ is a λ-r.e. linear ordering with no λ-recursive descending sequence then R is a well-ordering. This ordinal is much smaller than the first non-projectible.

[1] P. Aczel and W. Richter, Inductive definitions and analogues of large cardinals, these Proceedings pp. 1-9.

2. J. L. BELL and F. JELLETT: An effective implication in functional analysis

Let ZF be Zermelo-Fraenkel set theory without choice, let BPI stand for the Boolean prime ideal theorem, and HB for the Hahn-Banach theorem. It is well known that $ZF \vdash BPI \to HB$. Whether the converse holds is still an open question. In this paper we provide a partial solution to this problem by showing that $ZF \vdash HB \wedge KM \to BPI$, where KM is (a version of) the Krein-Milman theorem on the existence of extreme points in compact convex sets. The proof uses (1) a theorem of Phelps which characterizes homomorphisms between real function algebras as the extreme points of a certain set of linear maps between the algebras, and (2) a result of Luxemburg which demonstrates the effective equivalence between the Hahn-Banach theorem and a weak form of Alaoglu's theorem on weak* compactness of closed unit spheres. Unfortunately attempts to prove $ZF \vdash BPI \to KM$ have so far proved abortive.

3. ROGER CUSIN: Quasi-complet theories

We introduce a notion which generalizes both the notions of complete theory and model-complet theory. We say that a theory T in a first-order language with equality is quasi-complet if:

1) T is consistent,

2) If $\mathfrak{A}$, $\mathfrak{B}$ are models of T and if $\mathfrak{A} \subset \mathfrak{B}$, then $\mathfrak{A} \equiv \mathfrak{B}$.

Elementary properties.

1) Every complete theory is quasi-complet.

2) Every model-complet theory is quasi-complet.

3) If T' is a consistent theory such that $T' \vdash T$ (i.e. $T' \supset T$), then T' is quasi-complet if T is quasi-complet.

4) If $\mathfrak{A}_0$ is a prime-model of T and T is quasi-complet, then T is complete.

Remark. The properties 1) and 2) show that the notion of quasi-complet theory is strictly weaker than the notion of complete theory and the notion of model-complet theory.

If B_0 is the boolean algebra of sentences (mod. the theorems of predicate calculus), a theory is a filter of B_0. The following theorem gives a necessary and sufficient condition for a theory to be quasi-complet. A sentence ψ of $\mathcal{L}(T)$ (language of T) is primitive, if it has the form $\exists x_1 \ldots x_k (\psi_1 \wedge \ldots \wedge \psi_n)$, where ψ_i are elementary formulas or negations of such formulas of $\mathcal{L}(T)$ (this notion is more restrictive than the notion of primitive formula in A. Robinson).

THEOREM. T is quasi-complet iff for all complete theories T' of $\mathcal{L}(T)$ with $T' \supset T$, the ultrafilter T' is generated by $\{\phi \wedge \psi \mid \phi$ and ψ sentences, $\phi \in T$ and ψ primitive sentence of $T'\}$.

4. PAUL EKLOF and GABRIEL SABBAGH: Definability problems for modules and rings

We are concerned with questions of the following kind: Let L be a language of the form $L_{\alpha\omega}$ and let C be a class of modules over a fixed ring or a class of rings; is it possible to define C in L?

I. C is a class of modules over a fixed ring Λ. Sample of results:

a) We characterize the rings Λ such that the class of free (respectively projective, respectively flat) left Λ-modules is elementary (in the wider sense).

b) The class of injective modules is definable in $L_{\infty\omega}$ if and only if it is definable in $L_{\omega\omega}$ if and only if Λ is noetherian.

Some of the results are obtained by investigating the equivalence with respect to $L_{\infty\omega}$ of the direct sum and the direct product of a family of modules. A typical result is:

c) If Λ is right noetherian, the left Λ-modules $\Lambda^{(I)}$ and Λ^{J} are $\infty\omega$-equivalent if I and J are infinite sets.

II. C is a class of rings. Sample of results:

a) The class of artinian rings is definable in $L_{\omega_1\omega}$.

b) The class of noetherian rings is not definable in $L_{\infty\omega}$.

c) The class of commutative principal ideal domains is definable in $L_{\omega_1\omega}$.

d) The class of (not necessarily commutative) principal ideal domains is not definable in $L_{\infty\omega}$.

5. PAUL FOULKES: The logic of "And" in the anatomy of proof: a new model for entailment

The connection between premisses and conclusions of a multi-stage deductive argument is not the same as that between propositions within each step: the formal (non-truthfunctional) relation between adjacent elements, here called strict entailment, is intransitive. This accompanies the logical constant that links the premisses of a deductive step. A forthcoming article in MIND shows that the logical constant "and" represents two radically different meanings in English, corresponding to two logical functions:

1) <u>syntopic</u> "and", grammatically a conjunction putting conjuncts in the same place; 2) <u>syndetic</u> "and", a preposition governing the instrumental case, binding elements together. Contrary to tradition, "and" between joint premisses of a deductive step is <u>syndetic</u>, non-truthfunctional and not subject to De Morgan's rules.

The fundamental formula is

$$(p \circ q) \rightharpoonup r$$

where p, q, r are propositions, all different and subject to a set of formal restrictions; in particular, the premisses must be compatible, operative, independent of each other and of the contradictory of the conclusion. $\circ$ represents <u>syndetic</u> "and", $\rightharpoonup$ represents <u>strict entailment</u>. We further have $[(p \circ q) \rightharpoonup r] \rightleftharpoons [(p \circ \sim r) \rightharpoonup \sim q]$, the antilogism.

Strict entailment can be displayed in a three-dimensional topological model (just as class relations between terms can in the two-dimensional model of Euler's circles): a vector looped at each end represents a proposition p; and, read in reverse, ~p (fig. 1). $p \rightharpoonup q$ shows as two vectors interlinked (fig. 2), while in the fundamental formula (fig. 3) each ring at the centre engages the other two. All the formal restrictions can be read off these diagrams.

Reductio ad absurdum is revealed as an antilogism, involving no self-contradiction: assumption p with admitted premiss q strictly entails r, whereas ~r obtains; then ~r with q strictly entails ~p. Thus represented, the argument is immune to intuitionist objections.

The topology of double-loop vectors will throw light on the detailed structure of deductive systems.

fig. 1:

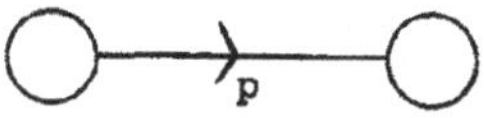

fig. 2:

fig. 3:

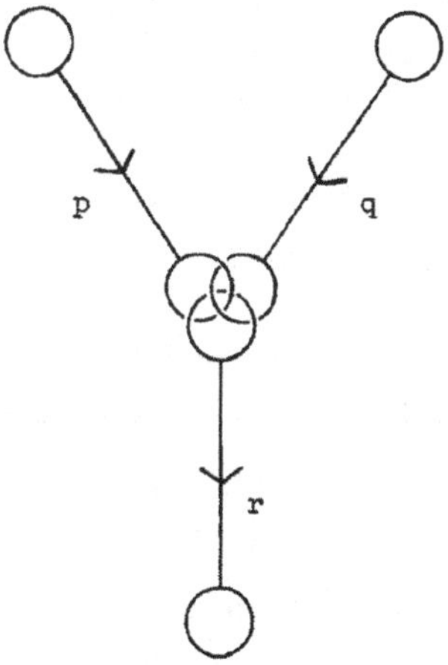

17th July, 1970

6. DOV M. GABBAY: <u>Selective filtration in modal logics I, II, III</u>

Let K be the modal logic with the axiom

$$\Box(\phi \rightarrow \psi) \rightarrow (\Box\phi \rightarrow \Box\psi)$$

and the rule $\vdash \phi \Rightarrow \vdash \Box\phi$. Then the following extensions of K have the finite model property:

$(a)_m$ $\qquad \phi \rightarrow \Box^m \phi$

$(b)_{m,n}$ $\qquad \Diamond^m \phi \rightarrow \Box^n \phi$

$(c)_{m,n}$ $\qquad \phi \wedge \Diamond^n \phi \rightarrow \Box^m \phi$

$(d)_m$	$\Box\phi \to \Box^{m+1}\phi$
$(e)_m$	$\Diamond^m\Box\phi \to \Box\phi$
(f)	$\Diamond\Box\phi \to \Box^2\phi.$

That is, for each fixed m, n $\geq$ 1, we get an extension of K which has the finite model property and hence is decidable.

7. J. HEIDEMA: Metamathematical representation of radicals in universal algebra

The result announced is that two approaches to the problem of developing a theory of radicals in universal algebra, which have been initiated independently, are in fact equivalent. Consider a fixed (usually primitive) class of algebras of the same type. A radical according to Hoehnke ([2], [3]) is a function R which assigns to every algebra A of the class a congruence R(A) on A, satisfying the following conditions:

a) if $\phi : A \to B$ is an epimorphism, then $R(A)\phi \subseteq R(B)$;
(if $S \subseteq A\times A$, $S\phi = \{(a\phi, b\phi) : (a,b) \in S\}$);

b) $R(A/R(A)) = 1_{A/R(A)}$, the identity congruence on $A/R(A)$.

If C is any congruence on A one can now define the radical R(C) of C as the congruence induced on A by the canonical homomorphism $A \to A/C \to (A/C)/R(A/C)$. Then $C \subseteq R(C)$, $R(A) = R(1_A)$, and R is a closure operator on the lattice of congruences on A.

The present author initiated another approach, employing Abraham Robinson's metamathematical theory of ideals ([4], ch. vii). Let L be a language of the first order predicate calculus, while J, K and K^* are sets of sentences in L, such that $K \subseteq K^*$. If I is any metamathematical ideal in J over K, the ideal in J over K^* generated by I will be denoted by I^*. Then $I \subseteq I^*$ and * is a closure operation on the lattice of ideals in J over K, [1].

Now again consider a class of algebras of the same type. Then there is a language L, and, for every algebra A, sets of sentences J_A and K_A in L, such that there is a 1-1 correspondence between the sets AxA and J_A which at the same time establishes a lattice isomorphism between the congruence lattice of A and the lattice of all metamathematical ideals in J_A over K_A. Let K_A^* be any set of sentences of L containing K_A. A congruence C on A corresponds to an ideal I in J_A over K_A, which has a closure I^* over K_A^*, which corresponds to a congruence C^* on A, containing C. In this approach, C^* is the radical of C.

These two approaches are equivalent: every radical * has Hoehnke's two properties, and every Hoehnke-radical R can be represented by prescribing a language L, and, for every A of the class of algebras, sets of sentences J_A, K_A, K_A^*, in a metamathematically uniform way.

[1] Heidema, J., Metamathematical prime ideals and radicals, Indag. Math. 30 (1968), 280-285.

[2] Hoehnke, H.-J., Einige neue Resultate über Abstrakte Halbgruppen, Coll. Math. 14 (1966), 329-348.

[3] Hoehnke, H.-J., Radikale in allgemeinen Algebren, Math. Nachr. 32 (1966), 347-383.

[4] Robinson, A., Introduction to Model Theory and to the Metamathematics of Algebra, North-Holland Publ. Co., Amsterdam, 1963.

8. P. HENRARD: Forcing with infinite conditions

I. α-topological spaces. An α-topological space is a topological space in which the intersection of a family of less than $\aleph_\alpha$ open sets is an open set.

A Baire's α-topological space is an α-topological space in which the intersection of $\aleph_\alpha$ dense open sets is a dense set.

II. Application to forcing. Let $\aleph_\gamma$ be a fixed regular cardinal; L a language of the lower predicate calculus without function symbols whose cardinal is less than $\aleph_\gamma$; M a set containing as elements the individual constants of L (if any) and $\aleph_\gamma$ other elements.

If α is an ordinal $\leq \gamma$ we call α-condition every consistent set of less than $\aleph_\alpha$ basis sentences (i.e. atomic or negation of an atomic) of the language L(M).

We define the notion of "the α-condition P α-forces the sentence θ of L(M)" (in symbols $P \Vdash_\alpha \theta$) by induction on the length of θ.

i) $P \Vdash_\alpha \theta$ iff $\theta \in P$ if θ is an atomic sentence.

ii) $P \Vdash_\alpha \chi \wedge \psi$ iff $P \Vdash_\alpha \chi$ and $P \Vdash_\alpha \psi$.

iii) $P \Vdash_\alpha \chi \vee \psi$ iff $P \Vdash_\alpha \chi$ or $P \Vdash_\alpha \psi$.

iv) $P \Vdash_\alpha \sim\chi$ iff no α-condition containing P α-forces χ.

v) $P \Vdash_\alpha \exists x\ \theta(x)$ iff there exists an element m of M such that $P \Vdash_\alpha \theta(m)$.

Let $\mathfrak{M}$ be the set of relational structures for L(M) whose universe is M. If A(P) is the set of structures in $\mathfrak{M}$ which satisfies the γ-condition P, the family of the sets A(P), P a γ-condition, is a basis of a Baire's γ-topology on $\mathfrak{M}$.

We say that a sentence θ of L(M) is α-forced in a structure Ⓜ of $\mathfrak{M}$ (in symbols Ⓜ $\in B_\alpha(\theta)$) if Ⓜ satisfies an α-condition which α-forces θ. A structure Ⓜ is complete if every sentence θ of L(M) or its negation is 0-forced in Ⓜ.

THEOREM 1. For each $\alpha \leq \gamma$, $B_0(\theta) = B_\alpha(\theta)$.

THEOREM 2. The set $B_\gamma(\theta)$ is an open set of $\mathfrak{M}$ and $B_\gamma(\sim\theta)$ is the interior of the complement of $B_\gamma(\theta)$.

THEOREM 3. The set of complete structures is dense in $\mathfrak{M}$.

COROLLARY. For each condition P, there exists a complete structure which satisfies P.

9. TH. LUCAS: Equations in the theory of monadic algebras

A monadic algebra is considered as a structure $\langle A,+,.,-,C\rangle$ where $\langle A,+,.,-\rangle$ is a boolean algebra and C is a quantifier on $\langle A,+,.,-\rangle$. Let CA_1 be the class of monadic algebras. An identity refers to the universal closure of an equation (in the first-order language with equality of CA_1). We use a notion of standard identity to prove the following two theorems in a straightforward manner:

THEOREM 1. (D. Monk) 1) The lattice of equational classes of monadic algebras is a chain:

$$H_0 \subset H_1 \subset \ldots \subset H_\omega = CA_1.$$

2) For every $0 \leqslant n \leqslant \omega$, H_n is finitely axiomatizable and its equational theory is decidable.

THEOREM 2. For $0 < n < \omega$, the minimum number of bound variables needed in an identity characterizing H_n (relatively to CA_1) is the smallest p such that $2^p \geqslant n + 1$.

The same notion is also used to prove that equational classes are closed under certain types of extension among which completions and canonical embeddings.

10. G. P. MONRO: The possible pattern of cardinals

Let ZF(K) be the theory derived from ZF by adding a one-place predicate K and the axioms of replacement for formulas involving K. A model N of ZF(K) is described in which

(i) K is a proper class, K can be mapped onto the universe and there is no injection of ω into K

(ii) every linearly ordered set can be embedded into the pattern of cardinals.

These two results provide answers to the questions "How large can a Dedekind-finite set be?" and "How bad can the pattern of cardinals be?". The construction is as follows. Let M be a countable model of ZF + V = L. A model M[G] of ZFC is constructed by a method similar to Easton's, in which for each regular cardinal λ, a set K_λ containing λ generic subsets of λ is introduced. N is the inner model of M[G] consisting of sets constructible from the sequence (K_λ); $K = S_\omega(\cup_\lambda K_\lambda)$.

Note: The result stated at the conference was stronger than that stated here and the author is indebted to Professor R. M. Solovay for pointing out an error in the original proof.

11. B. P. MOSS: A picture of a Kripke model for S4

A Kripke model for a finite consistent set S of sentences of modal propositional calculus can be displayed in a picture involving a finite number of distinct possible worlds. Since S may contain sentences of the form $\Box\Diamond P$ & $\Box\Diamond{\sim}P$, the actual model is infinite, in the sense that every world has a successor distinct from itself.

To combine comprehensiveness with simplicity, consider

$$S = \{X, \Box A, \Diamond B, \Box\Diamond C, \Diamond\Box D, \Box\Diamond\Box E, \Diamond\Box\Diamond F\}$$

where X, A, B, C, D, E, F are formulae of propositional calculus. For any sentences P, Q, R of propositional calculus, let <P,Q,R> be a model for the sentences. Δ is said to be an <u>immediate successor</u> of Γ if (i) $\Gamma \mathcal{R} \Delta$; (ii) $\Gamma \neq \Delta$; (iii) $\mathcal{B} \mathcal{R} \Delta$ & $\mathcal{B} \neq \Delta \rightarrow \mathcal{B} \mathcal{R} \Gamma$. Then there is a Kripke model having $\Gamma = \langle X,A \rangle$ as its first world, in the sense that every world Δ satisfies $\Gamma \mathcal{R} \Delta$; and in this model, S is valid in Γ. Γ has immediate successors $\langle A \rangle$, $\langle A,B \rangle$, $\langle A,C \rangle$, $\langle A,D \rangle$, $\langle A,E \rangle$ and $\langle A,F \rangle$, to ensure that all the formulae beginning with $\Diamond$ are satisfied in some immediate successor of $\langle X,A \rangle$. This is the second row of the picture.

The set of immediate successors of any world in the second and subsequent rows is one of six tableaux, each containing two, three or four worlds; four of these tableaux appear in the third row, and the other two in the fourth. For instance, $\langle A,B \rangle$ has immediate successors $\langle A \rangle$, $\langle A,C \rangle$ and $\langle A,E \rangle$, forming tableau α. $\langle A \rangle$ and $\langle A,C \rangle$ are always succeeded by tableau α, and $\langle A,E \rangle$ by $\langle A,E \rangle$ and $\langle A,C,E \rangle$, forming tableau γ. These tableaux have the property that the section of the next row formed by immediate successors of the worlds in one tableau is determined solely by the tableau. Thus the first five rows delineate the entire model (the fifth row being used solely to show the development of the two new tableaux in row four).

The model for these seven sentences contains twelve distinct worlds. Extensions to larger sets are easy, but the extension to predicate logic uses a countable picture, and relies on the Barcan formula.

12. GEROLD STAHL: <u>Temporal terms in functional systems</u>

After the distinction between temporality de re (Charles was travelling yesterday) and temporality de dicto (It was true yesterday "Charles is travelling"), a second order functional system for temporality de re (FTR) is constructed, whose universe of discourse has a subclass <u>P</u> of the temporal positions and an individual <u>ac</u>, the actual position. An infinite list of two-place second order functional symbols

"$\underline{Cen}_{(\ldots,\ldots)}$" will be used, where the points separated by the comma correspond to two successive numerals beginning with "1,2". With the symbols "$\underline{Cen}$" (which represent what is called "central relations") we introduce by definition for each n-place first order propositional function $\underline{F}$ a n+1-place function $\underline{F}^*$: $\underline{F}^* =_{df} \imath\underline{G}(\underline{F}\ \underline{Cen}_{(n,n+1)}\underline{G})$. Besides the usual axioms and axiomatic rules for the second-order functional systems, FTR has the following (independent) axioms:

$$\vdash \underline{Fy}_1,\ldots,\underline{y}_n \equiv \underline{F}^*\underline{ac},\underline{y}_1,\ldots,\underline{y}_n;$$
$$\vdash \underline{F}^* \cap \underline{G}^* = (\underline{F} \cap \underline{G})^*;$$
$$\vdash -_r\underline{F}^* = (-\underline{F})^*,$$

where "$-_r\underline{F}^*$" represents the complement of $\underline{F}^*$ with respect to $(\underline{P} \times \underline{F}) \cup (\underline{P} \times -\underline{F})$. Some models for FTR are indicated. If we construct FTD (a second order functional system for temporality de dicto) in analogy to FTR over the sentences of FTR and over the class $\underline{P}$ and if we formulate semantical definitions not only for "$\underline{T}$" (the class of the true sentences of FTR), but also for "$\underline{T}^*$" then we get results like: $\underline{T}$"$(\underline{y})\underline{F}^*\underline{ky}$" $\equiv \underline{T}^*\underline{k}$"$(\underline{y})\underline{Fy}$".

Casilla 9733, Santiago, Chile

13. JOHN TUCKER: Algorithmic unsolvability in biological contexts

The investigation of algorithmic unsolvability in cell automata is used as a model for biological processes at the molecular level, and the question which is to be raised here is this: Which features of unsolvability in the automata-theoretical model can be expected to be exemplified in one or both of two ways: (i) as a heterological procedure (ii) as a diagonal procedure. Both of these possibilities will be discussed with reference to the literature, and an account of the distinction will be given.

(i) <u>The computer simulation of algorithmic unsolvability in the form of heterological procedures</u>

The simulation consists in the computer continuing to operate without halting, on account of an 'inherent logical loop'. [1] Now what is simulated here is a faulty procedure and the unsolvability is detected *via* the fault. The appearance of the heterological 'loop' is a *sign* of algorithmic unsolvability but is not identical with it. Such loops need not appear, and moreover, in a biological cell which was free of defects would not appear, even though there are tasks which the cell cannot in principle carry out, and even though our knowledge that this is so depends upon the derivation of 'loops' in the model. Thus, the heterological 'loop' is a fault in the model which gives information about the biological context but which need not be found in that context. The heterological 'loop' is a particular type of non-constructive defect, the characteristics of which will be discussed, which would arise if a hypothetical test automaton *had* to answer the question 'true or false?' in a certain situation. But if the test automaton were free of this defect it would reject the question instead of going into a 'loop'. Only if the automaton were defective in a very special way would it in fact go into a 'loop'.

(ii) <u>Diagonal procedures in models of adaptation</u>

By contrast with the heterological 'loop', which is non-constructive, diagonal procedures proper are constructive, and when present in a model might therefore be reasonably expected to be exemplified by some constructive process in the biological context at the molecular level. So for example in Myhill's model we have a reiteration of a diagonal process which could, in principle, correspond to adaptation. [2]

(iii) Diagonal procedures, on account of their constructive character, can correspond to constructive biological processes at the molecular level. Heterological procedures, if exemplified at that level, would give rise to nothing. Normally functioning biological systems would be expected to be constructive and free of non-constructive 'loops'.

[1] W. R. Stahl, Algorithmically unsolvable problems for a cell automaton, Journal of Theoretical Biology 8 (1965), pp. 371-394.

[2] J. Myhill, The abstract theory of self-reproduction, in Views on General Systems Theory, edited M. D. Mesarovic (1964), pp. 106-118.

Department of Pure Mathematics
University of Waterloo
Waterloo, Ontario
Canada

14. S. S. WAINER: A subrecursive hierarchy over the predicative ordinals

Feferman (Systems of Predicative Analysis, J.S.L. 29) has considered various (autonomous) progressions of formal systems intended to correspond to the notion of predicative proof, and has shown that each of these progressions "closes off" at a certain classical ordinal Γ_0. We present here a recursion-theoretic analogue of these results.

In (Systems of Predicative Analysis II, J.S.L. 33) Feferman defines for each limit ordinal $\lambda \leq \Gamma_0$, a natural fundamental sequence λ_i $(i = 0,1,2,\ldots)$. On the basis of these fundamental sequences we construct a Grzegorczyk-type hierarchy $\{\mathcal{E}_\alpha\}_{\alpha\leq\Gamma_0}$ of classes of recursive (number-theoretic) functions such that $\mathcal{E}_\alpha \subset \mathcal{E}_\beta$ whenever $\alpha < \beta$.

We say that an ordinal $\eta \leq \Gamma_0$ is recognized by a class $\mathcal{E}_\alpha$ if there are functions $f(n,x)$, $g(i) \in \mathcal{E}_\alpha$ and a constant c such that as n ranges over a certain (primitive recursive) set of notations for the ordinals $< \eta$, $f(n,x)$ enumerates a sequence $\{f_\nu\}_{\nu<\eta}$ of unary functions with the property that f_β is majorized by f_γ whenever $\beta < \gamma < \eta$, where

(i) if $\beta + 1 < \eta$ then for all $x > c$, $f_\beta(x) < f_{\beta+1}(x)$,

(ii) if λ is a limit ordinal $< \eta$ then for each i and all $x > g(i)$, $f_{\lambda_i}(x) < f_\lambda(x)$.

THEOREM I. If $0 < \eta < \Gamma_0$ then there is an $\alpha < \eta$ such that η is recognized by $\mathcal{E}_\alpha$, but there is no $\beta < \Gamma_0$ such that Γ_0 is recognized by $\mathcal{E}_\beta$.

The functions definable from the primitive recursive functions by means of explicit definitions and nested recursions over certain standard well-orderings of non-negative integers of order-types $< \Gamma_0$ form a large and interesting class, which we denote by $\mathcal{N}_{\Gamma_0}$. The second result provides a classification of $\mathcal{N}_{\Gamma_0}$ in terms of computational complexity.

THEOREM II. $\mathcal{N}_{\Gamma_0} = \cup_{\alpha<\Gamma_0} \mathcal{E}_\alpha$.

Leeds University

ɜcture Notes in Mathematics

nprehensive leaflet on request

74: A. Fröhlich, Formal Groups. IV, 140 pages. 1968. DM 16,-

75: G. Lumer, Algèbres de fonctions et espaces de Hardy. VI, ages. 1968. DM 16,-

76: R. G. Swan, Algebraic K-Theory. IV, 262 pages. 1968. 18,-

77: P.-A. Meyer, Processus de Markov: la frontière de Martin. IV, pages. 1968. DM 16,-

78: H. Herrlich, Topologische Reflexionen und Coreflexionen. XVI, Seiten. 1968. DM 16,-

79: A. Grothendieck, Categories Cofibrées Additives et Complexe angent Relatif. IV, 167 pages. 1968. DM 16,-

80: Seminar on Triples and Categorical Homology Theory. Edited . Eckmann. IV, 398 pages. 1969. DM 20,-

81: J.-P. Eckmann et M. Guenin, Méthodes Algébriques en Mécae Statistique. VI, 131 pages. 1969. DM 16,-

82: J. Wloka, Grundräume und verallgemeinerte Funktionen. VIII, eiten. 1969. DM 16,-

83: O. Zariski, An Introduction to the Theory of Algebraic Surfaces. 00 pages. 1969. DM 16,-

84: H. Lüneburg, Transitive Erweiterungen endlicher Permutationspen. IV, 119 Seiten. 1969. DM 16,-

85: P. Cartier et D. Foata, Problèmes combinatoires de commun et réarrangements. IV, 88 pages. 1969. DM 16,-

86: Category Theory, Homology Theory and their Applications I. ed by P. Hilton. VI, 216 pages. 1969. DM 16,

87: M. Tierney, Categorical Constructions in Stable Homotopy ry. IV, 65 pages. 1969. DM 16,-

88: Séminaire de Probabilités III. IV, 229 pages. 1969. 8,-

89: Probability and Information Theory. Edited by M. Behara, ickeberg and J. Wolfowitz. IV, 256 pages. 1969. DM 18,-

90: N. P. Bhatia and O. Hajek, Local Semi-Dynamical Systems. 7 pages. 1969. DM 16,-

91: N. N. Janenko, Die Zwischenschrittmethode zur Lösung mehrnsionaler Probleme der mathematischen Physik. VIII, 194 Seiten. . DM 16,80

92: Category Theory, Homology Theory and their Applications II. ed by P. Hilton. V, 308 pages. 1969. DM 20,-

93: K. R. Parthasarathy, Multipliers on Locally Compact Groups. 4 pages. 1969. DM 16,-

94: M. Machover and J. Hirschfeld, Lectures on Non-Standard ysis. VI, 79 pages. 1969. DM 16,-

95: A. S. Troelstra, Principles of Intuitionism. II, 111 pages. 1969 16,-

96: H.-B. Brinkmann und D. Puppe, Abelsche und exakte Kateen, Korrespondenzen. V, 141 Seiten. 1969. DM 16,-

97: S. O. Chase and M. E. Sweedler, Hopf Algebras and Galois ry. II, 133 pages. 1969. DM 16,-

98: M. Heins, Hardy Classes on Riemann Surfaces. III, 106 pages. . DM 16,-

99: Category Theory, Homology Theory and their Applications III. d by P. Hilton. IV, 489 pages. 1969. DM 24,

00: M. Artin and B. Mazur, Etale Homotopy. II, 196 Seiten. 1969. 6,-

01: G. P. Szegö et G. Treccani, Semigruppi di Trasformazioni voche. VI, 177 pages. 1969. DM 16,-

02: F. Stummel, Rand- und Eigenwertaufgaben in Sobolewschen en. VIII, 386 Seiten. 1969. DM 20,-

03: Lectures in Modern Analysis and Applications I. Edited by Taam. VII, 162 pages. 1969. DM 16,-

04: G. H. Pimbley, Jr., Eigenfunction Branches of Nonlinear ators and their Bifurcations. II, 128 pages. 1969. DM 16,-

05: R. Larsen, The Multiplier Problem. VII, 284 pages. 1969. 3,-

06: Reports of the Midwest Category Seminar III. Edited by S. Mac II, 247 pages. 1969. DM 16,-

07: A. Peyerimhoff, Lectures on Summability. III, 111 pages. DM 16,-

08: Algebraic K-Theory and its Geometric Applications. Edited M. F. Moss and C. B. Thomas. IV, 86 pages. 1969. DM 16,-

09: Conference on the Numerical Solution of Differential Equa- Edited by J. Ll. Morris. VI, 275 pages. 1969. DM 18,

10: The Many Facets of Graph Theory. Edited by G. Chartrand . F. Kapoor. VIII, 290 pages. 1969. DM 18,-

Vol. 111: K. H. Mayer, Relationen zwischen charakteristischen Zahlen. III, 99 Seiten. 1969. DM 16,-

Vol. 112: Colloquium on Methods of Optimization. Edited by N. N. Moiseev. IV, 293 pages. 1970. DM 18,-

Vol. 113: R. Wille, Kongruenzklassengeometrien. III, 99 Seiten. 1970. DM 16,-

Vol. 114: H. Jacquet and R. P. Langlands, Automorphic Forms on GL (2). VII, 548 pages. 1970. DM 24,-

Vol. 115: K. H. Roggenkamp and V. Huber-Dyson, Lattices over Orders I. XIX, 290 pages. 1970. DM 18,-

Vol. 116: Séminaire Pierre Lelong (Analyse) Année 1969. IV, 195 pages. 1970. DM 16,-

Vol. 117: Y. Meyer, Nombres de Pisot, Nombres de Salem et Analyse Harmonique. 63 pages. 1970. DM 16,-

Vol. 118: Proceedings of the 15th Scandinavian Congress, Oslo 1968. Edited by K. E. Aubert and W. Ljunggren. IV, 162 pages. 1970. DM 16,-

Vol. 119: M. Raynaud, Faisceaux amples sur les schémas en groupes et les espaces homogènes. III, 219 pages. 1970. DM 16,-

Vol. 120: D. Siefkes, Büchi's Monadic Second Order Successor Arithmetic. XII, 130 Seiten. 1970. DM 16,-

Vol. 121: H. S. Bear, Lectures on Gleason Parts. III, 47 pages. 1970. DM 16,-

Vol. 122: H. Zieschang, E. Vogt und H.-D. Coldewey, Flächen und ebene diskontinuierliche Gruppen. VIII, 203 Seiten. 1970. DM 16,-

Vol. 123: A. V. Jategaonkar, Left Principal Ideal Rings. VI, 145 pages. 1970. DM 16,-

Vol. 124: Séminare de Probabilités IV. Edited by P. A. Meyer. IV, 282 pages. 1970. DM 20,-

Vol. 125: Symposium on Automatic Demonstration. V, 310 pages. 1970. DM 20,-

Vol. 126: P. Schapira, Théorie des Hyperfonctions. XI, 157 pages. 1970. DM 16,-

Vol. 127: I. Stewart, Lie Algebras. IV, 97 pages. 1970. DM 16,-

Vol. 128: M. Takesaki, Tomita's Theory of Modular Hilbert Algebras and its Applications. II, 123 pages. 1970. DM 16,-

Vol. 129: K. H. Hofmann, The Duality of Compact Semigroups and C*-Bigebras. XII, 142 pages. 1970. DM 16,-

Vol. 130: F. Lorenz, Quadratische Formen über Körpern. II, 77 Seiten. 1970. DM 16,-

Vol. 131: A Borel et al., Seminar on Algebraic Groups and Related Finite Groups. VII, 321 pages. 1970. DM 22,-

Vol. 132: Symposium on Optimization. III, 348 pages. 1970. DM 22,-

Vol. 133: F. Topsøe, Topology and Measure. XIV, 79 pages. 1970. DM 16,-

Vol. 134: L. Smith, Lectures on the Eilenberg-Moore Spectral Sequence. VII, 142 pages. 1970. DM 16,-

Vol. 135: W. Stoll, Value Distribution of Holomorphic Maps into Compact Complex Manifolds. II, 267 pages. 1970. DM 18,-

Vol. 136: M. Karoubi et al., Séminaire Heidelberg-Saarbrücken-Strasbuorg sur la K-Théorie. IV, 264 pages. 1970. DM 18,

Vol. 137: Reports of the Midwest Category Seminar IV. Edited by S. MacLane. III, 139 pages. 1970. DM 16,-

Vol. 138: D. Foata et M. Schützenberger, Théorie Géométrique des Polynômes Eulériens. V, 94 pages. 1970. DM 16,-

Vol. 139: A. Badrikian, Séminaire sur les Fonctions Aléatoires Linéaires et les Mesures Cylindriques. VII, 221 pages. 1970. DM 18,-

Vol. 140: Lectures in Modern Analysis and Applications II. Edited by C. T. Taam. VI, 119 pages. 1970. DM 16,-

Vol. 141: G. Jameson, Ordered Linear Spaces. XV, 194 pages. 1970. DM 16,-

Vol. 142: K. W. Roggenkamp, Lattices over Orders II. V, 388 pages. 1970. DM 22,-

Vol. 143: K. W. Gruenberg, Cohomological Topics in Group Theory. XIV, 275 pages. 1970. DM 20,-

Vol. 144: Seminar on Differential Equations and Dynamical Systems, II. Edited by J. A. Yorke. VIII, 268 pages. 1970. DM 20,-

Vol. 145: E. J. Dubuc, Kan Extensions in Enriched Category Theory. XVI, 173 pages. 1970. DM 16,-

Please turn over

Vol. 146: A. B. Altman and S. Kleiman, Introduction to Grothendieck Duality Theory. II, 192 pages. 1970. DM 18,–

Vol. 147: D. E. Dobbs, Cech Cohomological Dimensions for Commutative Rings. VI, 176 pages. 1970. DM 16,–

Vol. 148: R. Azencott, Espaces de Poisson des Groupes Localement Compacts. IX, 141 pages. 1970. DM 16,–

Vol. 149: R. G. Swan and E. G. Evans, K-Theory of Finite Groups and Orders. IV, 237 pages. 1970. DM 20,–

Vol. 150: Heyer, Dualität lokalkompakter Gruppen. XIII, 372 Seiten. 1970. DM 20,–

Vol. 151: M. Demazure et A. Grothendieck, Schémas en Groupes I. (SGA 3). XV, 562 pages. 1970. DM 24,–

Vol. 152: M. Demazure et A. Grothendieck, Schémas en Groupes II. (SGA 3). IX, 654 pages. 1970. DM 24,–

Vol. 153: M. Demazure et A. Grothendieck, Schémas en Groupes III. (SGA 3). VIII, 529 pages. 1970. DM 24,–

Vol. 154: A. Lascoux et M. Berger, Variétés Kähleriennes Compactes. VII, 83 pages. 1970. DM 16,–

Vol. 155: Several Complex Variables I, Maryland 1970. Edited by J. Horváth. IV, 214 pages. 1970. DM 18,–

Vol. 156: R. Hartshorne, Ample Subvarieties of Algebraic Varieties. XIV, 256 pages. 1970. DM 20,–

Vol. 157: T. tom Dieck, K. H. Kamps und D. Puppe, Homotopietheorie. VI, 265 Seiten. 1970. DM 20,–

Vol. 158: T. G. Ostrom, Finite Translation Planes. IV. 112 pages. 1970. DM 16,–

Vol. 159: R. Ansorge und R. Hass. Konvergenz von Differenzenverfahren für lineare und nichtlineare Anfangswertaufgaben. VIII, 145 Seiten. 1970. DM 16,–

Vol. 160: L. Sucheston, Constributions to Ergodic Theory and Probability. VII, 277 pages. 1970. DM 20,–

Vol. 161: J. Stasheff, H-Spaces from a Homotopy Point of View. VI, 95 pages. 1970. DM 16,–

Vol. 162: Harish-Chandra and van Dijk, Harmonic Analysis on Reductive p-adic Groups. IV, 125 pages. 1970. DM 16,–

Vol. 163: P. Deligne, Equations Différentielles à Points Singuliers Reguliers. III, 133 pages. 1970. DM 16,–

Vol. 164: J. P. Ferrier, Seminaire sur les Algebres Complètes. II, 69 pages. 1970. DM 16,–

Vol. 165: J. M. Cohen, Stable Homotopy. V, 194 pages. 1970. DM 16,–

Vol. 166: A. J. Silberger, PGL_2 over the p-adics: its Representations, Spherical Functions, and Fourier Analysis. VII, 202 pages. 1970. DM 18,–

Vol. 167: Lavrentiev, Romanov and Vasiliev, Multidimensional Inverse Problems for Differential Equations. V, 59 pages. 1970. DM 16,–

Vol. 168: F. P. Peterson, The Steenrod Algebra and its Applications: A conference to Celebrate N. E. Steenrod's Sixtieth Birthday. VII, 317 pages. 1970. DM 22,–

Vol. 169: M. Raynaud, Anneaux Locaux Henséliens. V, 129 pages. 1970. DM 16,–

Vol. 170: Lectures in Modern Analysis and Applications III. Edited by C. T. Taam. VI, 213 pages. 1970. DM 18,–

Vol. 171: Set-Valued Mappings, Selections and Topological Properties of 2^X. Edited by W. M. Fleischman. X, 110 pages. 1970. DM 16,–

Vol. 172: Y.-T. Siu and G. Trautmann, Gap-Sheaves and Extension of Coherent Analytic Subsheaves. V, 172 pages. 1971. DM 16,–

Vol. 173: J. N. Mordeson and B. Vinograde, Structure of Arbitrary Purely Inseparable Extension Fields. IV, 138 pages. 1970. DM 16,–

Vol. 174: B. Iversen, Linear Determinants with Applications to the Picard Scheme of a Family of Algebraic Curves. VI, 69 pages. 1970. DM 16,–

Vol. 175: M. Brelot, On Topologies and Boundaries in Potential Theory. VI, 176 pages. 1971. DM 18,–

Vol. 176: H. Popp, Fundamentalgruppen algebraischer Mannigfaltigkeiten. IV, 154 Seiten. 1970. DM 16,–

Vol. 177: J. Lambek, Torsion Theories, Additive Semantics and Rings of Quotients. VI, 94 pages. 1971. DM 16,–

Vol. 178: Th. Bröcker und T. tom Dieck, Kobordismentheorie. XVI, 191 Seiten. 1970. DM 18,–

Vol. 179: Seminaire Bourbaki – vol. 1968/69. Exposés 347-363. IV, 295 pages. 1971. DM 22,–

Vol. 180: Séminaire Bourbaki – vol. 1969/70. Exposés 364-381. IV, 310 pages. 1971. DM 22,–

Vol. 181: F. DeMeyer and E. Ingraham, Separable Algebras over Commutative Rings. V, 157 pages. 1971. DM 16,–

Vol. 182: L. D. Baumert. Cyclic Difference Sets. VI, 166 pages. 1971. DM 16,–

Vol. 183: Analytic Theory of Differential Equations. Edited by P. F. Hsieh and A. W. J. Stoddart. VI, 225 pages. 1971. DM 20,–

Vol. 184: Symposium on Several Complex Variables, Park City, Utah, 1970. Edited by R. M. Brooks. V, 234 pages. 1971. DM 20,–

Vol. 185: Several Complex Variables II, Maryland 1970. Edited by J. Horváth. III, 287 pages. 1971. DM 24,–

Vol. 186: Recent Trends in Graph Theory. Edited by M. Capobianco/J. B. Frechen/M. Krolik. VI, 219 pages. 1971. DM 18,–

Vol. 187: H. S. Shapiro, Topics in Approximation Theory. VIII, 275 pages. 1971. DM 22,–

Vol. 188: Symposium on Semantics of Algorithmic Languages. Edited by E. Engeler. VI, 372 pages. 1971. DM 26,–

Vol. 189: A. Weil, Dirichlet Series and Automorphic Forms. V, 164 pages. 1971. DM 16,–

Vol. 190: Martingales. A Report on a Meeting at Oberwolfach, May 17-23, 1970. Edited by H. Dinges. V, 75 pages. 1971. DM 16,–

Vol. 191: Séminaire de Probabilités V. Edited by P. A. Meyer. IV, 372 pages. 1971. DM 26,–

Vol. 192: Proceedings of Liverpool Singularities – Symposium I. Edited by C. T. C. Wall. V, 319 pages. 1971. DM 24,–

Vol. 193: Symposium on the Theory of Numerical Analysis. Edited by J. Ll. Morris. VI, 152 pages. 1971. DM 16,–

Vol. 194: M. Berger, P. Gauduchon et E. Mazet. Le Spectre d'une Variété Riemannienne. VII, 251 pages. 1971. DM 22,–

Vol. 195: Reports of the Midwest Category Seminar V. Edited by J.W. Gray and S. Mac Lane. III, 255 pages. 1971. DM 22,–

Vol. 196: H-spaces – Neuchâtel (Suisse)- Août 1970. Edited by F. Sigrist, V, 156 pages. 1971. DM 16,–

Vol. 197: Manifolds – Amsterdam 1970. Edited by N. H. Kuiper. V, 231 pages. 1971. DM 20,–

Vol. 198: M. Hervé, Analytic and Plurisubharmonic Functions in Finite and Infinite Dimensional Spaces. VI, 90 pages. 1971. DM 16,–

Vol. 199: Ch. J. Mozzochi, On the Pointwise Convergence of Fourier Series. VII, 87 pages. 1971. DM 16,–

Vol. 200: U. Neri, Singular Integrals. VII, 272 pages. 1971. DM 22,–

Vol. 201: J. H. van Lint, Coding Theory. VII, 136 pages. 1971. DM 16,–

Vol. 202: J. Benedetto, Harmonic Analysis on Totally Disconnected Sets. VIII, 261 pages. 1971. DM 22,–

Vol. 203: D. Knutson, Algebraic Spaces. VI, 261 pages. 1971. DM 22,–

Vol. 204: A. Zygmund, Intégrales Singulières. IV, 53 pages. 1971. DM 16,–

Vol. 205: Séminaire Pierre Lelong (Analyse) Année 1970. VI, 243 pages. 1971. DM 20,–

Vol. 206: Symposium on Differential Equations and Dynamical Systems. Edited by D. Chillingworth. XI, 173 pages. 1971. DM 16,–

Vol. 207: L. Bernstein, The Jacobi-Perron Algorithm – Its Theory and Application. IV, 161 pages. 1971. DM 16,–

Vol. 208: A. Grothendieck and J. P. Murre, The Tame Fundamental Group of a Formal Neighbourhood of a Divisor with Normal Crossings on a Scheme. VIII, 133 pages. 1971. DM 16,–

Vol. 209: Proceedings of Liverpool Singularities Symposium II. Edited by C. T. C. Wall. V, 280 pages. 1971. DM 22,–

Vol. 210: M. Eichler, Projective Varieties and Modular Forms. III, 118 pages. 1971. DM 16,–

Vol. 211: Théorie des Matroïdes. Edité par C. P. Bruter. III, 108 pages. 1971. DM 16,–

Vol. 212: B. Scarpellini, Proof Theory and Intuitionistic Systems. VII, 291 pages. 1971. DM 24,–

Vol. 213: H. Hogbe-Nlend, Théorie des Bornologies et Applications. V, 168 pages. 1971. DM 18,–

Vol. 214: M. Smorodinsky, Ergodic Theory, Entropy. V, 64 pages. 1971. DM 16,–